Statistical Tableau

How to Use Statistical Models and Decision Science in Tableau

Ethan Lang

Beijing • Boston • Farnham • Sebastopol • Tokyo

Statistical Tableau

by Ethan Lang

Published by O'Reilly Media, Inc., 1005 Gravenstein Highway North, Sebastopol, CA 95472.

O'Reilly books may be purchased for educational, business, or sales promotional use. Online editions are also available for most titles (*http://oreilly.com*). For more information, contact our corporate/institutional sales department: 800-998-9938 or *corporate@oreilly.com*.

Acquisitions Editor: Michelle Smith
Development Editor: Sara Hunter
Production Editor: Beth Kelly
Copyeditor: nSight, Inc
Proofreader: Krsta Technology Solutions
Indexer: Sue Klefstad
Interior Designer: David Futato
Cover Designer: Karen Montgomery
Illustrator: Kate Dullea

May 2024: First Edition

Revision History for the First Edition

2024-05-02: First Release

See *http://oreilly.com/catalog/errata.csp?isbn=9781098151799* for release details.

978-1-098-15179-9

[LSI]

Table of Contents

Preface

I was first introduced to statistics and Tableau as an undergrad at the University of Kansas. I was enrolled in the engineering school, studying computer science, and I had to take a statistics course. That was really my first exposure to data and the power statistical analysis has. I told my guidance counselor in the semester after that course that I was interested in statistics and data in general. I asked if there was a degree that would lead to a career doing the types of analysis we conducted during the previous semester. At the time, the University of Kansas was just starting a brand new program called Business Analytics, and my counselor introduced me to the director of that program. I scheduled a call with him, and he told me all about the data industry and the growing career field available. I switched my major that afternoon and started my journey in data analytics.

Throughout my coursework, I was eventually introduced to Tableau and, just like statistics, I immediately saw the impact it could have. I dove into the tool and eventually became the go-to resource for Tableau at work and in class. After that, I began finding ways to bring statistical analysis into Tableau. It is these strategies and tactics that inspired this book.

Throughout my time in this career, I have had the opportunity to help solve some of the world's biggest challenges using Tableau and statistical analysis. I have worked for brands all over the world and across every industry you can think of. I have had the honor to be named Tableau Ambassador, join the Veterans Advocacy Tableau User Group as colead, receive Tableau's Viz of the Day several times, and become a member of the Tableau Speaker Bureau. All this has been made possible through my own curiosity and the amazing resources that the Tableau community has published.

This Book's Purpose

Statistical analysis and data visualization are often considered two separate things. However, both disciplines rely on one another. To understand your data and make accurate predictions or assumptions through statistical analysis, you must visualize your data. On the other hand, to deliver actionable insights and enable your audience to get the most from data visualizations, we must back up our assumptions with some statistical analysis.

This book is to help you bring statistics into your visualizations in Tableau. You will learn how to read and interpret statistical models, implement them in Tableau, and ultimately draw out actionable insights to present to your stakeholders. Most decisions using data come with risks; it is your responsibility to arm the decision maker with as much information as possible and to mitigate those risks.

This Book's Audience

This book is best for readers who are looking to build on foundational knowledge about Tableau and statistics. As I mentioned before, it is your responsibility to arm your stakeholders with as much information as possible to help them mitigate the risks that come with decision making. That being said, I have two big caveats:

- Chapter 1 provides some of the foundational methods, definitions, and familiarization you need to get up to speed. For more advanced users, this chapter will likely be a review of core concepts that will be used in subsequent chapters.
- "With great power comes great responsibility." This quote (attributed to Stan Lee) holds true for this book. Statistics is a very deep discipline. In this book, I will give you foundational knowledge to make predictions and assumptions about data. However, be careful when applying these tactics and always approach every analysis with caution. You need to mitigate risks, not increase them with incorrect assumptions. Do your research and conduct your analysis in an ethical manner.

This Book's Structure

Chapter 1, "Introduction"
: Newer users of Tableau will get up to speed with the basics. I also introduce some definitions and foundational statistical concepts that we will build on as we progress through each chapter.

Chapter 2, "Overview of the Analytics Pane"
: The Analytics pane is introduced in Tableau and gives you steps on how to access it from the authoring interface.

Chapter 3, "Benchmarking in Tableau"
: I discuss what benchmarking is, how to apply benchmarking in your visualizations, and best practices when incorporating it in Tableau.

Chapter 4, "Understanding Normal Distribution Using Histograms"
: In statistics, you often need to understand the distribution of your data to apply the appropriate method. In this chapter, I discuss how to quickly test data for normal distribution.

Chapter 5, "Understanding Confidence Intervals"
: Confidence intervals are described, along with how to use them in Tableau and how to calculate them using custom calculated fields.

Chapter 6, "Anomaly Detection on Normally Distributed Data"
: I introduce three methods you can implement to visually detect anomalies in your data.

Chapter 7, "Anomaly Detection on Nonnormalized Data"
: Here are three more methods you can implement to visually detect anomalies in your data, even if it fails the normalization assumption.

Chapter 8, "Linear Regression in Tableau"
: Linear regression is introduced, as is how to implement it in Tableau and how to understand the results of the model.

Chapter 9, "Polynomial Regression in Tableau"
: Polynomial regression is discussed, along with how to implement it in Tableau and interpret the results of the model.

Chapter 10, "Forecasting in Tableau"
: Exponential smoothing is discussed, as is how to implement this forecasting method in Tableau and how to interpret the results of the model.

Chapter 11, "Clustering in Tableau"
: K-means clustering is introduced, along with how to implement this clustering method in Tableau and understand the results of the model.

Chapter 12, "Creating an External Connection to R Using Tableau"
: I will show you how to download the appropriate software needed to make an external connection to R from Tableau.

Chapter 13, "Creating an External Connection to Python Using Tableau"
: Here I will talk about Python and how to download the appropriate software needed to make an external connection from Tableau.

Chapter 14, "Understanding Multiple Linear Regression in R and Python"
: Multiple linear regression is introduced, along with how to implement it in R and Python and interpret the results of the model.

Chapter 15, "Using External Connections in Tableau"
: Here are several examples of using external connections in both R and Python to implement new modeling methods in Tableau.

Conventions Used in This Book

The following typographical conventions are used in this book:

Italic
: Indicates new terms, URLs, email addresses, filenames, and file extensions.

`Constant width`
: Used for program listings, as well as within paragraphs to refer to program elements such as variable or function names, databases, data types, environment variables, statements, and keywords.

`Constant width bold`
: Shows commands or other text that should be typed literally by the user.

`Constant width italic`
: Shows text that should be replaced with user-supplied values or by values determined by context.

O'Reilly Online Learning

For more than 40 years, *O'Reilly Media* has provided technology and business training, knowledge, and insight to help companies succeed.

Our unique network of experts and innovators share their knowledge and expertise through books, articles, and our online learning platform. O'Reilly's online learning platform gives you on-demand access to live training courses, in-depth learning paths, interactive coding environments, and a vast collection of text and video from O'Reilly and 200+ other publishers. For more information, visit *https://oreilly.com*.

How to Contact Us

Please address comments and questions concerning this book to the publisher:

O'Reilly Media, Inc.
1005 Gravenstein Highway North
Sebastopol, CA 95472
800-889-8969 (in the United States or Canada)

707-827-7019 (international or local)
707-829-0104 (fax)
support@oreilly.com
https://www.oreilly.com/about/contact.html

We have a web page for this book, where we list errata, examples, and any additional information. You can access this page at *https://oreil.ly/statistical-tableau*.

For news and information about our books and courses, visit *https://oreilly.com*.

Find us on LinkedIn: *https://linkedin.com/company/oreilly-media*

Watch us on YouTube: *https://youtube.com/oreillymedia*

Acknowledgments

Thank you to the data community for always being supportive and encouraging. When getting started, I was lucky enough to have had so many fantastic resources I could turn to. Thank you to all the content creators and the folks who have supported me.

Thank you to my mentors, Ryan Sleeper, Kaleb Gilliland, and many others. Without your support and guidance, it would have taken me years to get where I am professionally and personally. Thank you for taking the time and endowing me with your leadership and knowledge.

Thank you to the technical reviewers of this book, Maddie Dierkes, Lorna Brown, Ann Jackson, and Christopher Gardner. Your feedback helped make this book 100 times better.

Thank you Sara Hunter, development editor from O'Reilly. Your words of encouragement and continuous feedback helped make all this possible. Thank you so much; I couldn't have done it without you.

Thank you to my mom, dad, and brothers for believing in me and pushing me along the right path. I often needed a course correction, and you were always there to help steer me in the right direction, especially in the early days.

Special thanks to my wife Sandra Lang and my kids Jameson, Ophelia, and Edalyn. Without your support and encouragement, I would never be able to create anything like this. Thank you guys for putting up with my crazy ideas, putting off home projects, and disappearing into my office for hours at a time to write. I love you all so much!

CHAPTER 1

Introduction

It is estimated that 70%–80% of job postings for a data analyst mention statistics as a desired skill or requirement. I haven't found a way to prove those numbers myself, but looking at job postings, I would argue in favor of that estimate. With ever-increasing amounts of data, businesses are looking for ways to interpret and understand that data. Statistics is often the most scientific way to do that. However, I think many analysts and Tableau developers struggle to implement statistics into their analysis or data visualizations. There are many reasons for this, and I will be the first one to tell you that it is not for lack of trying. Statistics can be intimidating for both developers and the stakeholders who rely on their reports. Trying to explain and interpret complex statistical equations is tough without a firm understanding of the discipline.

That is the exact purpose of this book. I want to equip you with that firm understanding of statistics and give you the confidence to speak to the equations and implement them in your work. In this book, I will be focusing on bringing data visualization in Tableau together with statistical analysis so that you can support your insights with scientific evidence.

In this chapter, I will introduce you to some common Tableau terminology I will be using throughout the book. I am also going to introduce you to some basic statistical terms and ideas. Toward the end of the chapter, I will present you with a case study that ties both disciplines together, and I will discuss the importance of visualizing statistical results.

Introduction to Tableau

It is important to understand that Tableau is not simply a data visualization tool, but a company with a suite of tools to support data visualization and analytics at an enterprise level. There are many products within Tableau's ecosystem, including Tableau

Desktop, Tableau Cloud, Tableau Server, Tableau Prep Builder, Tableau Public, and more.

Some of these products require a license to use, while others, such as Tableau Public, do not require you to purchase a license, but there are certain limitations. With a license, you can publish your workbooks to Tableau Server or Tableau Cloud from Tableau Desktop. This allows your users to view and interact with your data visualizations from a browser. Checkout the Tableau website (*https://oreil.ly/1oQCa*) for a full list of all Tableau's products.

Common Terms of the Authoring Interface of Tableau Desktop

There are several common terms within Tableau Desktop that I want you to know and be familiar with. To begin with, when you open Tableau Desktop, you will land on the Start Page, as shown in Figure 1-1.

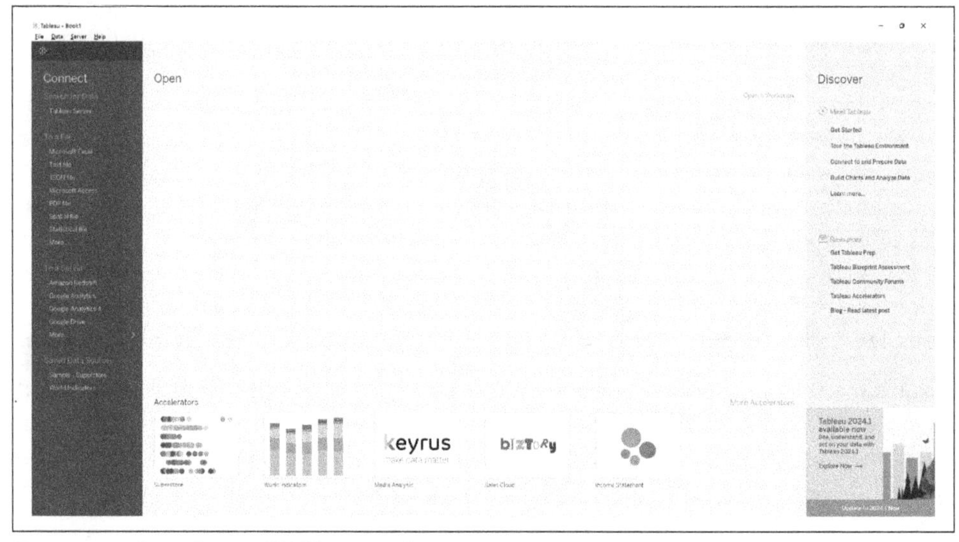

Figure 1-1. Start Page of Tableau Desktop

From the Start Page, you can connect to the data you want to visualize. Tableau has hundreds of connectors that you can use to access your data. A connector is basically like a built-in API that allows you to establish a connection to a database or file type to read that data into Tableau Desktop. On the lefthand side of the Start Page you can explore all the connectors that are available.

For all the demonstrations in this book, I will be using the Sample - Superstore dataset. To connect to this dataset, simply click on Sample - Superstore, as shown in Figure 1-2.

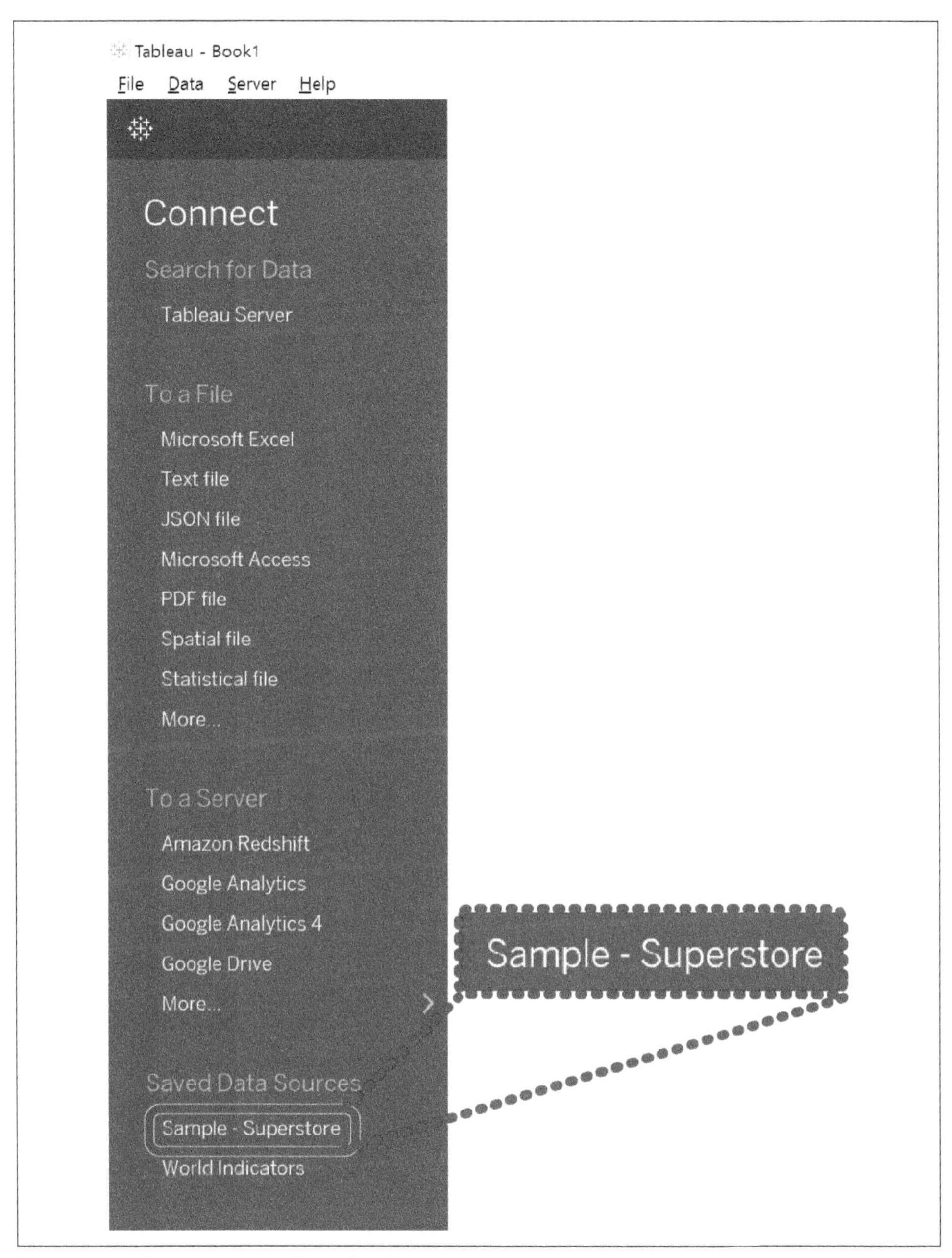

Figure 1-2. Connecting to Sample - Superstore

It is important to note that if you are using a different version of Tableau Desktop than I am, you may get different results. Tableau will occasionally update the Sample - Superstore dataset. I will be using version 2023.2 throughout this book. If you want to follow along exactly, you can download this version from Tableau's product support page.

After clicking on the sample dataset, you will be navigated from the Start Page to Tableau Desktop's authoring interface, as shown in Figure 1-3.

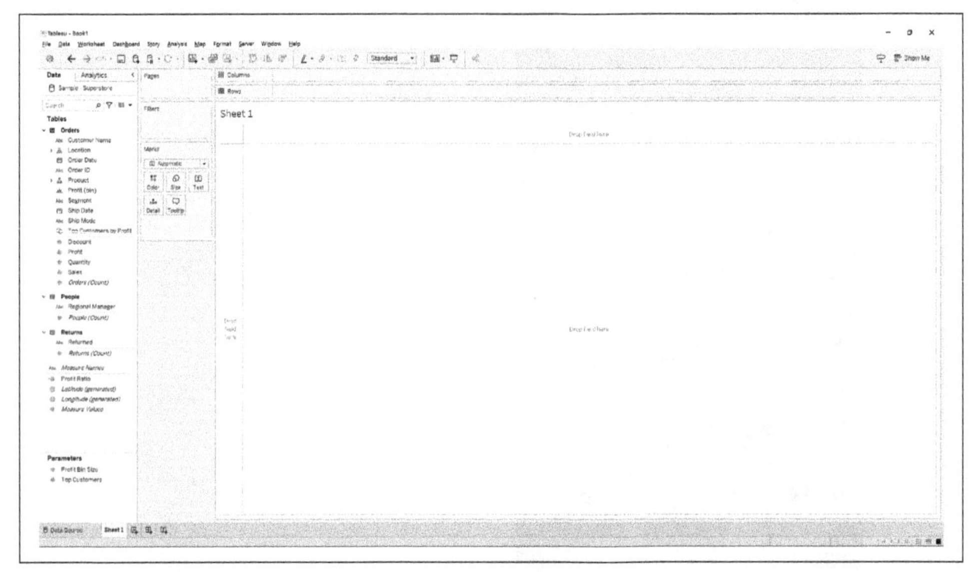

Figure 1-3. Tableau Desktop's authoring interface

To introduce you to the terms I will use throughout the book, on the lefthand side, you will find the Data pane, as shown in Figure 1-4.

At the top of the Data pane, you will see a list of the data sources you are connected to. Moving down, you will find a list of fields, including those that are calculated, separated by data source and whether Tableau believes that field is a measure or dimension.

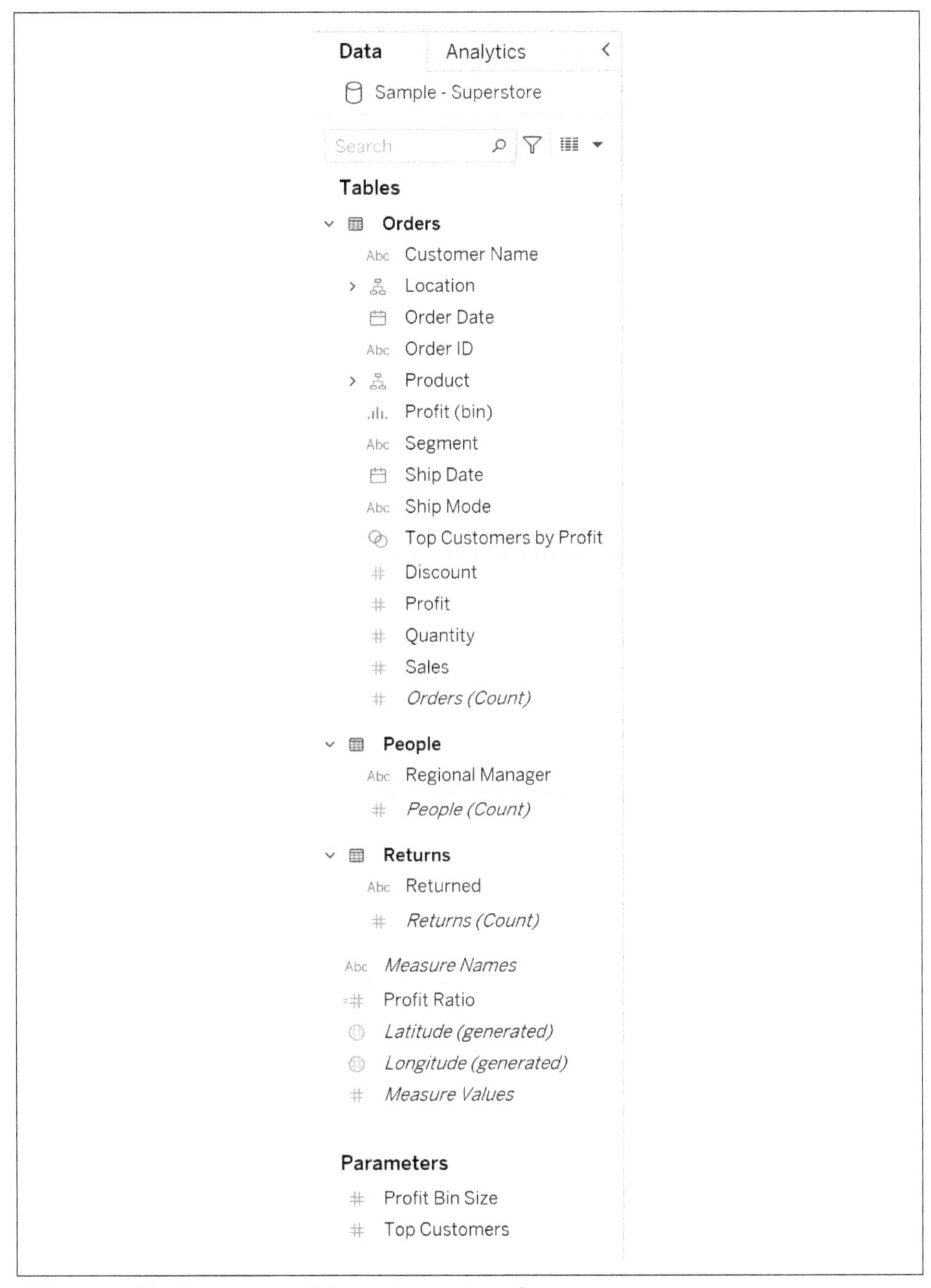

Figure 1-4. The Data pane of the authoring interface

To the right of the Data pane, you will find the different components used to create visualizations called shelves. There is the Marks shelf, Filters shelf, Pages shelf, Columns shelf, Rows shelf, and canvas, as shown in Figure 1-5.

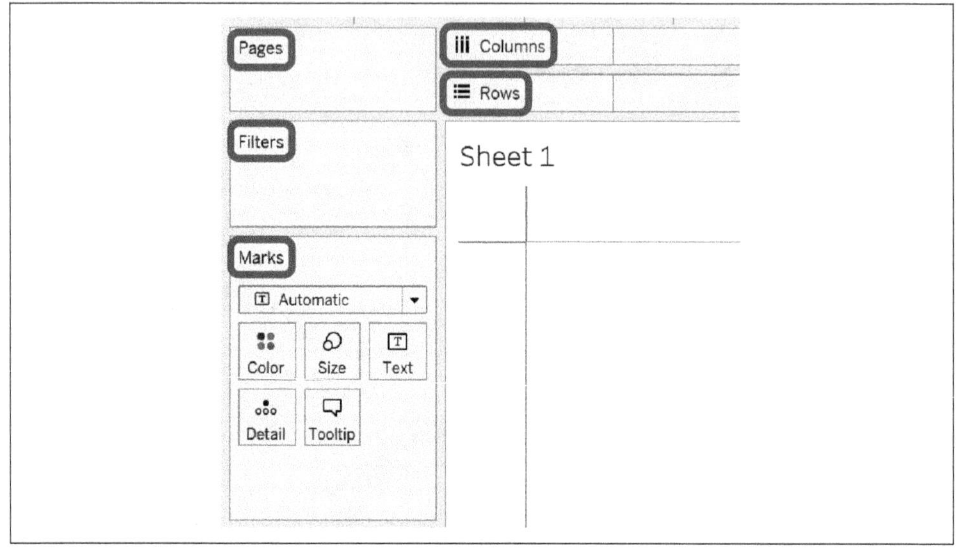

Figure 1-5. Key features of the authoring interface

To define each a little further, here is a brief explanation of each:

Marks shelf
: The Marks shelf is a key element on the authoring interface and allows you to drag fields into different properties that affect the view. The properties are Color, Size, Text, Detail, and Tooltip. There are different property options that will appear when certain conditions are met. For instance, in changing the Mark type to polygon, you will see a new property of angle in the Marks shelf.

Filters shelf
: The Filters shelf allows you to add different fields to filter the view on. There are eight different types of filters in Tableau that are processed at different times in Tableau's order of operations.

Pages shelf
: The Pages shelf lets you break the view up into pages so that you can analyze how a specific field affects the rest of the fields in the view. The most common use of this is to add a Date dimension and animate how things change over time.

Columns shelf
: The Columns shelf is where you can drag fields to create the columns of the visualization you are making. The Columns shelf will coordinate with the x-axis in the view.

Rows shelf
: The Rows shelf is where you can drag fields to create the rows of the visualization you are making. The Rows shelf coordinates with the y-axis in the view.

Canvas
: The canvas is where the data visualization will appear as you begin dragging fields to the various other shelfs. You can also drag different fields directly to the canvas when you are authoring a data visualization. Doing so will add the field to the appropriate shelf for you.

The last major feature I want to call out in this chapter is in the bottom-left corner of the authoring interface. There you will find a button to navigate to the data source page and three additional buttons. These buttons are used to create new worksheets, new dashboards, or new stories, as shown in Figure 1-6.

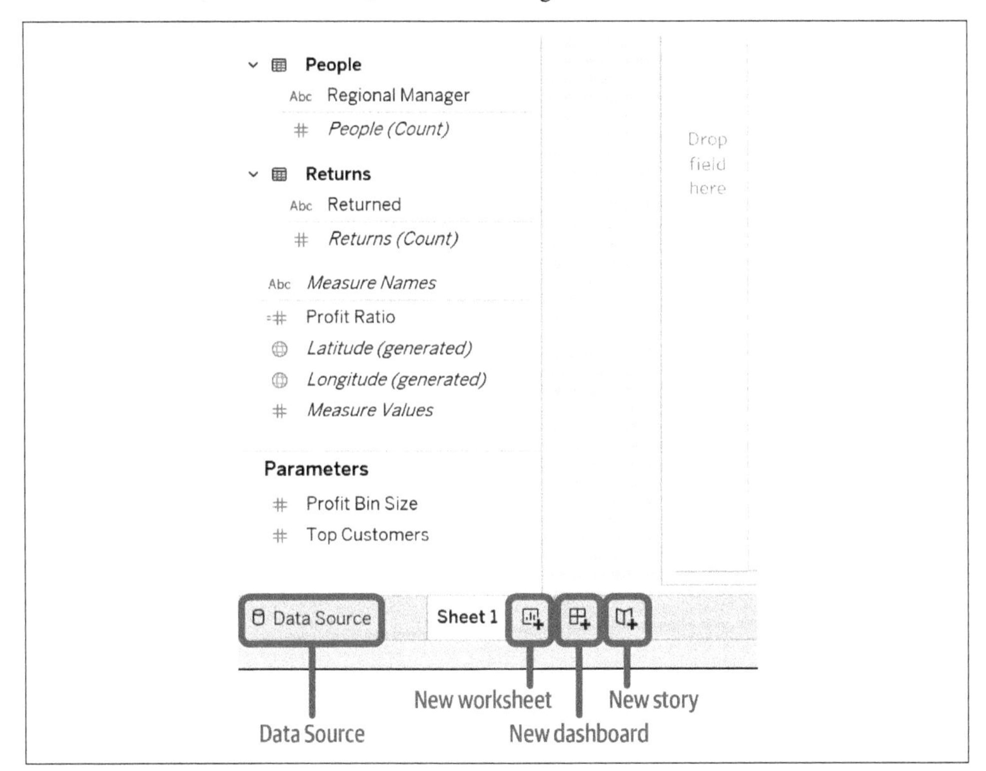

Figure 1-6. Navigation buttons on the authoring interface

To give you a little more context here is a brief description of each:

Data Source button
: This will navigate you to the Data Source page. From there, you can add new connections, create new data sources, and view the physical and logic layer for joins and blending.

New worksheet button
: Clicking this button will create a new worksheet and navigate you to that sheet's tab. From here you can author a new data visualization.

New dashboard button
: Selecting this button will create a new dashboard and navigate you to that dashboard's tab. From here you can drag sheets onto the canvas instead of fields to compile a new dashboard.

New story button
: Clicking the new story button will create a new story and navigate you to that story's tab. From here, you can compile a story using sheets or dashboards to create different pages within your story.

Example of the Step-by-Step Instructions Throughout This Book

To get you familiar with the instructions and writing style used in this book, this section gives a simple example that puts the common terms together. Using Tableau Desktop is very intuitive, and there are many different ways to do things. I am going to show you how to create two simple charts and add them to a dashboard using the Sample - Superstore dataset. Let's say you want to view the sales by order date. First, double-click on Sales in the Data pane, then double-click Order Date, as shown in Figure 1-7.

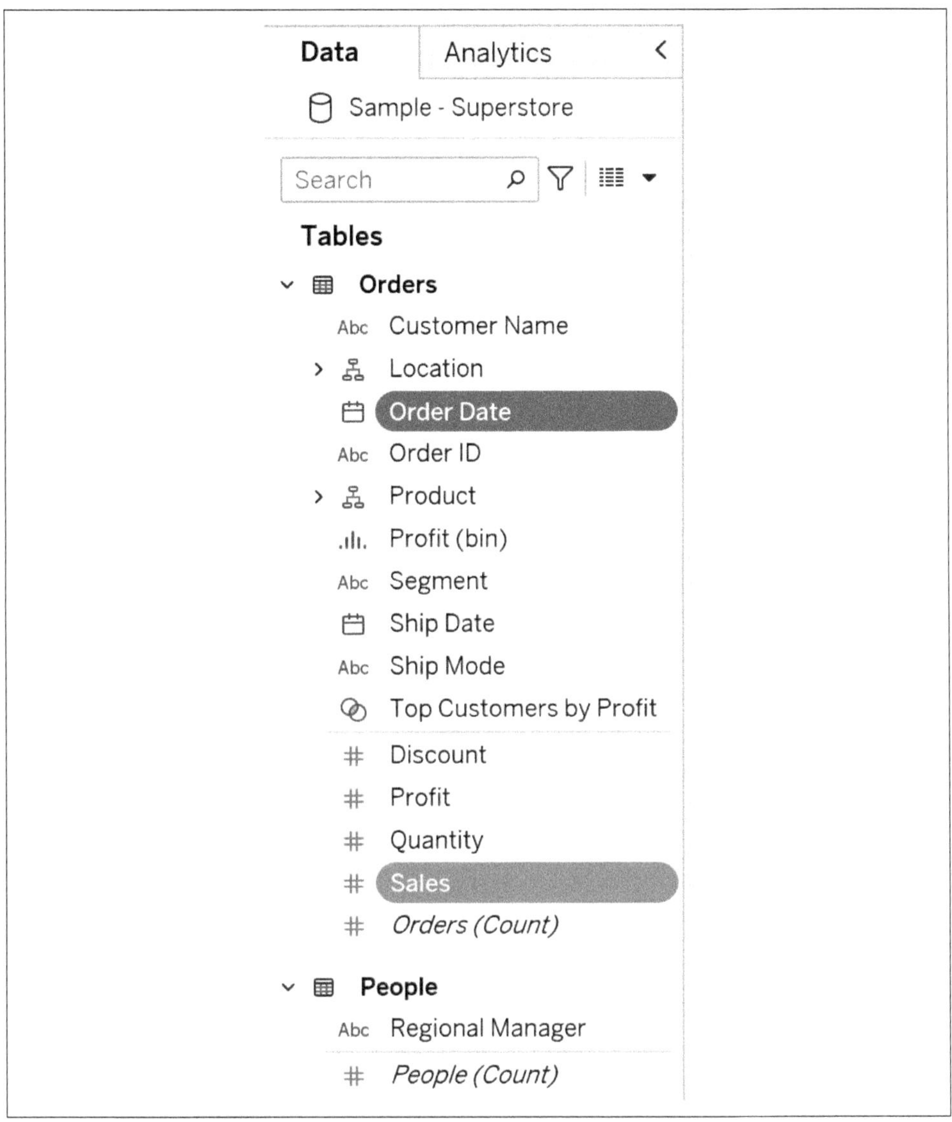

Figure 1-7. Clicking Sales and Order Date into the view

Tableau is intuitive enough to recognize that you likely want this data to trend over time, and it will automatically create a line chart, as shown in Figure 1-8.

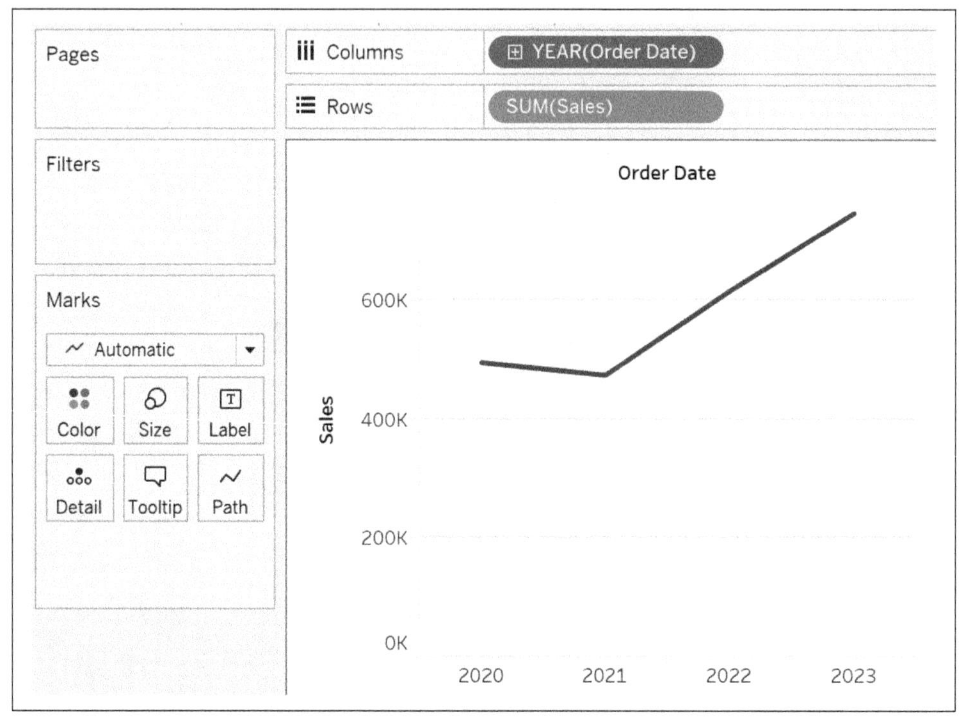

Figure 1-8. Creating a simple line chart in Tableau Desktop

Now let's say you also want to view your sales data by segment. Click on the "New worksheet" button at the bottom left of the authoring interface, as shown in Figure 1-9.

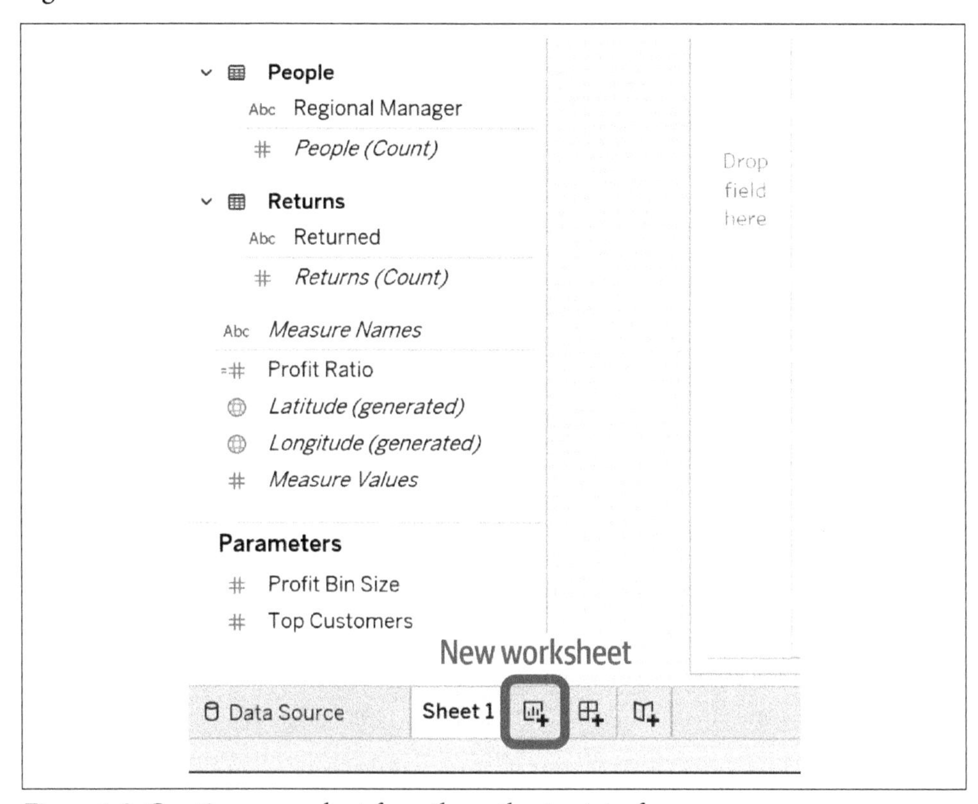

Figure 1-9. Creating a new sheet from the authoring interface

This will open Sheet 2; your first chart is still viewable by navigating back to Sheet 1. Double-click on Sales, then Segment in the Data pane, as shown in Figure 1-10.

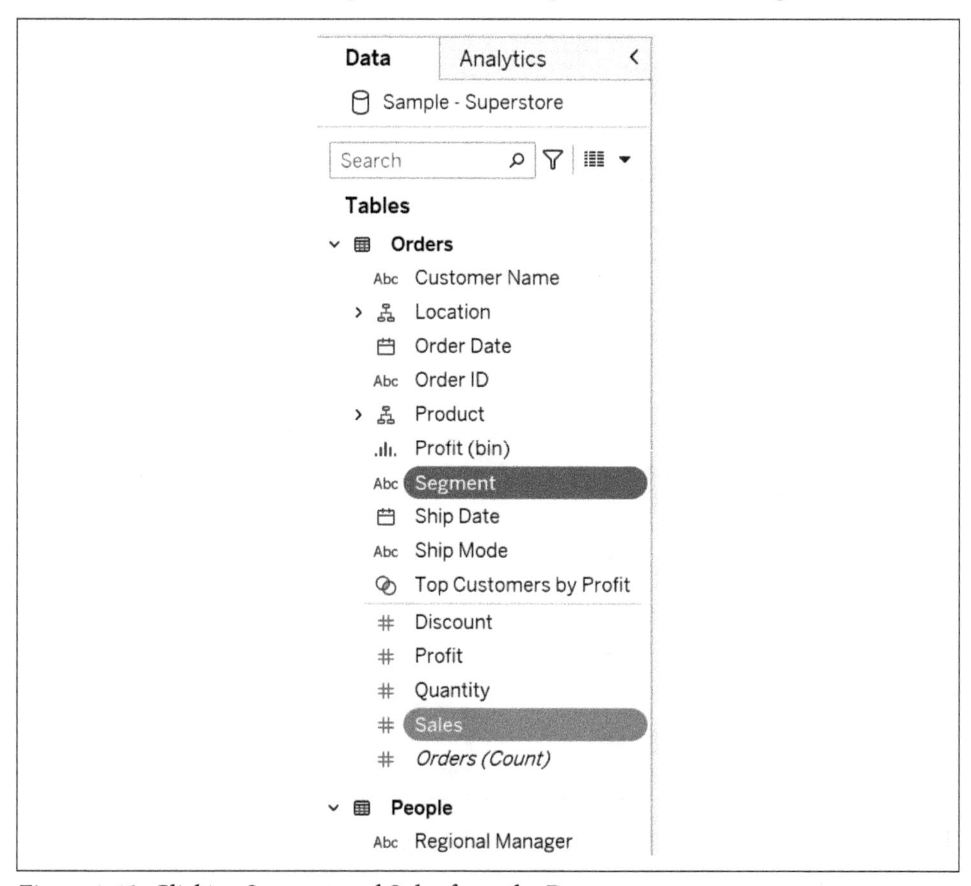

Figure 1-10. Clicking Segment and Sales from the Data pane

This will create a simple bar chart showing the SUM(Sales) (sum of sales) by Segment on the canvas, similar to Figure 1-11.

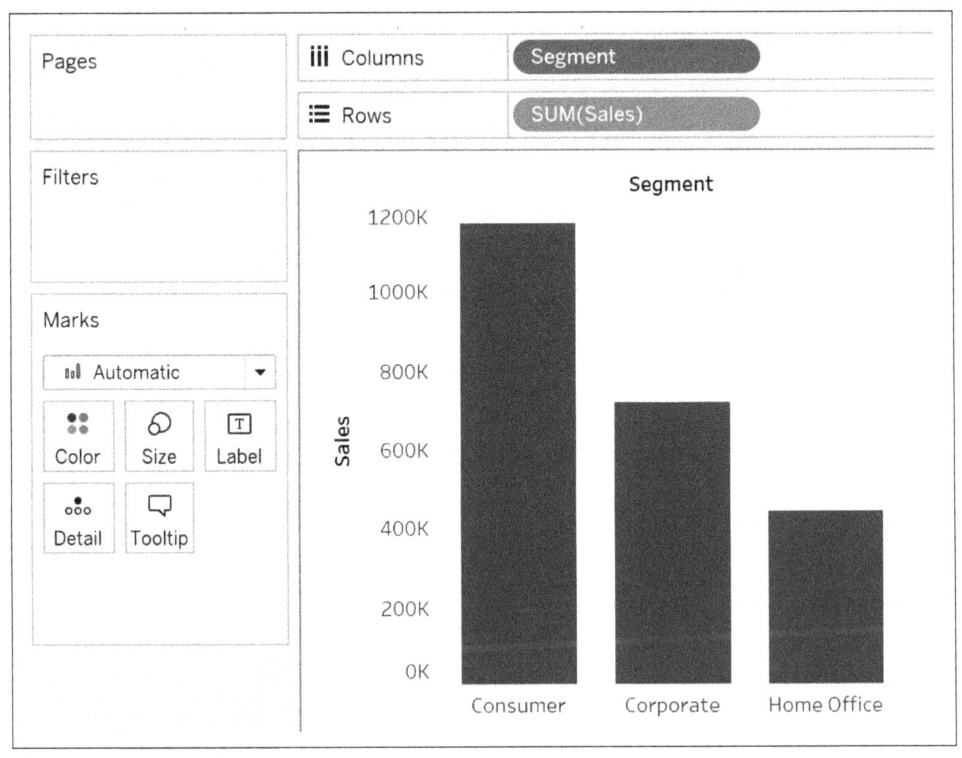

Figure 1-11. Creating a simple bar chart in Tableau Desktop

So far, you've been able to view these two charts in a working environment in Tableau. Let's say you want to share these charts with others in your organization. To begin that process, click on the "New dashboard" button in the bottom left of the authoring interface, as shown in Figure 1-12.

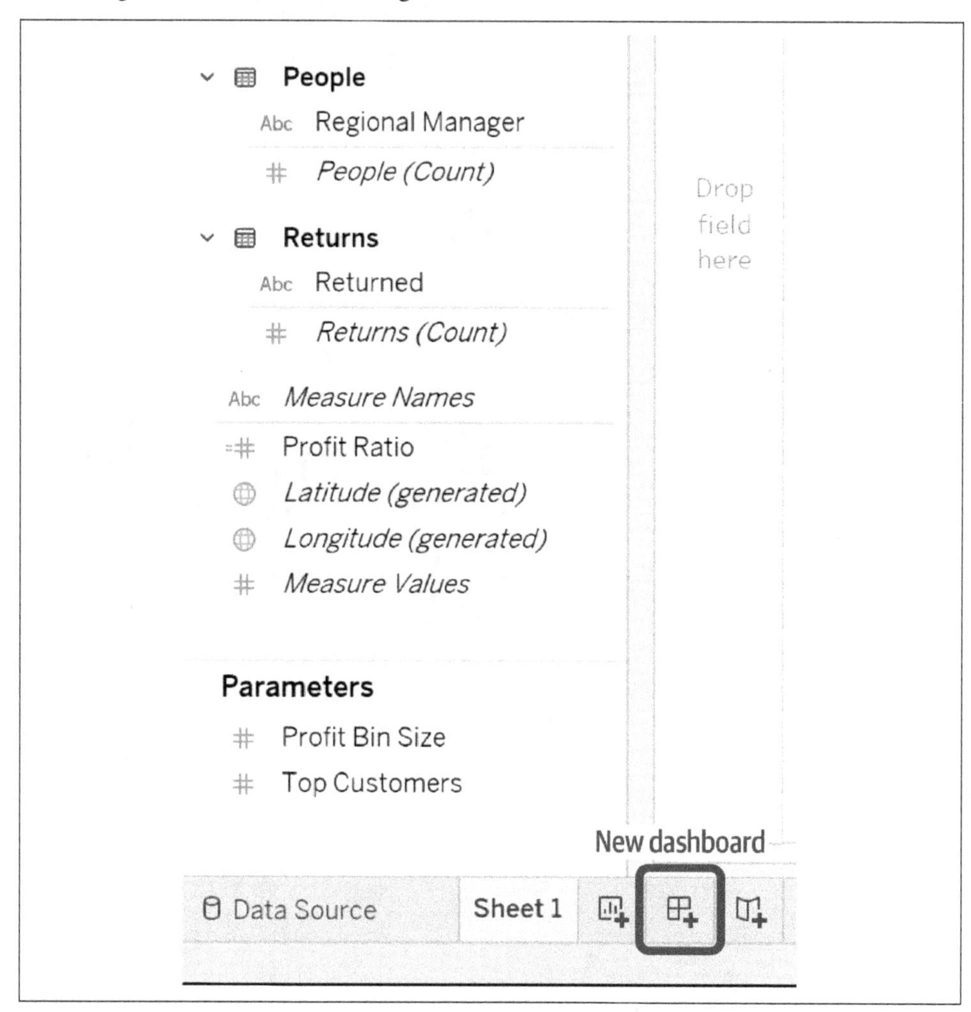

Figure 1-12. Creating a new dashboard from the authoring interface

This will open a new canvas where you can create dashboards, as shown in Figure 1-13. Dashboards are the bread and butter of Tableau and are ultimately what you will share for users to interact with.

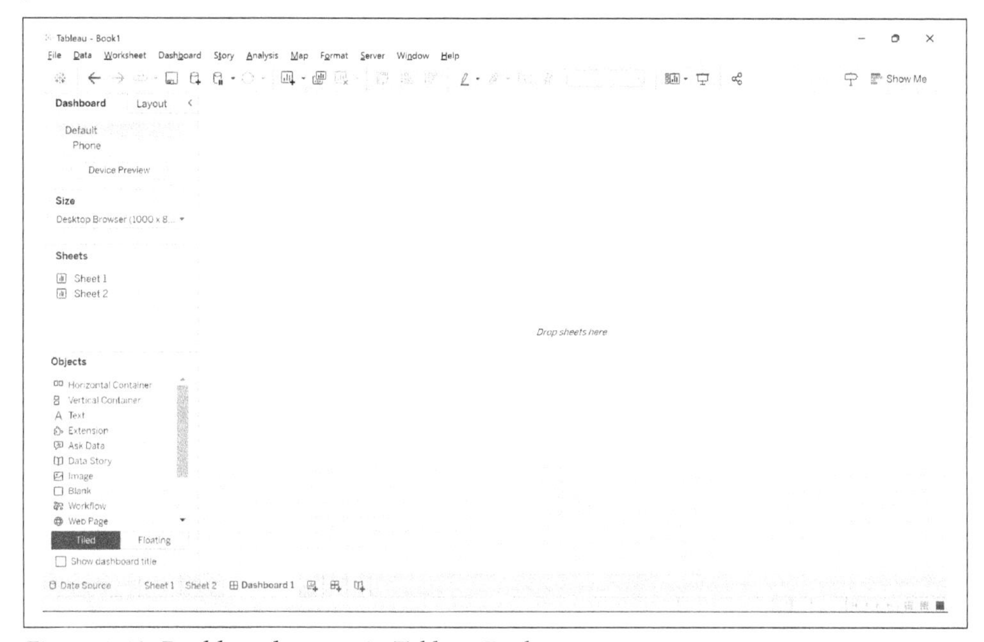

Figure 1-13. Dashboard canvas in Tableau Desktop

Now add your two sheets on the dashboard canvas. On the left, click and drag Sheet 1 onto the canvas. Then click and drag Sheet 2 onto the canvas. Your dashboard should now look similar to Figure 1-14.

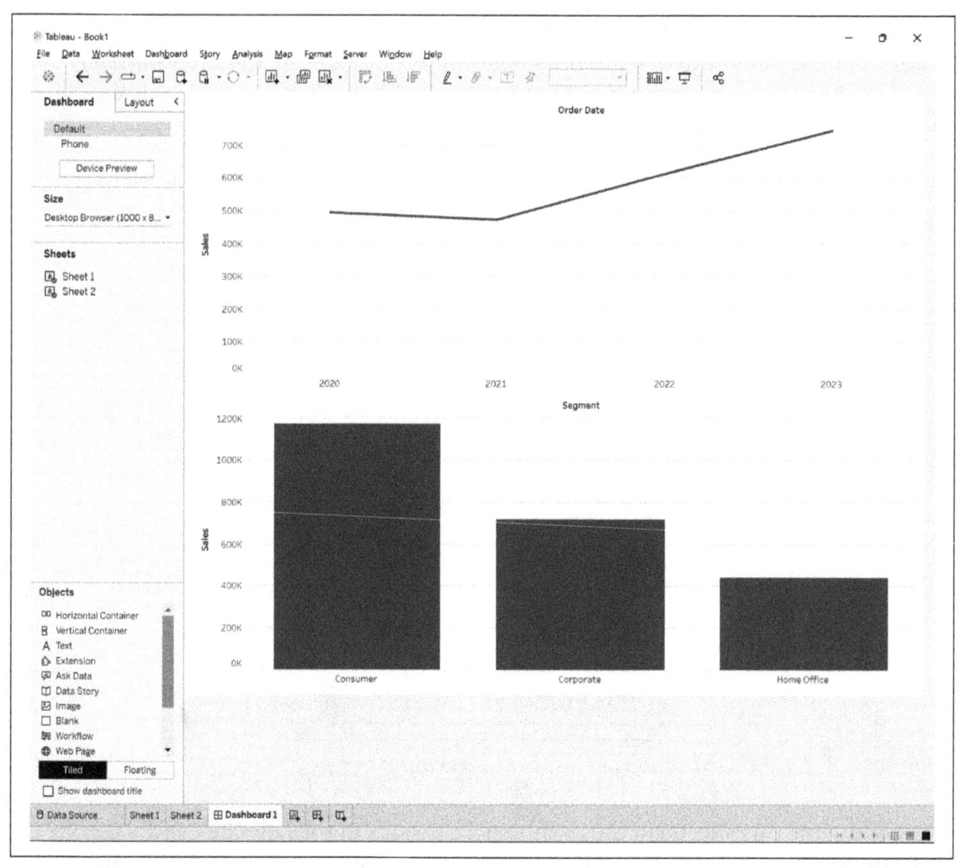

Figure 1-14. Creating a simple dashboard layout in Tableau Desktop

This example should help you see how Tableau's common terms will be used in the tutorials throughout this book. Knowing the layout of the tool and terms is the foundation for understanding Tableau Desktop as a whole. The content thus far has most likely been a review for you. From here on, I will be showing you how to tie statistics into your dashboards and giving you tangible examples of how to implement them into your work! In the next section, I will be introducing you to common statistical terms and showing you an example that ties everything together.

Introduction to Statistics

According to Merriam-Webster's online dictionary, statistics is defined as *a branch of mathematics dealing with the collection, analysis, interpretation, and presentation of masses of numerical data.* I personally think this definition nails it on the head, especially in today's business environment. To unlock deep insights in your data, you need to incorporate statistics into almost every aspect of the analytics process. This includes collecting data in an efficient and ethical way, understanding the data, finding deeper insights in the analysis, and presenting your findings so your stakeholders can make informed decisions.

In the next section, I will introduce you to some common statistical terms and ideas. I will also show you how powerful adding statistics to your analysis can be through a tangible case study example.

Common Statistical Terms

To level-set, I will briefly explain some of these terms and ideas. However, this is not a comprehensive list of everything there is to know about statistics. The purpose of this book is to get you more comfortable and familiar with foundational statistics so you can apply them to your own work. I will also go into more detail about some of these terms as you progress through each chapter, where applicable:

A statistic
: Throughout this book, you will see me refer to different things as a *statistic*. The definition of a statistic is *a fact or piece of data from a study of a large quantity of numerical data*. This means anything you can calculate from a large set of data could be referred to as a statistic. For instance, if we calculate the mean, median, or mode of a dataset, I would refer to each of those values as a statistic.

Hypothesis testing
: Setting up a hypothesis test is one of the most foundational steps in most statistical analysis. Without doing so, you will find yourself chasing some sort of statistical significance when simply proving there isn't a significant difference is just as powerful. Basically, a hypothesis test is when you create a *null hypothesis* and an *alternative hypothesis*. Then you lay out the conditions of what you consider a significant difference in favor of one hypothesis or the other by setting a significance level.

Significance level
: A *significance level* is the predetermined threshold used to determine statistical significance. The most common significance level is 0.05 (5%), but this is just an arbitrary number. There are times when you can be more or less significant. For instance, if you are in healthcare, you may want to show results that have a higher level of significance to ensure the most accurate interpretation of the results.

Statistical significance
: *Statistical significance* is a term used in statistics to determine whether an observed effect or relationship in data is likely to be genuine or if it could have occurred by chance. In other words, it helps analysts assess whether the results of an analysis are meaningful or if they could be attributed to random variation.

To tie that back to the hypothesis tests:

Null hypothesis (H_0)
: This is a statement that there is no significant difference or effect. It serves as the default assumption to be tested.

Alternative hypothesis (H_1 or H_a)
: This is the opposite of the null hypothesis and suggests that there is a significant difference or effect.

P-value
: The *p-value* (probability value) is a measure of the evidence against a null hypothesis. It represents the probability of obtaining the observed results (or more extreme results) if the null hypothesis is true. A low p-value (typically less than 0.05) is considered indicative of statistical significance.

: If the p-value is less than the chosen significance level (commonly 0.05), the null hypothesis is rejected in favor of the alternative hypothesis. If the p-value is greater than the significance level, there is not enough evidence to reject the null hypothesis.

In summary, you define what you are going to test by setting up a null hypothesis and alternative hypothesis. Then you decide what the significance level should be for your experiment. Then you test for a statistically significant difference and use the p-value as a unit of measure compared to your predetermined significance level.

To really drive these ideas home, I want to show you a practical example we can calculate by hand. This way, you can see how these terms all come together.

Practical Application Through a Case Study

Let's say that your company wants to test some new marketing in an email. However, they are worried that if the new marketing fails, it could significantly impact sales for this quarter. Therefore, they want to test the new marketing email by sending it to a subset of the total email list, then analyze the performance before deciding whether to move forward with the new marketing. Table 1-1 shows the results of the test displayed in a contingency table.

Table 1-1. Contingency table of marketing conversions

	Original email	New marketing email
Nonconversions	727	117
Conversions	23	8

A *contingency table* is a way to organize and display data in a table format, especially when studying the relationship between two categorical variables. Categorical variables are variables that represent categories or groups, such as colors, types of fruits, or responses to a yes-no question. In this example, we are displaying how many conversions the original email had compared to the new marketing email. *Conversion* has been defined as "the point at which a recipient of a marketing message performs a desired action."[1]

The marketing team has done a simple analysis looking at the conversion rates of the emails by taking the total sent/conversions of each campaign. Using this calculation, they found that the original email had a conversion rate of about 3% ($23 \div 750 = 0.030$) and the new marketing email had a conversion rate of about 6% ($8 \div 125 = 0.064$). They claim that the new email is an absolute success and that it will lead to double the amount of conversions when they send it out to their entire list next time.

Senior leaders at the business are thrilled with the idea of doubling the amount of sales and want to invest in several new salespeople to help with the increase. However, they come to you for a second opinion and ask if the analytics team could review the data and confirm the marketing team's assumptions.

Where do you begin? This is where statistical analysis will become your best friend. Armed with some basic statistics, you know that you can run a few simple tests to tell you if the new marketing email was statistically significant or not. Before I get too far in the weeds, let's set up the hypothesis and determine the significance level to test for.

Setting up the hypothesis test

The first thing you need to do in this situation is to set up a hypothesis test. In a standard hypothesis test, you set the two hypotheses: null and alternative. For this example the hypothesis will be as follows:

Null hypothesis
: The new marketing email *is not* statistically significant; therefore, email conversions will remain the same on average as the original.

Alternative hypothesis
: The new marketing email *is* statistically significant; therefore, email conversions will be higher on average than the original.

To prove the statistical significance, I will be looking for a p-value less than 0.05, which is my significance level.

In statistics, it's important to understand that you are always trying to validate your assumptions using mathematics. What do I mean by that? You always want to assume

---

1 See David Kirkpatrick's blog article on conversion (*https://oreil.ly/gfz7R*), "Marketing 101: What Is Conversion?," *MarketingSherpa*, March 15, 2021.

that the results aren't going to change when new things are introduced. Therefore, you want to assume that the null hypothesis is correct, and your test will determine if that is wrong. In statistics, you would say you have failed to reject the null hypothesis if the p-value is greater than your predetermined significance level. If the p-value is less than the significance level, then the test is statistically significant, and you would reject the null hypothesis in favor of the alternative.

Chi-square test

Now that you have your hypothesis set up, it's time to run a statistical analysis. In the spirit of providing you with a foundational understanding, I have decided to run a simple statistical test called a chi-square test. A *chi-square test* is a statistical test used to determine whether there is a significant association (or independence) between two categorical variables. It is particularly useful when working with data that can be organized into a contingency table.

This is a great option to run in this situation and very accessible, even if you're new to statistics. You don't have to have any special software or know any coding to calculate this test. You can do it by hand, run it in Excel, or look for a calculator online.

To begin, let's revisit the contingency table and add to it. As you can see in Table 1-2, I added totals for each column, row, and a grand total column.

Table 1-2. Adding totals to the contingency table

	Original email	New marketing email	Totals
Nonconversions	727	117	844
Conversions	23	8	31
Totals	750	125	875

Now you need to calculate *expected values* (*E*) for each of the cells in the table. The formula is very easy. Take the row total, multiply it by the column total for each cell, and then divide by the grand total. So for the top-left cell (original email by nonconversions) you would take $750 \times 844 \div 875 = 723.43$. I will calculate each of the expected values in the corresponding cells in Table 1-3.

Table 1-3. Calculating expected values

	Original email	New marketing email	Totals
Nonconversions	E_{11} $(750 \times 844) \div 875 = 723.43$	E_{12} $(125 \times 844) \div 875 = 120.57$	844
Conversions	E_{21} $(750 \times 31) \div 875 = 26.57$	E_{22} $(125 \times 31) \div 875 = 4.43$	31
Totals	750	125	875

You can see that I added some mathematical syntax for each cell (E_{11}, E_{12}, E_{21}, and E_{22}). This is referring to the expected value for the cell in row x and column y. So E_{11}

is the expected value in row 1/column 1. E_{12} is the expected value for row 1/column 2, and so on. I will continue to use mathematical expressions and syntax similar to this throughout the book and introduce you to mathematical syntax along the way.

With your expected values calculated, you need to finish by comparing those values to the values you observed. This step is expressed mathematically by the following formula:

$$X2 = \Sigma(\text{observed value} - \text{expected value})^2 \div \text{expected value}$$

Simply put, you need to take the original value minus the expected value you just calculated, square that, and then divide by the expected value. You will do this for each cell and then add up each of the values we get. Looking at E_{11}, we have the original value of 727 minus the expected value of 723.43, which equals 3.57. Take 3.57 and square it, which equals 12.7449. Then divide that by the expected value. So 12.7449 ÷ 723.43 = 0.017617. I'll round that number up to 0.018. You can follow along in Table 1-4 for each cell.

Table 1-4. Comparing expected values to observed values

	Original email	New marketing email	Totals
Nonconversions	$(727 - 723.43)^2 \div 723.43 = 0.018$	$(117 - 120.57)^2 \div 120.57 = 0.106$	844
Conversions	$(23 - 26.57)^2 \div 26.57 = 0.48$	$(8 - 4.43)^2 \div 4.43 = 2.877$	31
Totals	750	125	875

Now you take the values you got in each cell in Table 1-4 and add them up. Here are the values we got for each cell:

$E_{11} = 0.018$

$E_{12} = 0.106$

$E_{21} = 0.48$

$E_{22} = 2.877$

$X2 = (0.018 + 0.106 + 0.48 + 2.877) = 3.481$

That gives you an $X2$ observed value of 3.481. The decision rule for a chi-square test is as follows: if the $X2$ observed value is greater than the $X2$ critical value, you reject the null hypothesis. So far, I have calculated the $X2$ observed value, but I need to get the $X2$ critical value. Remember, for our hypothesis test, we set a significance level of 0.05. Using that significance level, you can determine the $X2$ critical value.

The best way to find the critical value is to look it up in a distribution table. A distribution table is a resource you can find online that is a large table of critical values that are precalculated for you. Using the significance level of 0.05, I found the $X2$ critical value to be 3.84.

Considering that the observed value versus the critical value 3.481 is not greater than 3.84, you would therefore fail to reject the null hypothesis. In simple terms, this means that the test proved that the new marketing email did not have a statistically significant increase in conversions. You can conclude that moving forward with this new email marketing campaign will yield similar results to the original, on average.

Conclusions drawn from statistical analysis

I chose this example for two reasons: (1) this is a common, real-world example that gives you a foundational understanding of statistics and how it's used, and (2) this example comes really close to being statistically significant. In statistics, one of the most important lessons is to understand the data and make some assumptions.

In this situation, I may go back and say that the results did not yield a significant increase in conversions. However, the data suggests that there is a slight improvement. My recommendation would be to hold off on hiring, run the test again next quarter, and split the total emails sent 50/50 versus 75/25. This would give the team a larger sample size to rerun the analysis. After all, you can make the assumption that while the new campaign did not yield statistically significant results to prove it increased conversions, the results did suggest that the new marketing email did not hurt conversions in any way.

Therefore, it's not always as black and white as it appears. Unlike traditional mathematics, when using statistics, you have to be able to think outside the box and make further recommendations after an analysis.

Data Visualization and Statistics

In closing, there is an obvious advantage of data visualization when trying to find quick insight in your data; from the previous example, you can see the power statistical analysis can have when making decisions. However, bringing them together is where you will truly unlock the most of any analytics tool or analysis.

I want to share a great example to drive home the importance of bringing data visualization together with statistical analysis. In Table 1-5, I have four statistical summaries from four different datasets.

Table 1-5. Statistical summary of Anscombe's quartet

	Dataset 1		Dataset 2		Dataset 3		Dataset 4	
	X	Y	X	Y	X	Y	X	Y
Obs	11	11	11	11	11	11	11	11
Mean	9.00	7.50	9.00	7.50090	9.00	7.50	9.00	7.50
SD	3.16	1.94	3.16	1.94	3.16	1.94	3.16	1.94
r	0.82		0.82		0.82		0.82	

Here you can see some statistics, such as the standard deviation, *r*, mean, and the number of observations in each dataset. I will explain each of these statistics in detail in upcoming chapters; however, notice they are the same across all four datasets. If you were to plot the datasets and visualize them, as shown in Figure 1-15, you could clearly see that each dataset is very different.

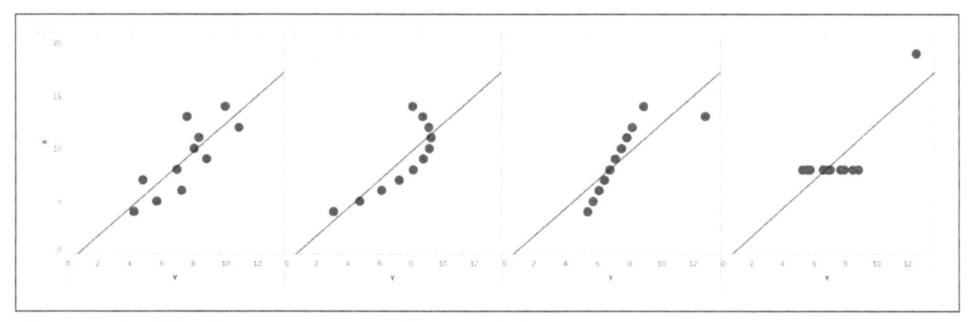

Figure 1-15. Visual representation of Anscombe's quartet

Figure 1-15 is an example of Anscombe's quartet, and it was constructed by the statistician Francis Anscombe in 1973 to demonstrate the importance of visualizing your data before and after modeling it. When building statistical models, you need to visualize the data to truly understand what the story is—if there are outliers, correlation, normalization; the list goes on. On the other hand, data visualization alone leaves a lot of assumptions and room for misinterpretation, so you need to back it up with statistics. The rest of this book will be about just that.

Summary

In this chapter, I discussed what Tableau is and listed several of its key products. Then I went over some key terms that I will use throughout the book when walking you through each tutorial. This foundational knowledge will be key in later chapters, especially if you are newer to Tableau.

Then I touched on some foundational statistical terms and ideas. After that, I tied those terms together with a practical case study. To introduce you to the idea of how statistics and data visualization come together, I showed you the Anscombe's quartet example.

In the following chapters, I will show you how to start incorporating statistical analysis into your data visualizations in Tableau. You will learn to visualize distribution of your data, detect outliers, forecast future values, create a cluster analysis, use regression to make predictions, and connect to external resources for more advanced statistical models.

If you have made it this far and still need some additional foundational practice, I would recommend the following books to become more familiar with Tableau Desktop and its capabilities:

- *Practical Tableau* by Ryan Sleeper (O'Reilly, 2018)
- *Tableau Desktop Cookbook* by Lorna Brown (O'Reilly, 2021)
- *Tableau Strategies* by Ann Jackson and Luke Stanke (O'Reilly, 2021)

CHAPTER 2

Overview of the Analytics Pane

In Chapter 1, I mentioned that implementing statistics into your work can be intimidating at first. Most developers think they have to master R and Python to effectively run statistical analysis on their data. While I do suggest picking up those skills (and cover briefly in later chapters of this book), I think the best place to start is directly within Tableau Desktop.

In this chapter, I will be introducing you to the Analytics pane in Tableau Desktop. The Analytics pane is a resource within Tableau that allows you to implement different statistical models to your work by simply dragging them into your view. I will show you the three sections available in the Analytics pane and go into all the available options in detail.

What Is the Analytics Pane?

Tableau Desktop makes bringing data visualization and statistics together extremely easy for you because it has incorporated statistical modeling directly in the tool. You can find these options accessible in the Analytics pane.

Accessing the Analytics pane in Tableau Desktop is pretty straightforward. From the authoring interface, navigate to the top left and you will find a toggle from the Data pane to the Analytics pane, which I have highlighted in Figure 2-1.

By clicking on Analytics from the toggle selection, the Data pane will switch to the Analytics pane, shown in Figure 2-2.

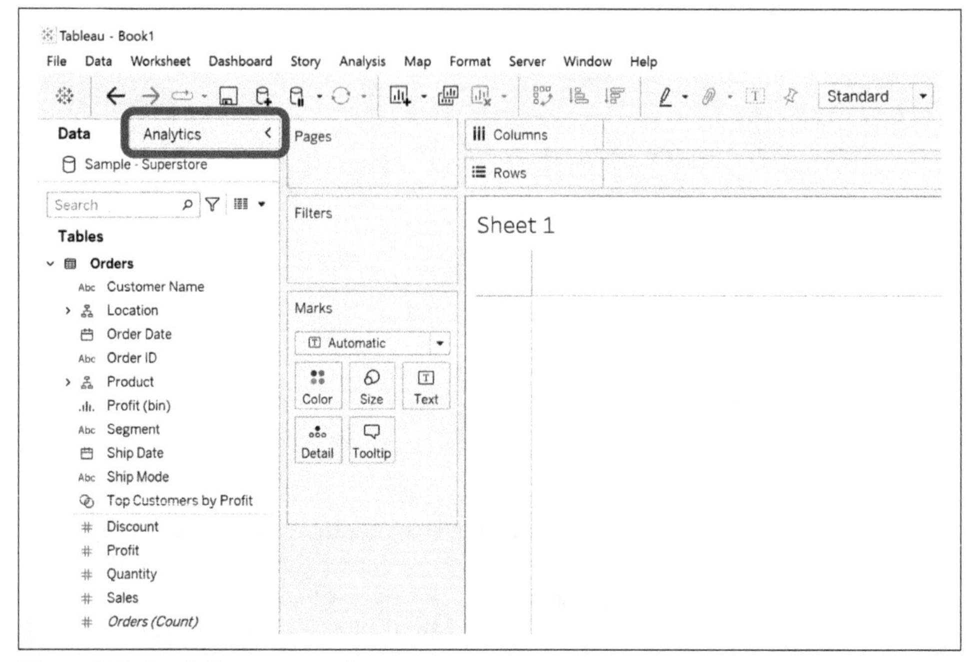

Figure 2-1. Analytics pane toggle

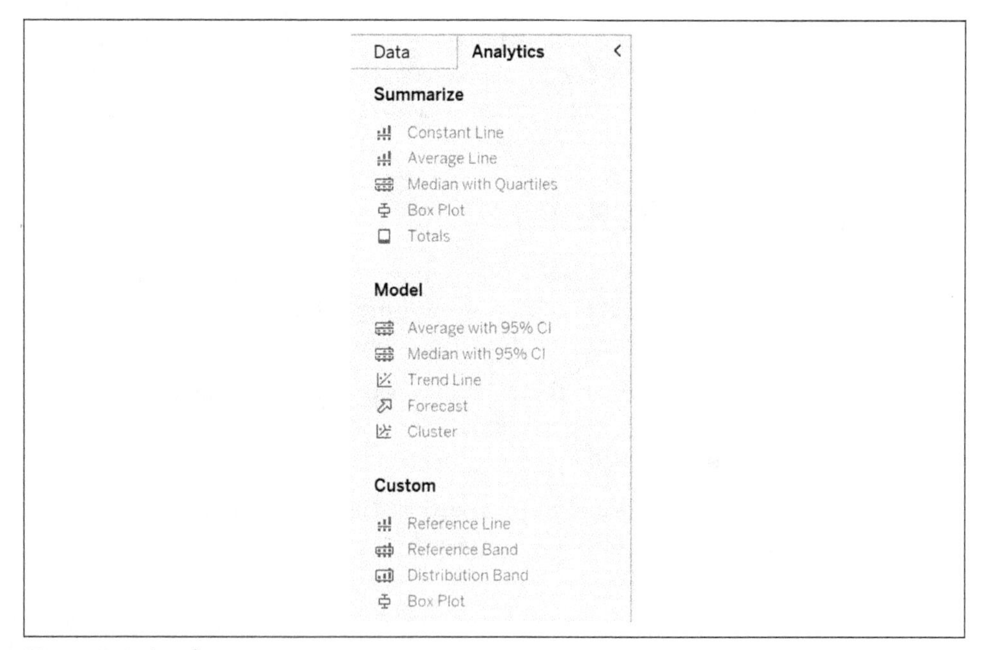

Figure 2-2. Analytics pane preview

Notice that all the options from the Analytics pane are grayed out in Figure 2-2 and not selectable. This is because the Analytics pane is designed to add more context to your visualization, and therefore we need to add measures and dimensions to the canvas first.

If you hover over any of the options that are grayed out, you will see a tooltip appear that will tell you what you need to have in the view to implement that option. As you can see in Figure 2-3, if I hover over "Median with Quartiles," it says I need at least one axis that uses a continuous measure or date.

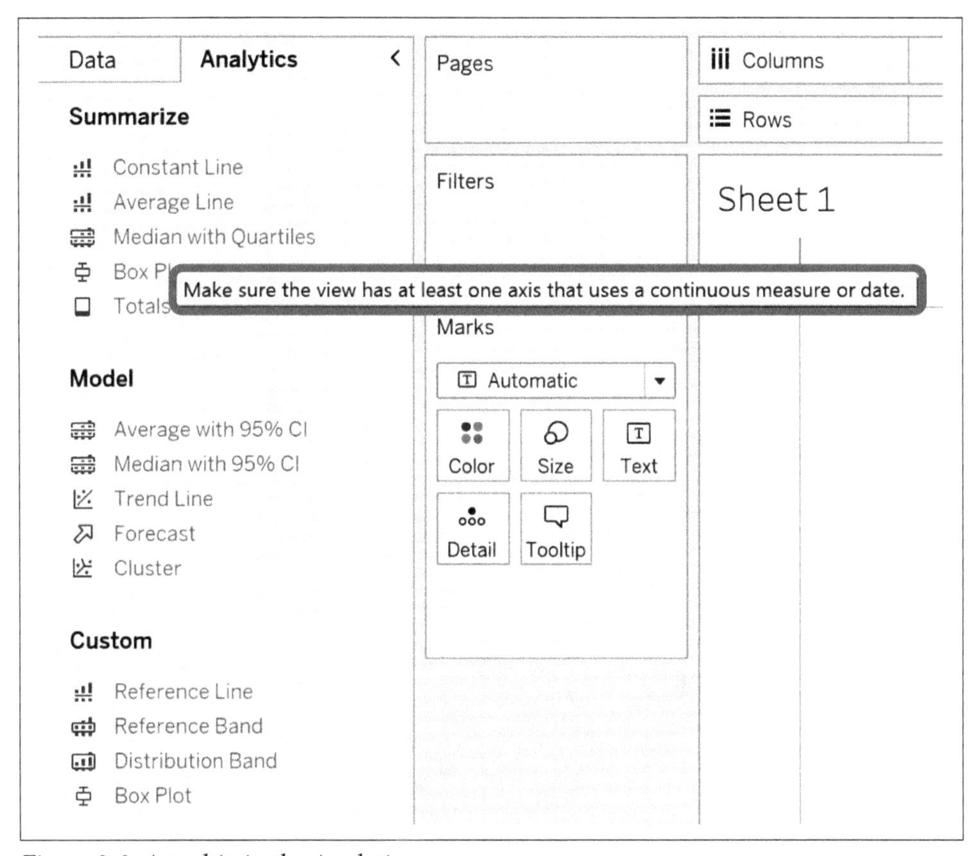

Figure 2-3. A tooltip in the Analytics pane

I am going to toggle back to the Data pane and double-click Sales. This will add SUM(Sales) to the Rows shelf and create a bar chart in the view, as shown in Figure 2-4.

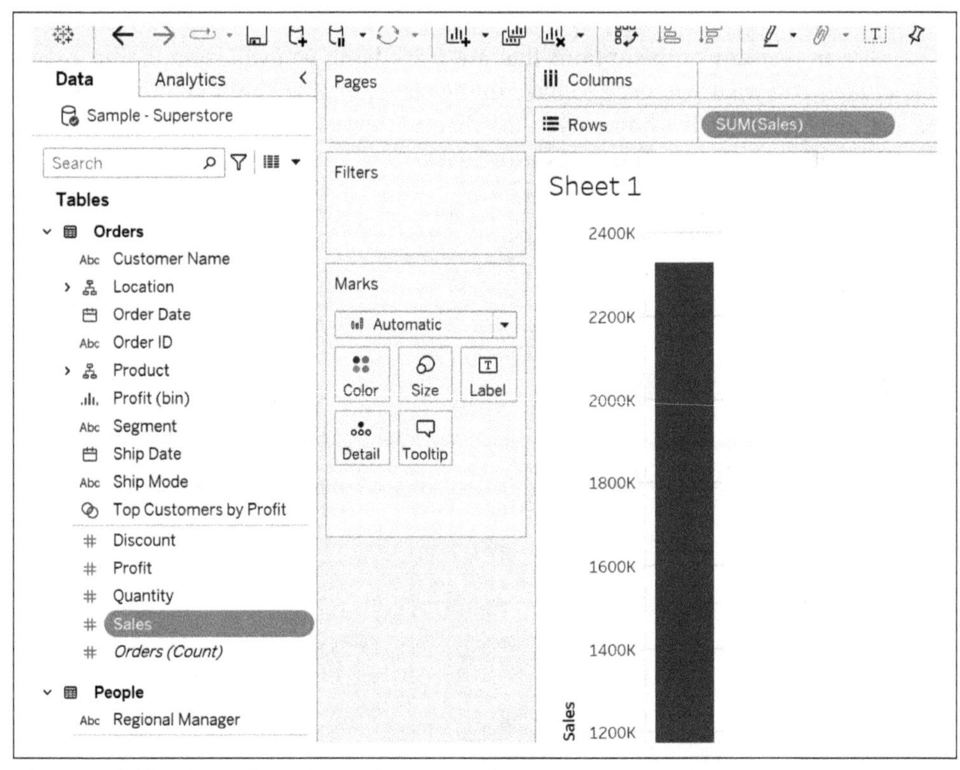

Figure 2-4. Creating a bar chart of the sum of sales

Now I will toggle back to the Analytics pane. You can see in Figure 2-5 that there are several options that are now bold. This means that the fields in the view meet the criteria needed to add those options from the Analytics pane to the view.

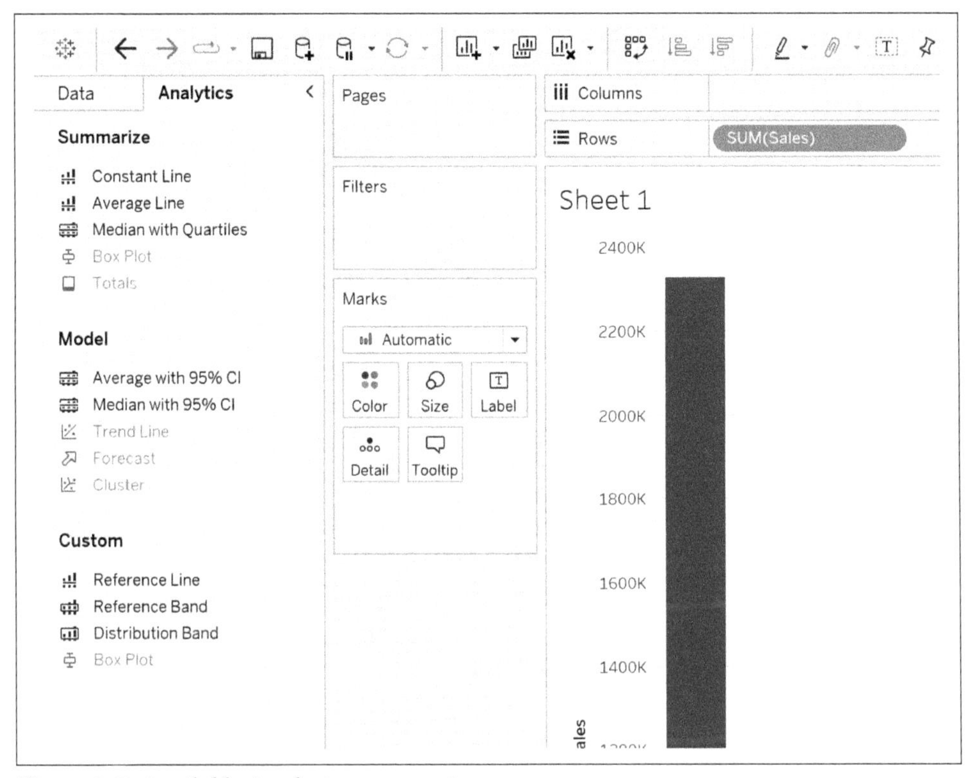

Figure 2-5. Available Analytics pane options

Before I take you through each option, let me first explain each section of the Analytics pane (as shown in Figure 2-6).

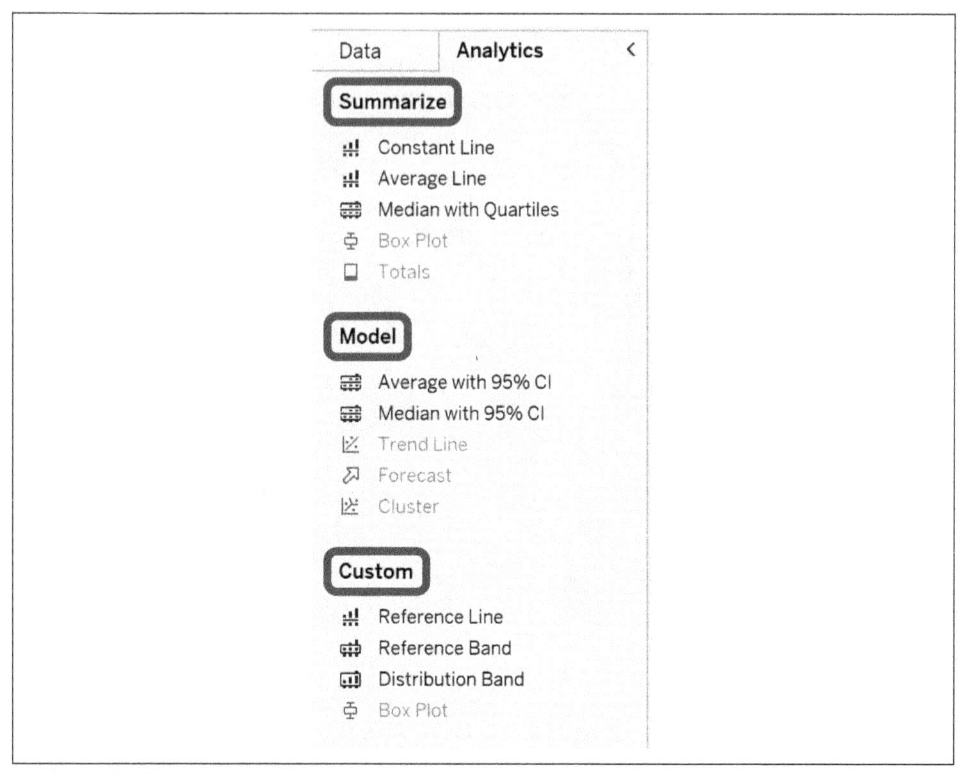

Figure 2-6. Three sections of the Analytics pane

Let's look at each section in more detail:

Summarize
: In this section, you will find quick ways to add reference lines, box plots, and totals to your views. As the name of the section suggests, these are great options to summarize the data in your view. However, there are many other uses for these options, which I will cover in the following chapters.

Model
: Within the Model section, you will find advanced statistical models that you can include in your data visualization. Regression models, forecasting, clustering, and confidence intervals can all be implemented from this section.

Custom
: This section gives you the ability to add reference lines, reference bands, and distribution bands into your view. Some of the options are similar to the Summarize section, but these selections are not predefined like the options in the Summarize section. That gives you the freedom to implement all kinds of other advanced statistical analysis into your view by customizing these options with calculated

fields and parameters. Later in this book, I will show you several practical use cases that you can implement using the Custom section.

As you can see, the Analytics pane offers you a lot of options to enhance your data visualizations with statistical analysis. In the next sections, I will go into more detail about each available option in the Analytics pane.

Implementing Summarization Options from the Analytics Pane

The Summarize section has some simple, predetermined selections that you can add to the view as reference lines. You can also create a box plot directly from this section. I will cover some important highlights in the chapter, but I will show you some more in-depth examples of how to use these options in other chapters.

To implement any of these options, simply click and drag the option into the view. You can see in Figure 2-7 that if I drag Constant Line into the view, there is a small window that appears at the top left of the canvas.

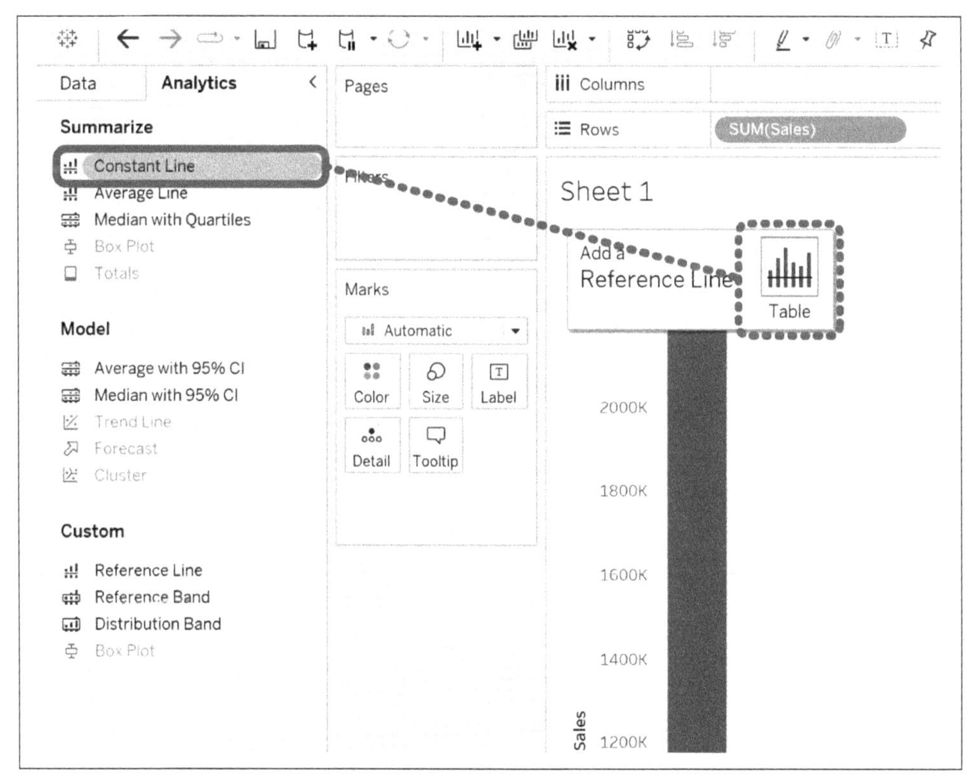

Figure 2-7. Adding Constant Line to the view

If I drop Constant Line on Table within that window, a new menu will appear and ask what value we want to use as our constant, as shown in Figure 2-8.

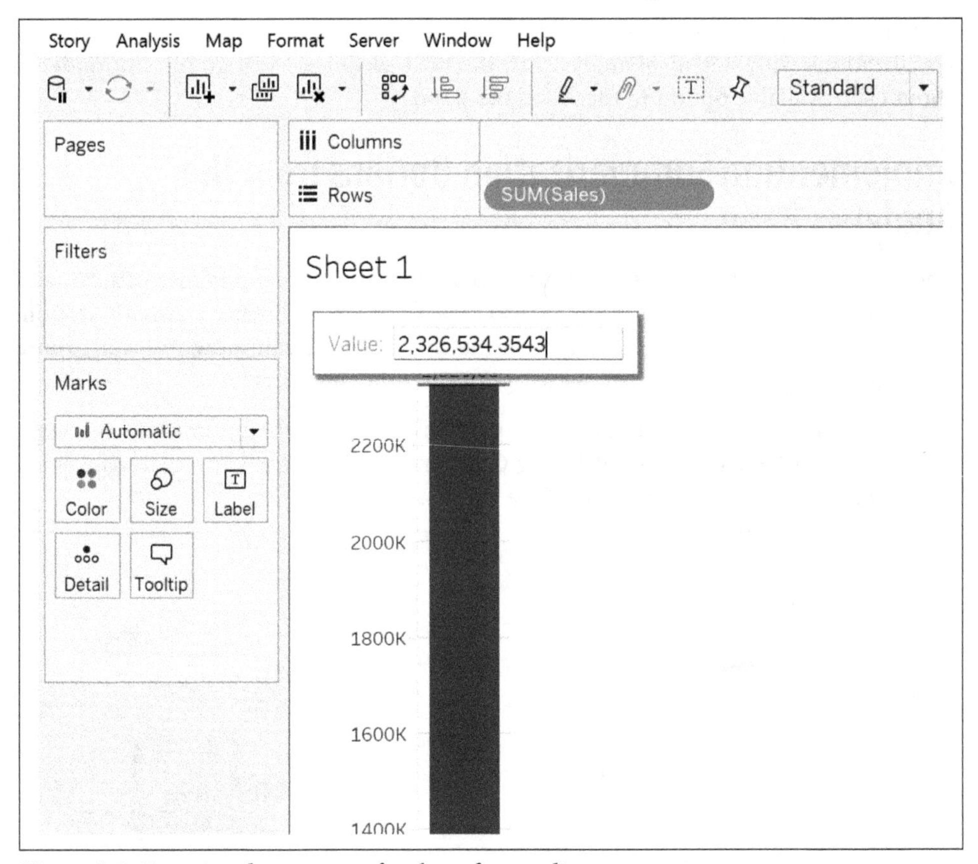

Figure 2-8. Entering the constant for the reference line

By default, Tableau Desktop uses the max value in the view as the constant. However, you can change this value to whatever you desire. As a demonstration, I will set the value to 2,000,000, as shown in Figure 2-9.

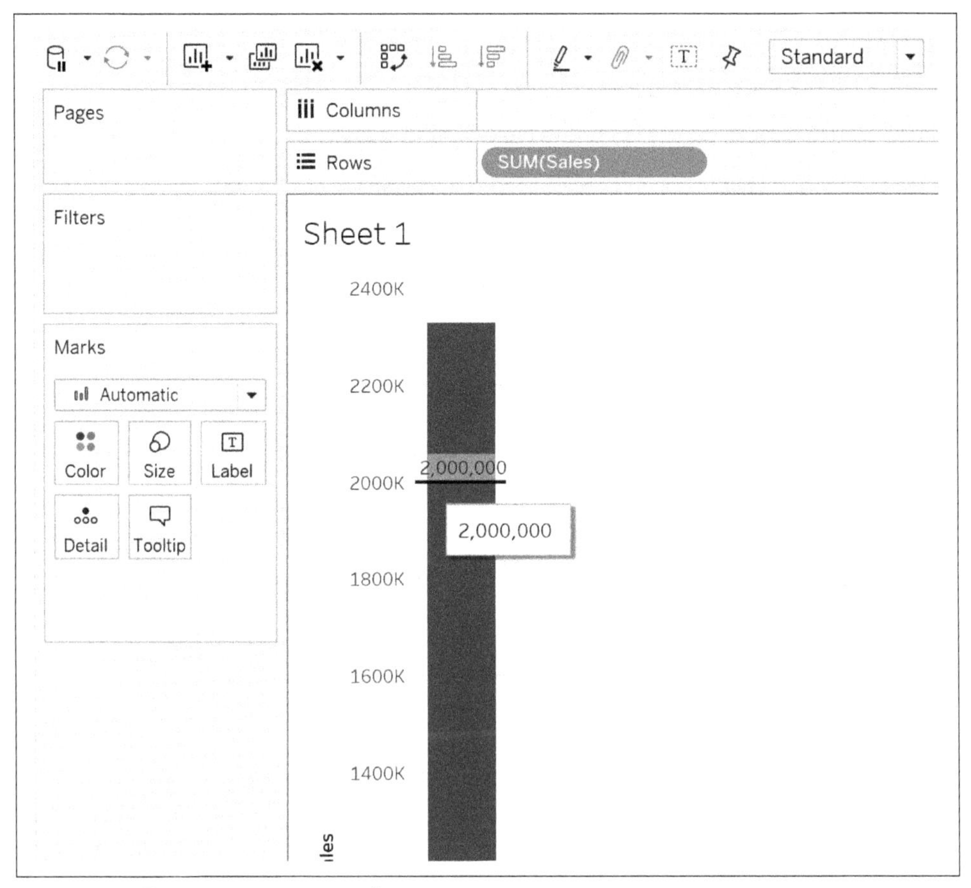

Figure 2-9. Entering a constant of 2,000,000

You can see that I now have a reference line that is drawn from the axis across the bar chart at that value. By default, Tableau Desktop formats that reference line to include a label and the value. You are able to format this label as well as the reference line itself. If you right-click on the reference line and select Edit from the list of options that appears, a new menu will open, as shown in Figure 2-10. If the reference line is too small to right-click, you can also right-click the axis and select Edit Reference Line from the menu.

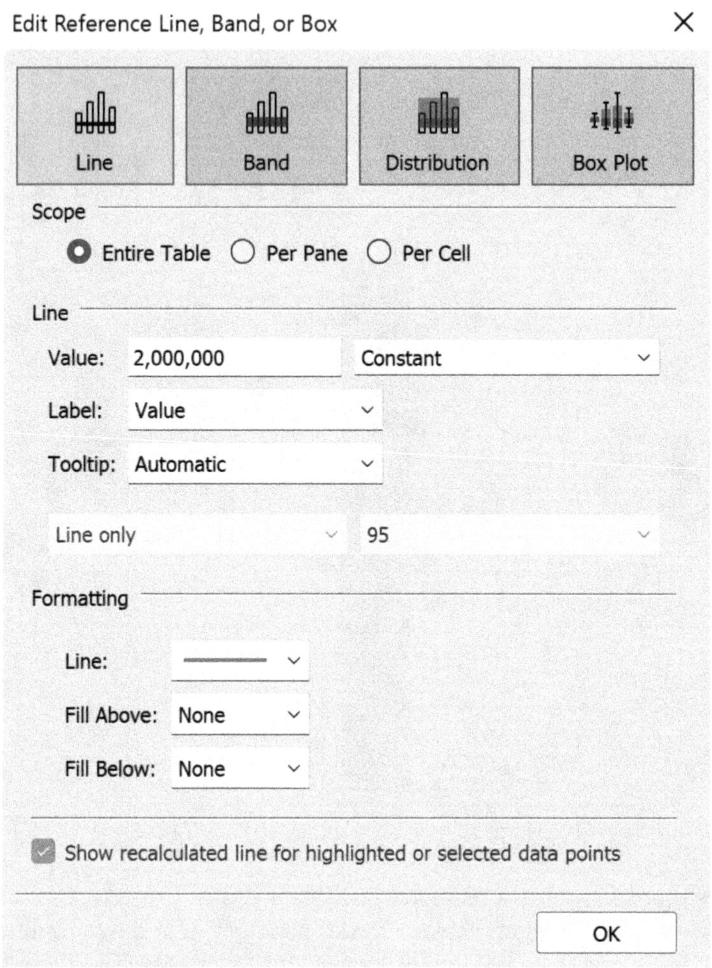

Figure 2-10. Format options for the reference line

You can see within the menu that there are three distinct sections from top to bottom—Scope, Line, and Formatting:

Scope
: You can see that there are three radio buttons here: Entire Table, Per Pane, and Per Cell. This is where you can change the level of detail the reference line will calculate at. For a constant line, this doesn't matter much, but when you are adding a reference line for the average or median, these options will change the

level of detail at which those statistics are calculated. Here is a brief description of each option:

Entire Table
: This will draw the reference line across the entire table.

Per Pane
: The line will be calculated at differing levels for each dimension value 1 lower than the whole table.

Per Cell
: The line will be calculated at each cell.

Line
: In this section, you will see three distinct options: Value, Label, and Tooltip. This section will affect what is displayed in the label. Here is a brief description of the available options in this section:

  Value
  : This is where you can change the value itself for the constant. You can also choose an aggregation from the drop-down to the right of the text box where you enter the value (shown in Figure 2-10) as Constant. The available aggregations are Average, Median, Minimum, and more.

  Label
  : The label option allows you to choose what you want displayed in the label. Currently, the Value option is selected on the drop-down menu, but there are other options, such as Custom Label, that allow you to type in a custom message for the label.

  Tooltip
  : This is what will be displayed in the tooltip if a user hovers over the reference line.

Formatting
: The formatting section is where you can change how thick the reference line is, its color, and its opacity. You can also add shading (fill) above or below the line. After the fill is incorporated, you have the option to edit how it looks as well:

  Line
  : This is where you can edit the way the reference line itself is formatted. You can change its color, opacity, and thickness from the menu.

  Fill Above
  : Under this option, you can add or edit the fill. As the name implies, this will add fill to the view above the reference line.

Fill Below

Under this option, you can add or edit the fill. This will add fill to the view below your reference line.

For now, it's good to know these options exist and what's available to you. In the following chapters, I will show you more in-depth tactics you can apply using these options that will help highlight your analysis.

To show you another example from the Summarize section of the Analytics pane, I will create a new sheet and then double-click Sales and Segment. Next, I will drag Customer Name onto the details property of the Marks shelf and change the mark type to circle. You can reference the way the view is set up in Figure 2-11. This visualization represents the sales by segment at the customer level of detail. This means that each mark or circle in the view represents the sales of a specific customer in that product segment.

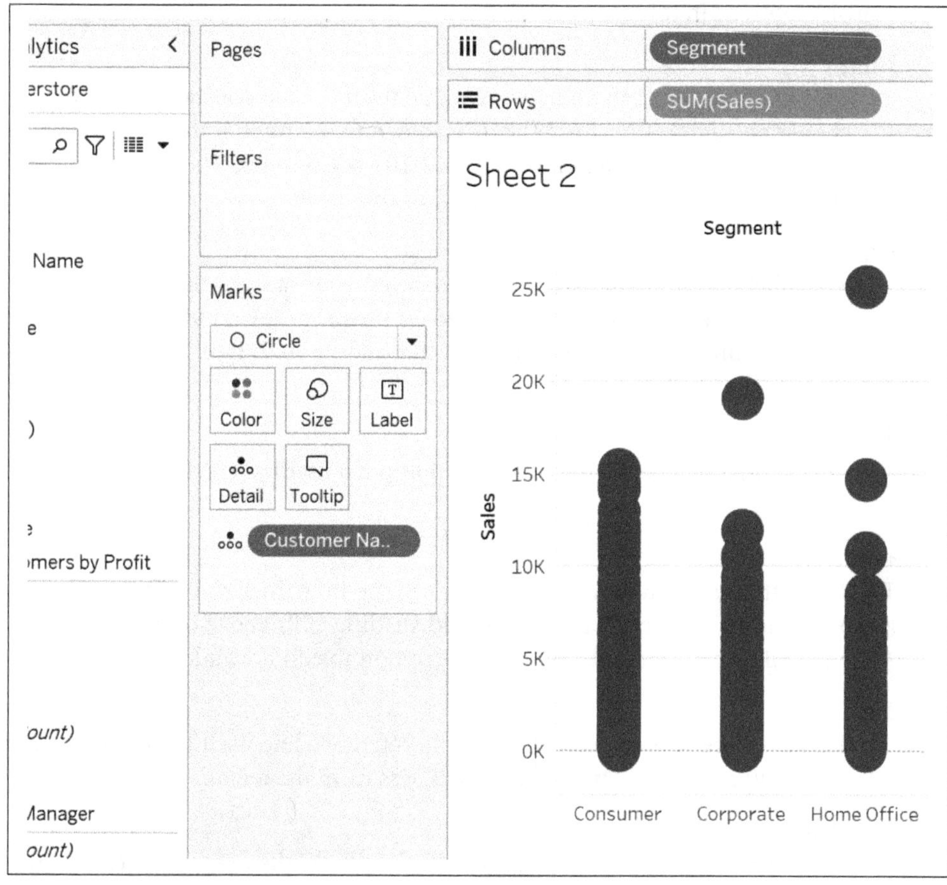

Figure 2-11. Scatter plot of customer sales by product segment

With this setup, I meet the criteria necessary to build a box plot using the Analytics pane—you will now see it in bold in the Analytics pane. A box plot can be used to find outliers within your data and analyze the distribution of your data by different categories.

I will toggle from the Data pane to the Analytics pane and drag Box Plot onto the view. You can see in Figure 2-12 that an option will appear in the top left of the canvas that says, "Add a Box Plot." Drop the Box Plot option onto Cell within this pane.

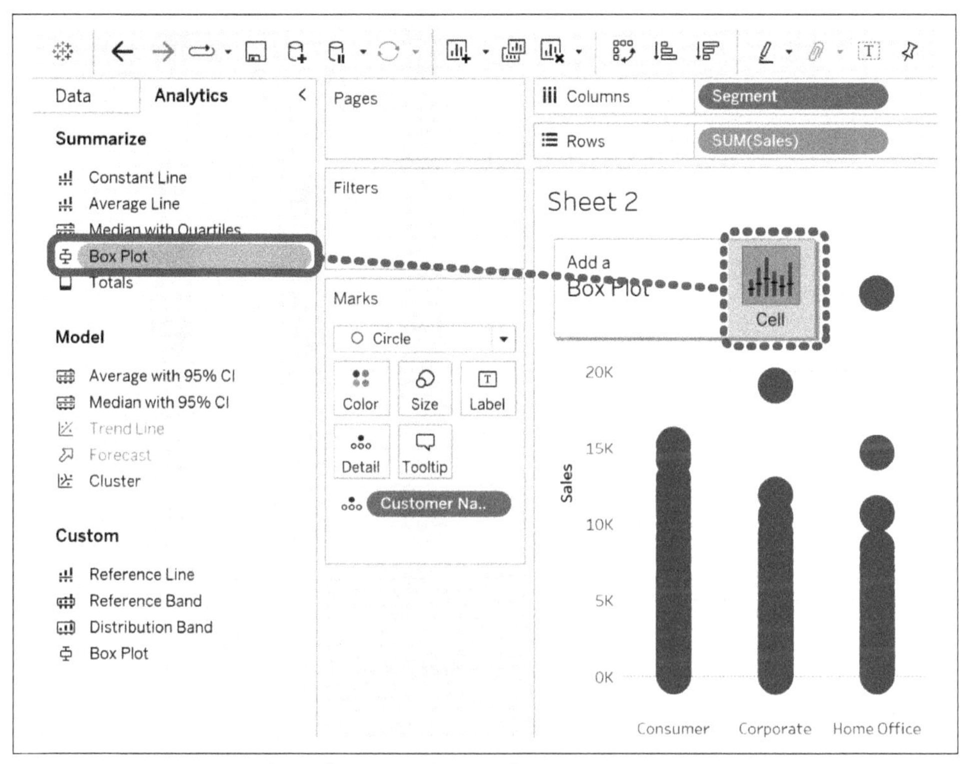

Figure 2-12. Creating a box plot using the Analytics pane

Once you incorporate the box plot on your view, you will end up with a data visualization shown in Figure 2-13.

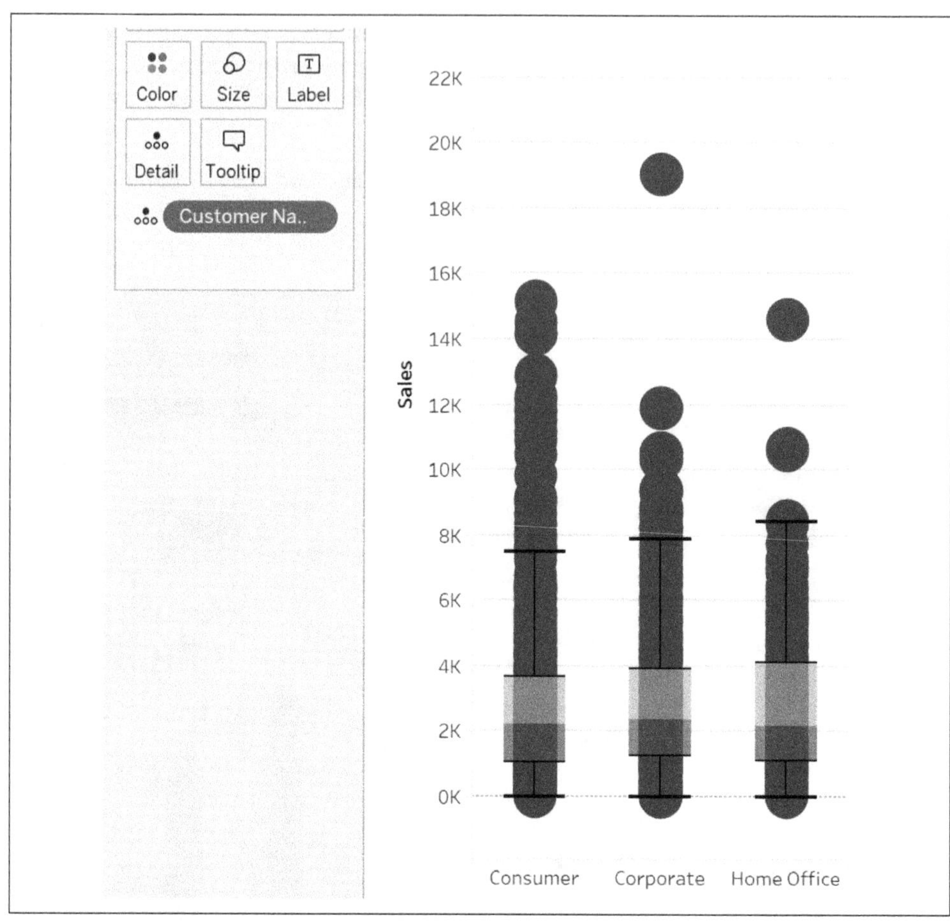

Figure 2-13. Box plot of customer sales by product segment

Let's improve the way this is displayed. Click the Size property in the Marks card and adjust the slider to the left. This will decrease the size of the marks in the view, allowing you to hover over each mark and see information about individual customers more easily (see Figure 2-14).

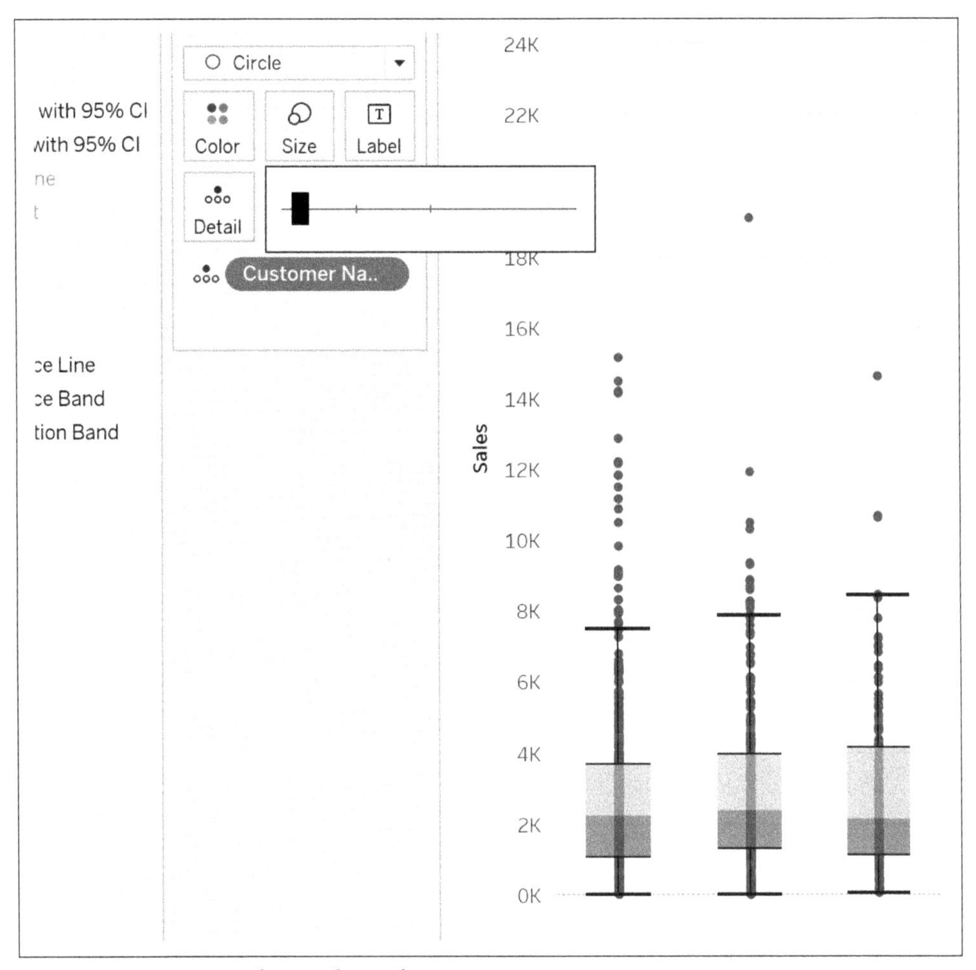

Figure 2-14. Resizing the marks in the view

This is another great view that gives you a powerful statistical analysis. For each segment in the chart, the box that is shaded is where ~50% of your data is. The dividing line between the shades is your median. You can see how each segment has a slightly different distribution using this chart. I will go more in depth into all the details of this chart and what each part of it represents in Chapter 6.

Implementing Model Options from the Analytics Pane

Within this section of the Analytics pane, you will find several powerful predictive models. This chapter covers the options you have available to you. However, in later chapters, I will circle back and cover these models in more detail so you have a better understanding of how they work.

Since these predictive models are used to predict what will happen in the future, most of them work best when you are visualizing time-bound data. With that said, I will create a new sheet and double-click Sales and Order Date. Just as in Chapter 1, this creates a line chart of sales by year of order date, as shown in Figure 2-15.

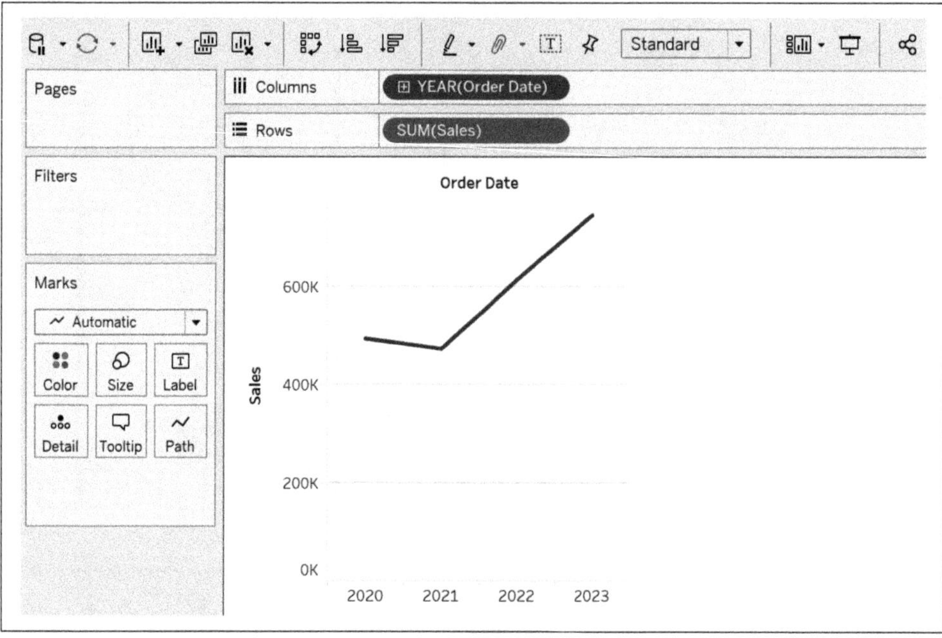

Figure 2-15. SUM of sales by year of order date line chart

Next, I will change the level of aggregation of Order Date to be a continuous month, allowing it to be displayed on an axis. To do this, right-click on YEAR(Order Date) in the Columns shelf and then select the continuous Month option, as shown in Figure 2-16. Note that there are two Month options in the menu. The first is a discrete Month that will show January, February, etc. The second is the continuous Month, which doesn't roll the months together; instead, it allows them to flow continuously on the axis as January 2023, February 2023, etc.

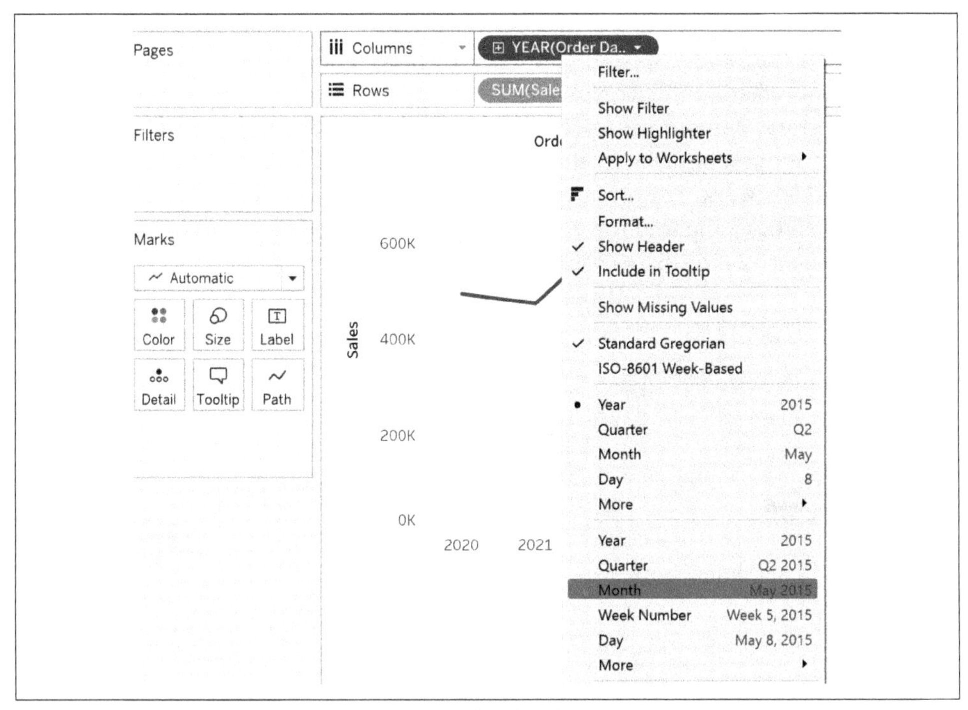

Figure 2-16. Selecting continuous Month of order date

This will give you a line chart of sales by month and year of order date. You should end up with a view that looks like Figure 2-17.

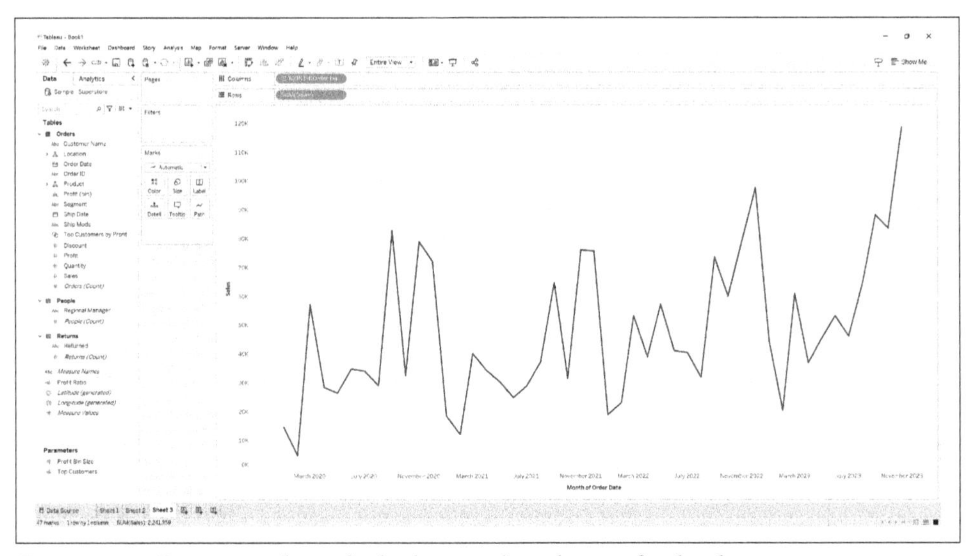

Figure 2-17. Continuous line of sales by month and year of order date

If we toggle to the Analytics pane now, we can see this view meets the criteria for each option within the Model section, and each of them is displayed in bold. To briefly explain to you what's possible, you can add confidence interval models to the view: linear regressions, polynomial regressions, forecasts, and clustering. I will be covering each of these models in detail in the following chapters so that I can explain how the statistical models work. However, to show you some capabilities, I will drag Forecast from the Analytics pane into the view. This will add a predictive model that forecasts sales for the next 12 months, as shown in Figure 2-18.

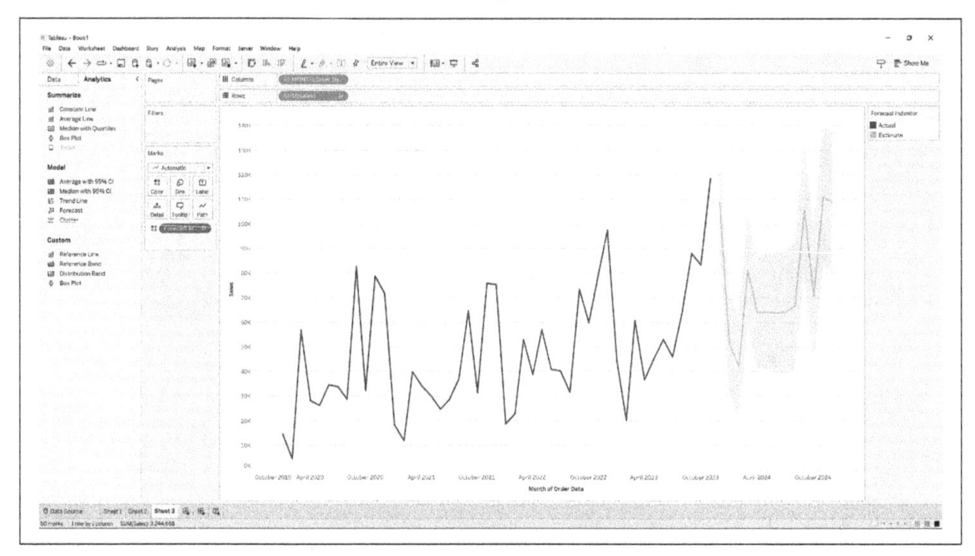

Figure 2-18. Adding the Forecast model in the view

As you can see, the model is highlighted using the Color property of the Marks shelf. Again, my intention with this chapter is to introduce you to the Analytics pane. I will cover each model in detail in the next chapters so that you have a better understanding of each of them.

Implementing Custom Options from the Analytics Pane

The options available in the Custom section of the Analytics pane are by far my favorites and nearly limitless in terms of what you could implement. To show you an example, I will use the line chart from the previous section and drag Reference Line from the Analytics pane. You will see a new menu appear at the top left of the canvas, as shown in Figure 2-19.

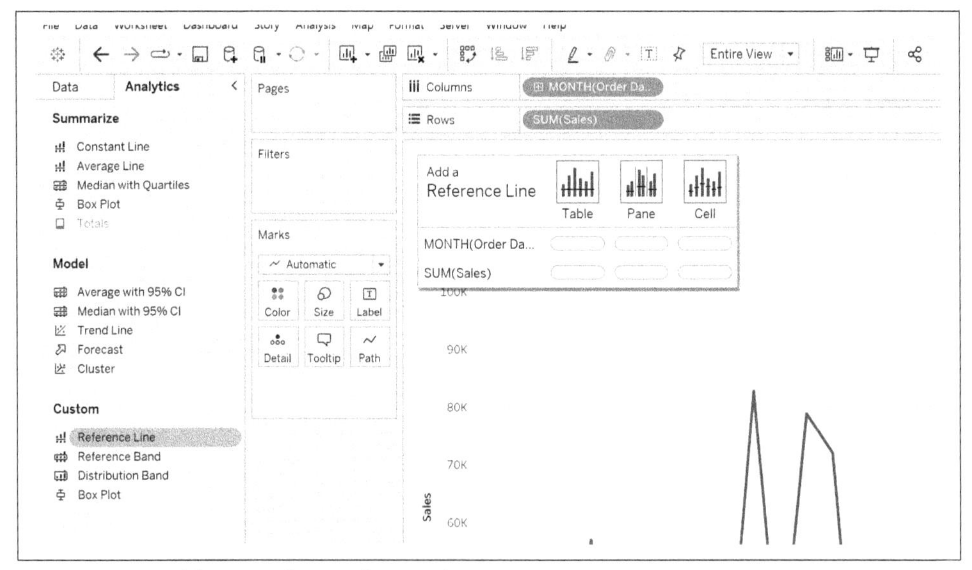

Figure 2-19. Adding a reference line to the view

The top options are the scope you want to use to compute the reference line. You can calculate the reference line across the entire view, by separate panes, or by cell—in the same way that we did for the Summarize section. Under the three options of scope are the measures and dimensions we have in the Columns and Rows shelves. You can see there are six rounded rectangles that create a grid. These rounded rectangles are referred to as pills, and if you drag the reference line over these pills, they will be highlighted. This gives you the ability to draw the reference line on the x-axis, y-axis, or both. This gives you maximum flexibility.

To demonstrate, I will drag the Reference Line option and drop it on the SUM(Sales) pill under the Table scope. You will see the menu appear that I covered in the Summarize section. Again, this menu gives you a full range of capabilities, such as adjusting formatting, choosing different values from your view, changing how the value is aggregated, and so on. This menu also allows you to switch between each of the Custom section options. At the top of the menu, you can see Reference Line, Band, Distribution, and Box Plot. Each of these options has its own menu. You can reference each menu and the options available to you in each from Figure 2-20.

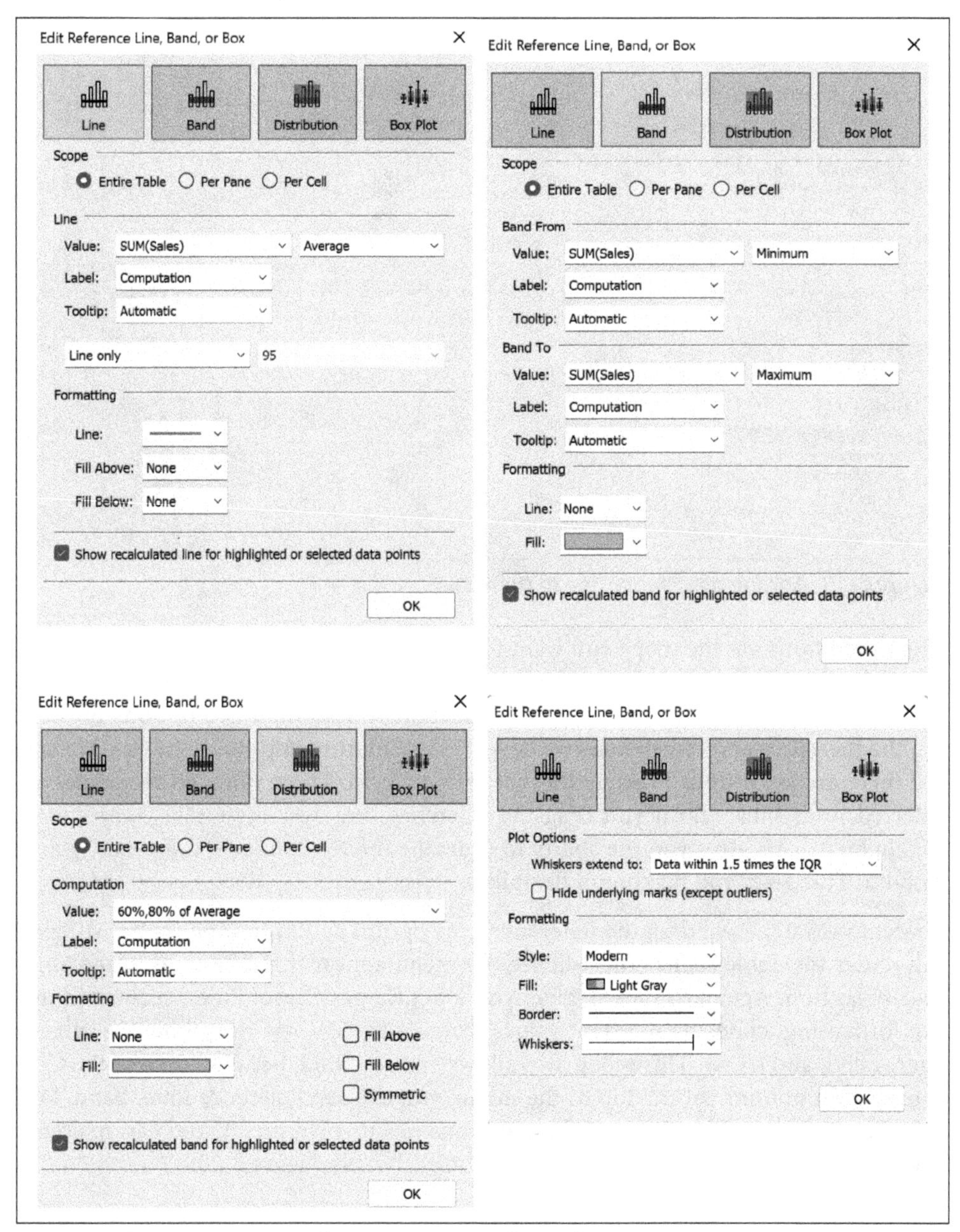

Figure 2-20. Custom section formatting menu for each option

I will cover these options in detail in Chapters 5, 6, and 7 but recommend you get familiar with them and what's available to you.

Summary

In this chapter, I introduced you to the Analytics pane within Tableau. I showed you the three sections within the Analytics pane: Summarize, Model, and Custom. I also showed you some basic examples of different functionalities within the three sections of the Analytics pane. I will be building upon this foundational material as you progress through the book.

CHAPTER 3

Benchmarking in Tableau

Oftentimes, your stakeholders will want you to incorporate benchmarks, targets, or goals into your dashboards. This helps give context to how the business is performing. While benchmarks can be a powerful variable to add to your data visualizations, they can also be a crutch for your stakeholders if implemented for the wrong reasons. As a developer, you have to think through the science behind a benchmark when implementing them. This is another opportunity for you to implement some simple statistics that can help guide the business in making better decisions.

In this chapter, I will explain what benchmarks are and show you some different types of benchmarks you can implement. I'll also discuss how to implement benchmarks in your data visualizations using Tableau.

What Is a Benchmark?

Benchmarking can be a business's North Star that helps them grow, adapt to changing conditions, and stay ahead of their competition. Benchmarking is the process of comparing a product, service, process, or performance against recognized industry standards or best practices. It involves measuring and evaluating performance against a set of predetermined criteria to determine strengths, weaknesses, and areas for improvement. There are two main ways to conduct benchmarking: internal and external.

Internal Benchmarking

Internal benchmarking is the process of comparing the performance of different departments, units, or teams within the same organization to identify best practices, areas of improvement, and opportunities for cost savings. It involves comparing key performance indicators (KPIs) such as productivity, quality, customer satisfaction, and efficiency across different departments or units.

Internal benchmarking can be a useful tool for organizations that have multiple departments or locations performing similar functions. By comparing the performance of these departments or locations, an organization can identify areas where improvements can be made and best practices can be shared across the organization.

Internal benchmarking typically involves collecting data on KPIs from different departments or units and analyzing the data to identify areas for improvement. The data can be collected through qualitative research such as surveys, questionnaires, focus groups, or interviews, or through quantitative data collection such as sales volume, customer satisfaction scores, or the number of products manufactured. These internal benchmarks can then be used as targets by business leaders to monitor through a performance dashboard.

Internal benchmarking can be a cost-effective way for organizations to improve their performance and efficiency, as it does not require the resources or expertise needed for external benchmarking. Additionally, it can create a culture of continuous improvement within the organization.

External Benchmarking

External benchmarking is the process of comparing an organization's performance, processes, or practices against those of other companies or organizations in the same industry or sector. The aim of external benchmarking is to identify best practices, areas for improvement, and opportunities for innovation.

External benchmarking typically involves gathering information about competitors or industry peers, analyzing their practices and performance, and then comparing them to one's own organization. This can be done through various methods such as surveys, interviews, market research, acquisition, and publicly available data.

External benchmarking can provide valuable insights and new ideas for an organization and can help to identify areas where the organization may be lagging behind competitors or industry best practices. By learning from others in the same industry, organizations can improve their own performance and better meet the needs of their customers.

External benchmarking can be a valuable tool for organizations to stay competitive and continuously improve their performance, as it provides an objective and data-driven way to identify areas of improvement and best practices.

Implementing Benchmarks in Tableau

Now that you understand what benchmarking is, let's dive into several ways you can add benchmarks to your data visualizations in Tableau.

Static Reference Line

Within the Analytics pane you will see an option to add a constant line. This is a simple approach to implementing a benchmark, but it is very effective. This tactic will work for both external and internal benchmarks.

To demonstrate, I have created a view that looks at the average discount by month of order date, shown in Figure 3-1.

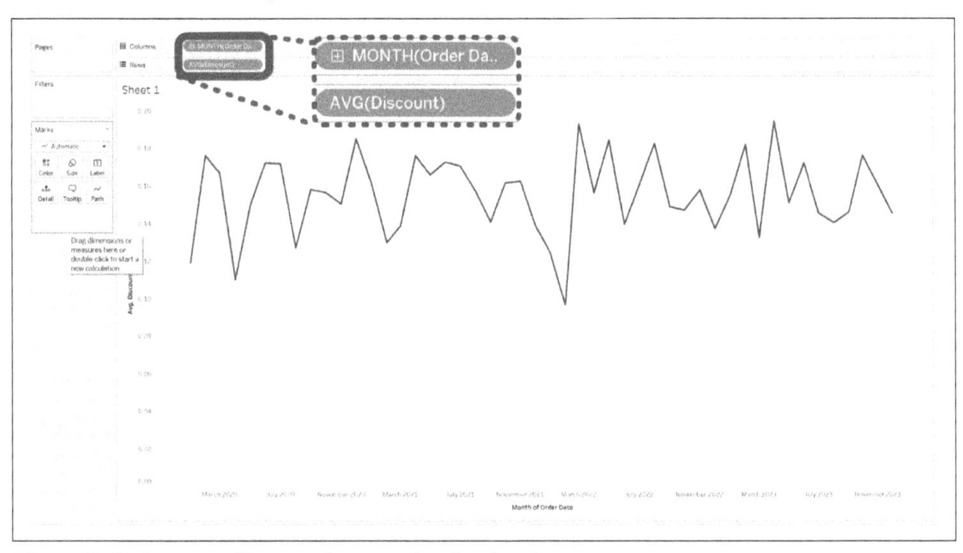

Figure 3-1. Average discount by month of order date

The goal here is to add a benchmark that will allow the user to see whether each period is above or below the benchmark. The first thing you'll do is toggle to the Analytics pane and drag Constant Line onto the view. You will be presented with one option in the upper left of the canvas that says, "Add a Reference Line." You have the option to incorporate that line in either the y-axis or x-axis, or both. For this example, you would add it to the y-axis, as shown in Figure 3-2.

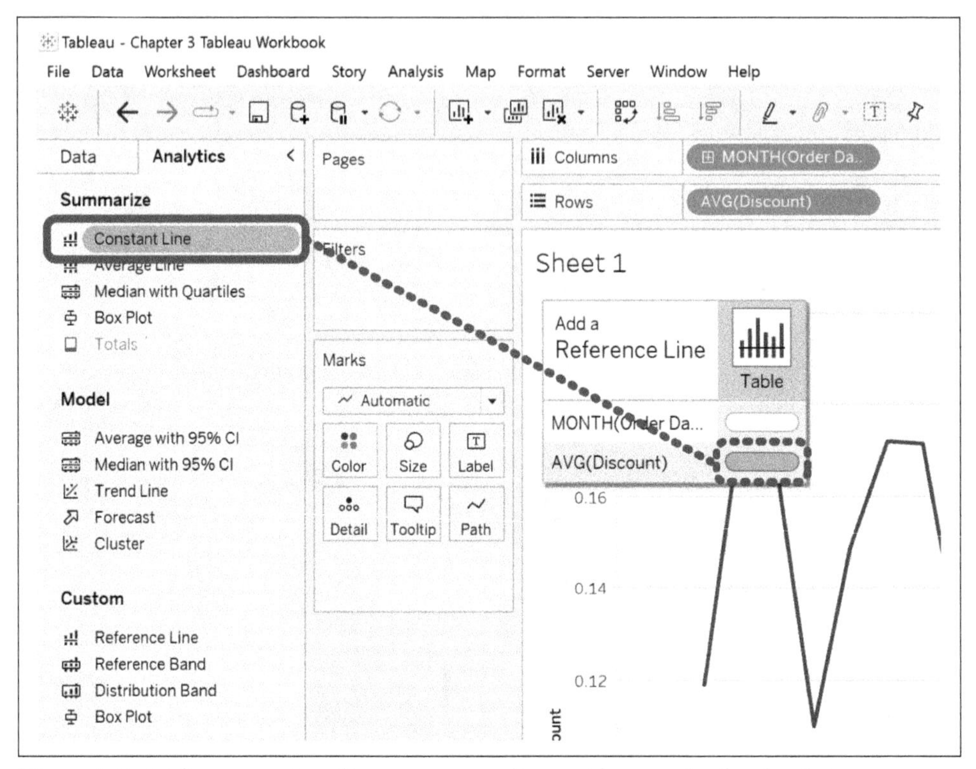

Figure 3-2. Adding a constant line to average discount by month of order date

When you drop this option onto that pill, you will see a new window appear that will let you type in the constant value of the reference line. Let's say that the business has attained an internal benchmark of 18%. You would type 0.18, since Tableau recognizes percentages as decimal numbers, into the box and press Enter. This will draw a reference line across your view at 18%. You can now easily see any period that was above or below this benchmark, as shown in Figure 3-3.

Once a constant line is added, the users of this tool will be able to compare each month's actuals versus the static benchmark.

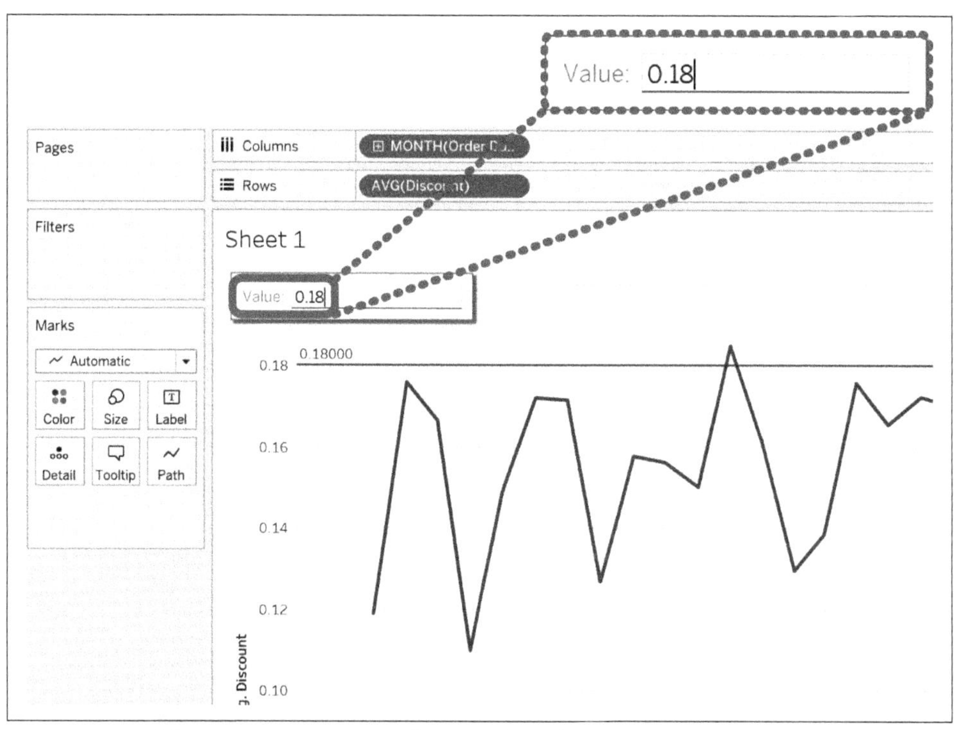

Figure 3-3. Adding a 0.18 constant value benchmark to the view

Dynamic Reference Line Using User-Driven Parameter

Using a static reference line to add in a benchmark is a simple and effective approach. However, what if you needed a more dynamic solution? In this section, I will show you how to create a dynamic reference line that the user can change on the fly. This gives them complete control of what value is displayed by the reference line.

To begin, I will create a line chart of average discount by month of order date, as shown in Figure 3-4.

Next, I will create a parameter by right-clicking in the Data pane and selecting Create Parameter from the drop-down menu that appears, as shown in Figure 3-5.

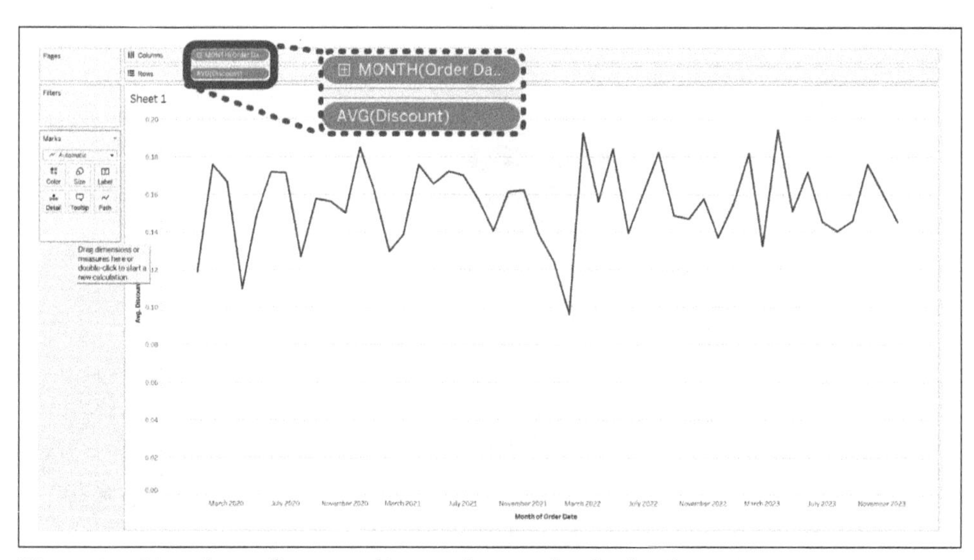

Figure 3-4. Line chart of average discount by month of order date

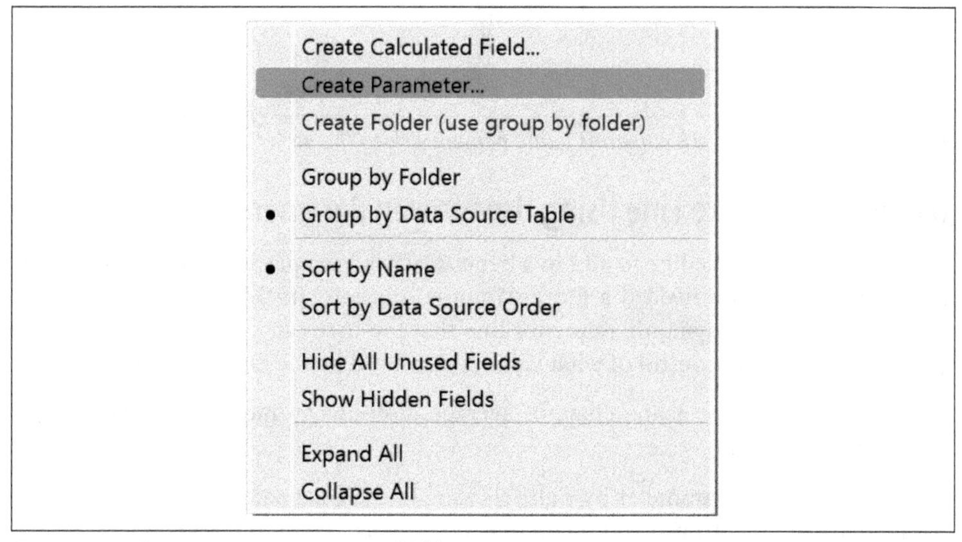

Figure 3-5. Creating a parameter in Tableau

In the Create Parameter menu, I will name it User Benchmark, keep the Data type as Float, and enter the current value at 0.18, as shown in Figure 3-6. In this example, I want the data type to be Float because Tableau recognizes decimals as percentages. Since we are analyzing a discount percent, I need the data type to be Float so the users can enter a decimal value. You can change the data type to whatever is best for your analysis.

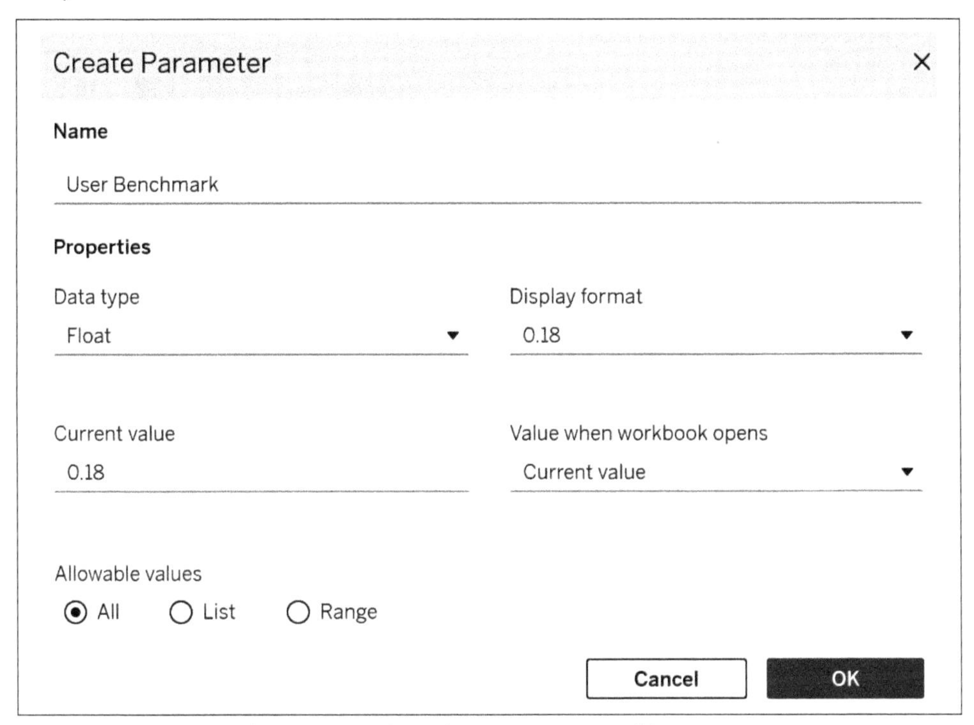

Figure 3-6. Creating the User Benchmark parameter in Tableau

With the parameter created, you need to add this benchmark to the view. To do this, toggle to the Analytics pane and drag Reference Line from the Custom section into the view, as shown in Figure 3-7. Since this reference line is a benchmark for average discount, you want to drop the reference line at the Table scope and in the y-axis pill next to AVG(Discount). Again, it is important to note that you may want your benchmark at a different scope or axis.

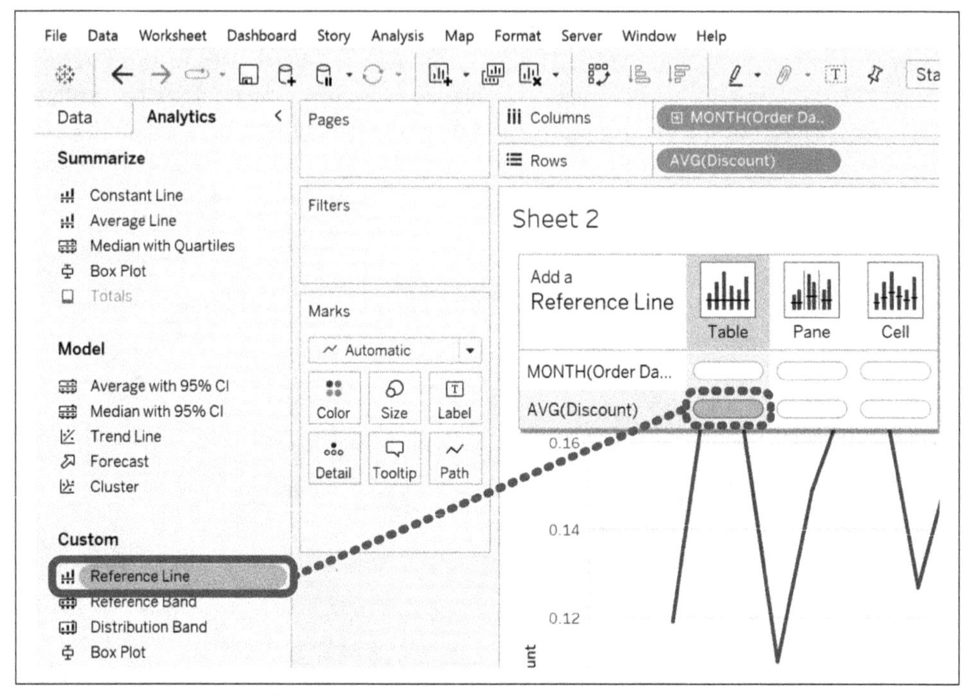

Figure 3-7. Dragging Reference Line onto the view from the Analytics pane

Once you drop the Reference Line option to the correct pill, you will see the "Edit Reference Line, Band, or Box" menu appear. In this menu, use the Value drop-down and select User Benchmark from the list of available options, as shown in Figure 3-8.

Figure 3-8. Selecting User Benchmark from the Value drop-down

As you can see in Figure 3-8, the view is already updated based on my selection. The User Benchmark parameter is a value that the users can change themselves if you add it to the view before publishing. To add the parameter to the view, simply right-click the parameter in the Data pane and select Show Parameter. You will see a text box appear to the right of the canvas that allows you to change the value of the parameter to anything you want, as shown in Figure 3-9.

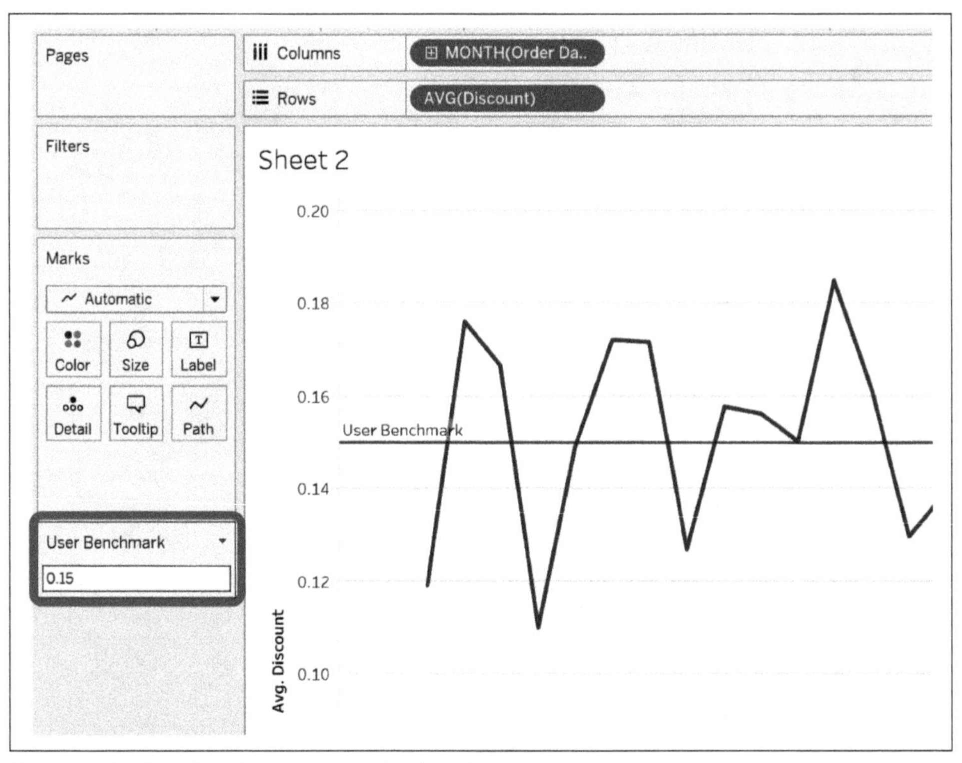

Figure 3-9. Showing the parameter in the view

You can see that I edited the User Benchmark parameter to 0.15 and the reference line dynamically changed along with it.

Dynamic Reference Line Using Level-of-Detail Calculation

In this section, I want to show you a more advanced feature that can be used to implement dynamic benchmarks. That feature is Tableau's Level of Detail (LOD) functions. LOD functions can be used to compute calculated fields at a different level-of-detail than what is in the view.

To set the stage, let's say your business wants to identify the region with the lowest average discount and add that statistic as a benchmark in our view. To achieve this, you would create a calculated field and enter the calculation from Figure 3-10.

In the calculation, I used a FIXED Level of Detail function to find the region that has the minimum average discount and fixed this calculation to that value. To show you what I mean, I created a small text table that shows average discount by region in Figure 3-11.

Figure 3-10. Calculated field for the lowest average discount by region

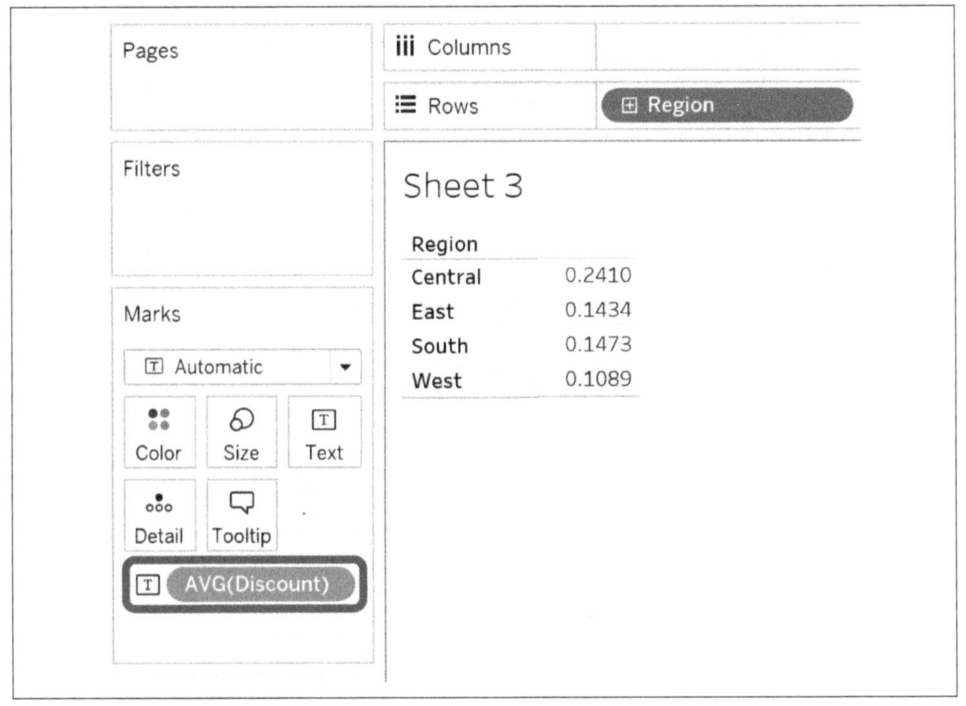

Figure 3-11. Average discount by region quality check sheet

You can see that the lowest average discount is 0.1089 or 10.89% in the West region. The calculated field will return that same value and appear as a benchmark in your view after you add the calculation as a reference line.

To continue, create the same line chart using average discount by month of order date. The first thing you would do is add the calculated field you just created to the Detail property in the Marks card. Then toggle to the Analytics pane and drag Reference Line from the Custom section to the y-axis at the Table scope, as shown in Figure 3-12.

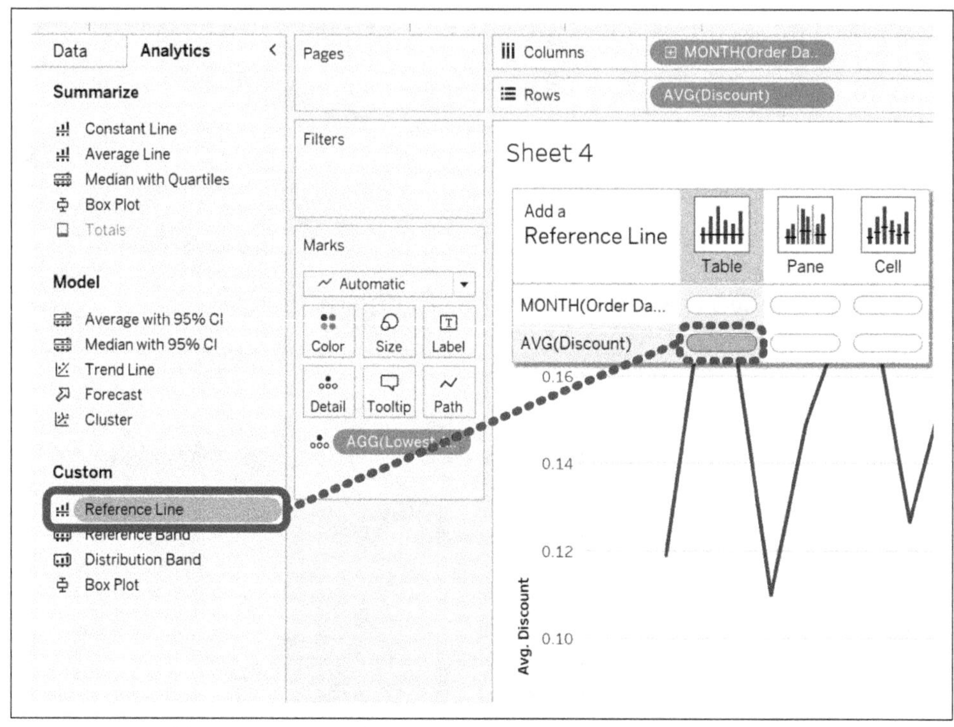

Figure 3-12. Adding a custom reference line to the view

When you add the reference line, you will see a new menu appear (see Figure 3-13). Within this menu, you need to select the calculated field you added to the Detail properties and select OK. The calculated field I created is named Lowest Average Discount by Region. I will choose that from the Value drop-down within the menu and click OK.

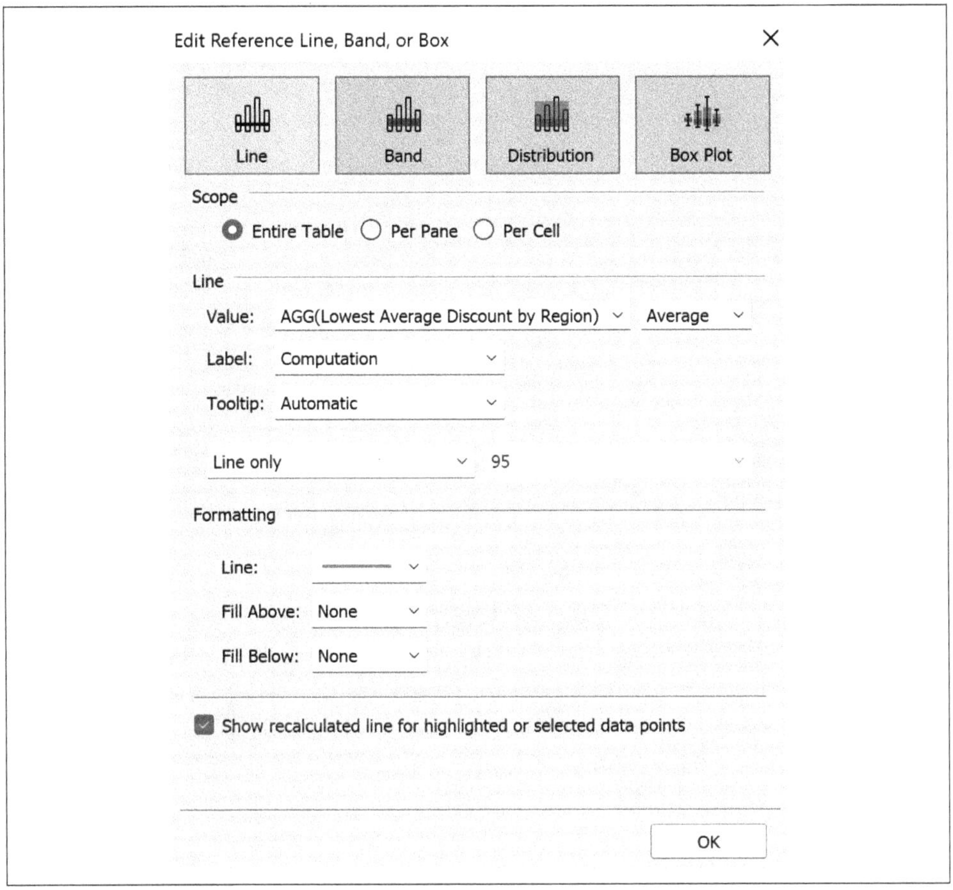

Figure 3-13. Setting the reference line parameters

You can see in Figure 3-14 that the lowest average discount by region is 10.89%. This means that the FIXED level-of-detail calculation is working like I want.

What's great about this tactic is that the reference line will automatically update as new data comes in. So if another region takes action and begins to outperform the West region, the calculation will update to the next lowest average discount by region.

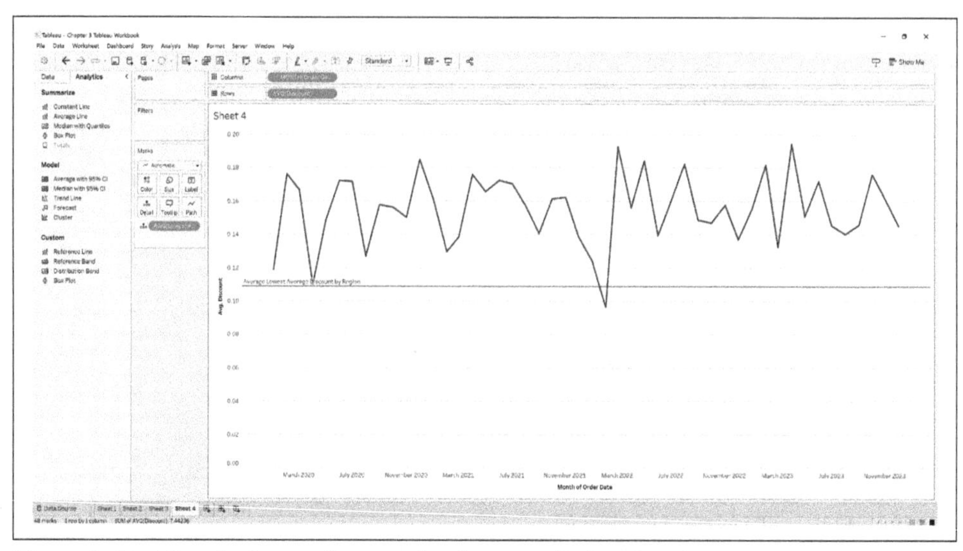

Figure 3-14. After the lowest discount has been applied to the view

Remember, this is only one example of how to apply this tactic using level-of-detail calculations. These functions within Tableau are extremely flexible, so you can manipulate them to fit your needs. Additionally, while this example in particular relates to an internal benchmark, you could also use this tactic with external benchmarks. If you had a dataset for external benchmarks, you could use level-of-detail calculations to fix the benchmark at comparable regions, for instance.

Median with Quartiles

Another great way to incorporate benchmarks using the Analytics pane is to use the Median with Quartiles option from the Analytics pane. This tactic adds a reference line that marks the median of your dataset as well as the upper and lower quartiles. The great thing about this tactic is that the interquartile range (IQR) will clearly denote where 50% of your data is in the view. In Figure 3-15 you can see I have plotted a bell curve that is normally distributed. You can see that the IQR makes up 50% of the dataset.

To implement this tactic in Tableau, you can create a new sheet with the same line chart you used in the previous examples and then drag "Median with Quartiles" from the Analytics pane, as shown in Figure 3-16.

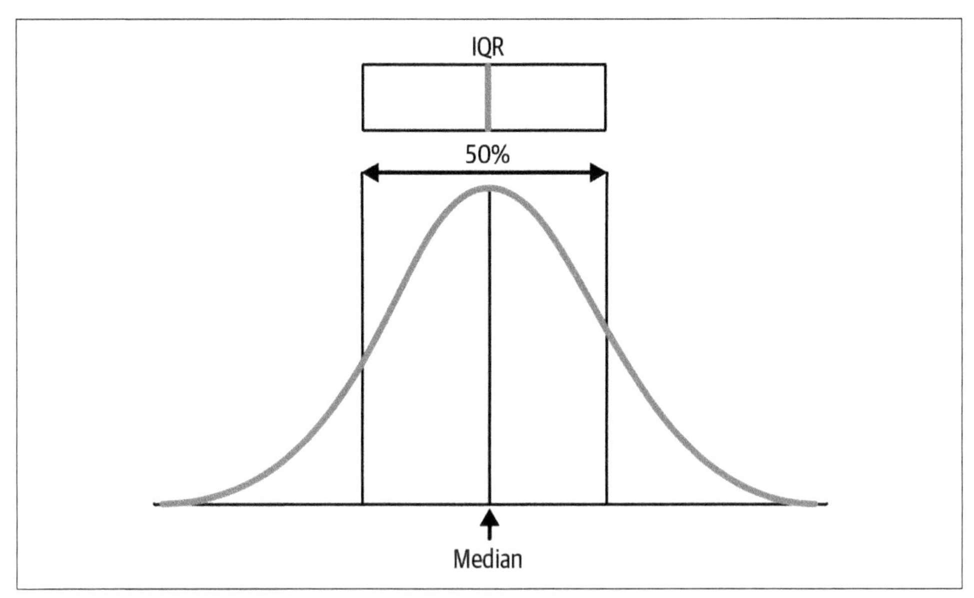

Figure 3-15. Median with Quartiles mapping the IQR

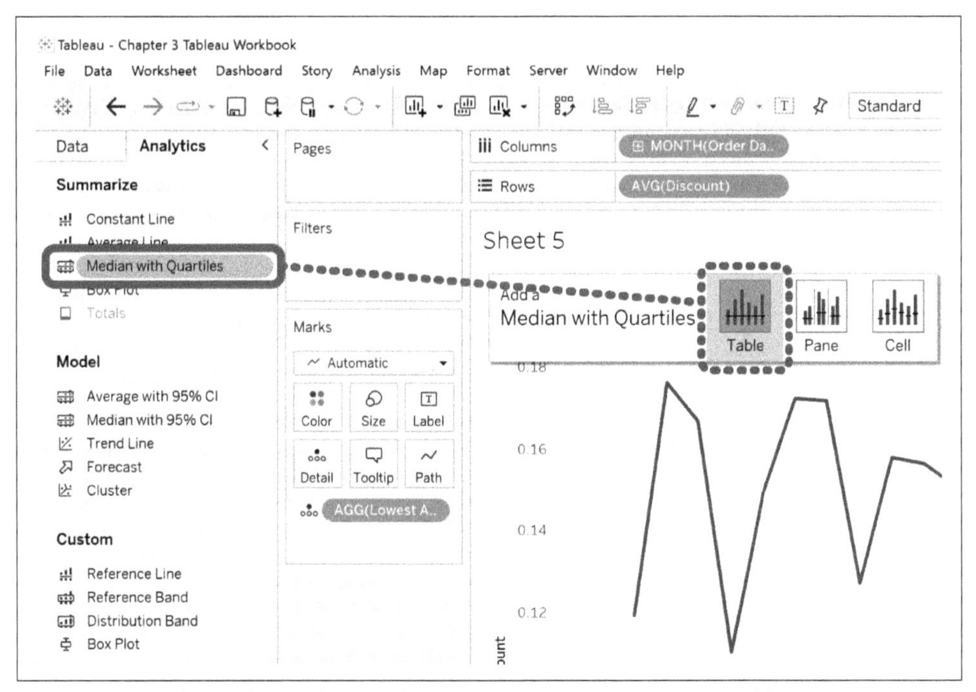

Figure 3-16. Adding the Median with Quartiles tactic to a view in Tableau

Drop the option on the Table scope, which will calculate the median across the entire table. Once it is added, you will end up with a view like Figure 3-17.

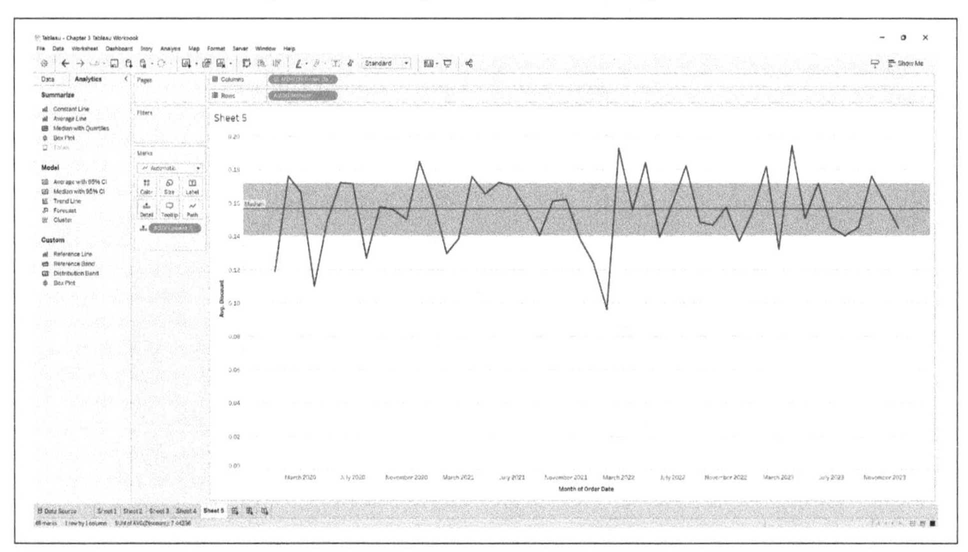

Figure 3-17. After the Median with Quartiles tactic is added to the view

Analyzing this view, you can see that 50% of the marks in the view fall within the interquartile range. Using this tactic, the business could use the upper and lower bounds of the IQR as the benchmark.

You can also see that the IQR does make sense, with 50% of the marks within that range. In terms of the example, the business could use the upper and lower bounds of the IQR as the benchmark.

Summary

There are many other ways to incorporate benchmarks, targets, or goals in Tableau. For the scope of this book, I have shown you several ways you can do it using Tableau's Analytics pane. You should now know what internal and external benchmarks are and how they can help support your business.

CHAPTER 4

Understanding Normal Distribution Using Histograms

When it comes to statistics, there are a few core concepts to know and understand. I've introduced you to some of these ideas in Chapter 1, including statistical significance, p-values, and hypothesis testing. However, one of the most important concepts to know and understand is the different ways data can be distributed. If you don't know how your data is distributed, you could be making some wrong assumptions in your analysis, which can lead to erroneous conclusions and false assumptions.

In this chapter, I will walk you through some ways your data can be distributed, provide examples of some different types of distribution, and then show you how to visualize distribution in Tableau using histograms.

Types of Distribution

In business or in most everyday analysis, you will run across different ways data is distributed. For example, if I flipped a coin 1,000 times, recorded the data, and visualized it, I would probably have two columns (heads and tails) that would be almost exactly evenly distributed because of the 50/50 chance to get either side. Another example: if I record the altitude of an airliner taking off and reaching 36,000 feet, the data would grow exponentially over time and slowly plateau at some point. And another example: if I recorded the height of every adult in a large lecture hall, I would probably end up with a normally distributed dataset.

All around us, we can record data and visualize it to reveal unique distributions. In business, it is just the same. If you visualize the distribution of profit by product, sales over time, orders by customer, etc., you will find and observe different distributions for each dataset. Here are some common examples.

Uniform Distribution

Uniform distribution is when your data is distributed equally across your dataset, as shown in Figure 4-1.

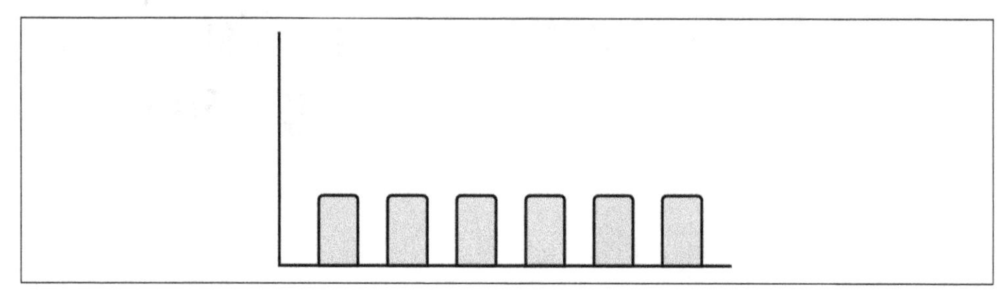

Figure 4-1. Uniform distribution

Most of the time, when data is collected, it will come with some sort of variance. That is why this form of distribution is unlikely to occur in most situations. However, to give you an example, imagine you were recording how many minutes occur every hour or plotting the probability of rolling a number 1 through 6 on a six-sided die. In these cases, you would end up with an equally distributed dataset.

Bernoulli Distribution

A Bernoulli distribution is when you collect data with only two possible outcomes, as shown in Figure 4-2. The Bernoulli distribution plays a pivotal role in probability theory and statistics. Named after the Swiss mathematician Jacob Bernoulli, this distribution forms the foundation of many statistical models and serves as a fundamental concept in various fields.

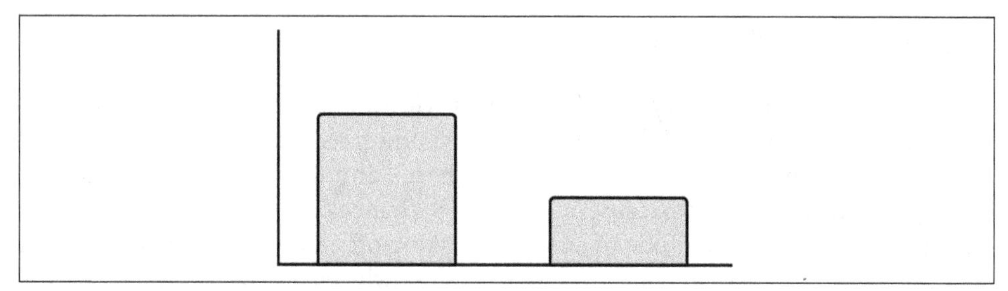

Figure 4-2. Bernoulli distribution

Data is often collected in very simple formats like this. Clinical trials, surveys, or server status can all be recorded in this Boolean format. Boolean means something has two outcomes: 0 or 1 in binary, yes or no, true or false, server is running or server is not running, etc.

Exponential Distribution

Exponential distribution occurs when your data grows exponentially as it's collected, then starts to taper off, as shown in Figure 4-3.

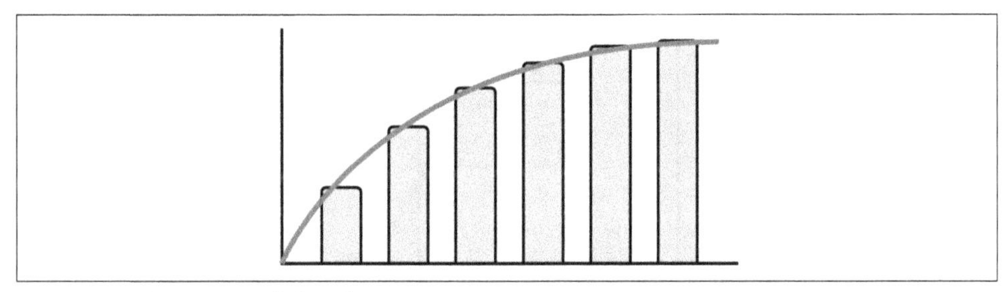

Figure 4-3. Exponential distribution

Exponential distribution can occur in a dataset when the data collected has a rapid increase or decrease and then stabilizes. Think of the altitude of an airplane as it climbs to its cruising altitude. The plane starts by climbing fast on takeoff and then tapers off as it levels out.

Normal Distribution

Normal distribution follows a bell-shaped curve, as shown in Figure 4-4. It is often referred to as a normal distribution, bell curve, or Gaussian distribution.

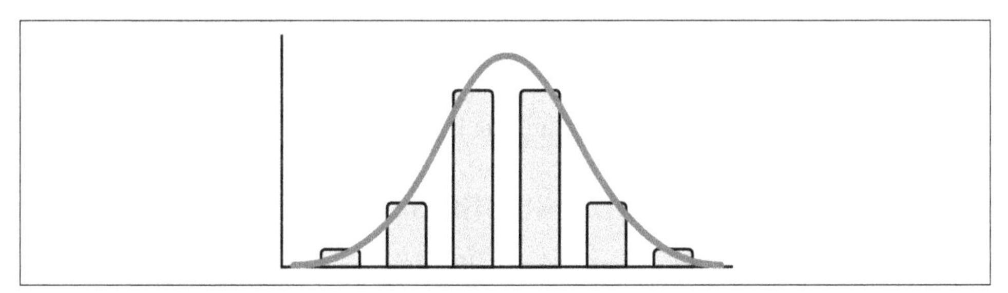

Figure 4-4. Normal distribution

This type of distribution occurs when the data is symmetrical in shape with no skew. To give you an example, if you measured the height of college students on a campus, you would see this pattern play out. There would be a few students that are shorter than average, but the majority would be around the national average, and some students would be taller than average.

To give you a mathematical explanation of a normal distribution, let's look at the empirical rule: 68–95–99.7. In a normal distribution, approximately 68% of the data falls within 1 standard deviation of the mean, 95% falls within 2 standard deviations, and 99.7% falls within three standard deviations, as illustrated in Figure 4-5.

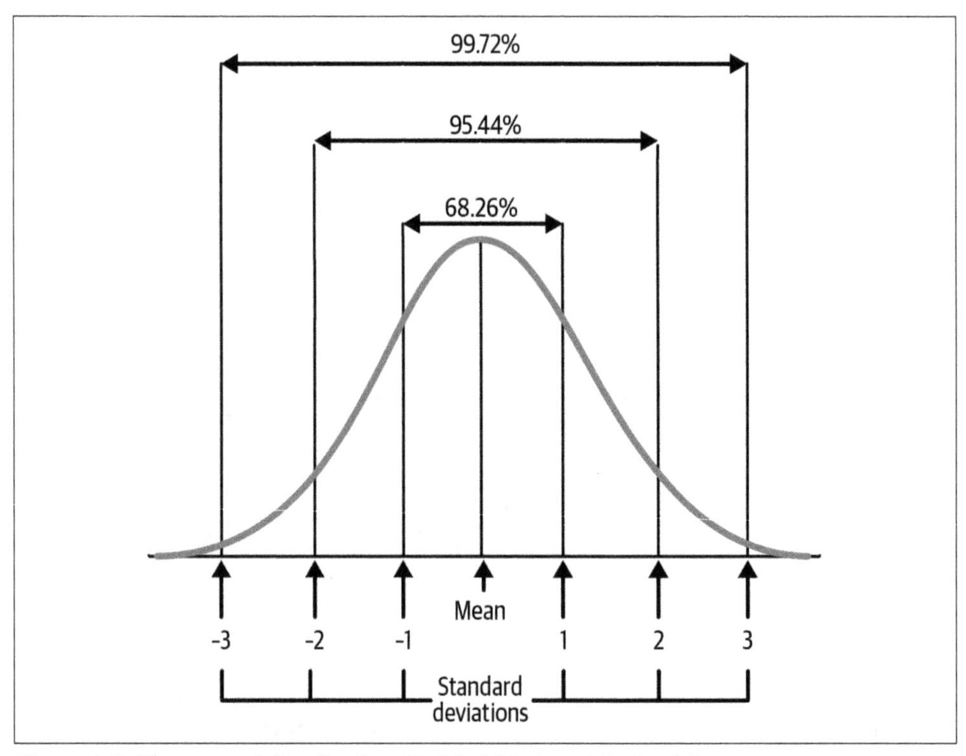

Figure 4-5. The normal distribution 68–95–99.7 (empirical rule)

This rule is very important to understand and will play a big part in outlier detection in Chapters 6 and 7. Also, many of the models that are built natively in Tableau are employed under the assumption that the data is normally distributed.

Normal Distribution and Skewness

To cover normal distribution in more depth, let's analyze what it means to have a normal distribution. If a dataset is perfectly normal, it means the mean, mode, and median are all equal to each other. Here are the definitions of those terms:

Mean
: The average of the numbers within our dataset

Mode
: The number that appears most frequently in our dataset

Median
: The middle number in a dataset when ordered lowest to highest

Consider this dataset: 3,4,5,5,6,6,7,7,7,8,8,9,9,10,11. Following the definitions of mean, mode, and median, you would have a perfectly distributed normal distribution, where the mean would equal 7, the mode would equal 7, and the median would equal 7.

If you visualize that dataset, you would have a normal distribution similar to Figure 4-6.

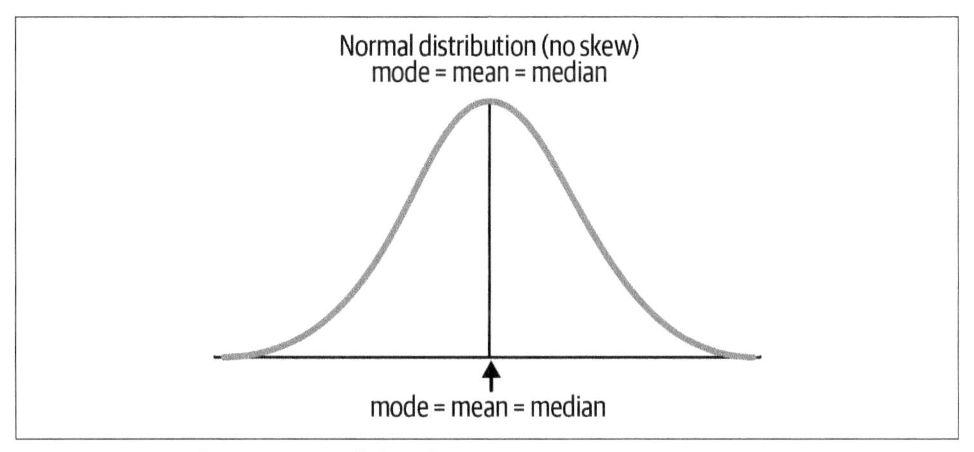

Figure 4-6. Visualizing a normal distribution

As you can see in Figure 4-6, mode = mean = median, so all of those values fall directly in the middle of the symmetrical curve. To make this concept clearer, think of the example I gave earlier about sampling the height of college students on a campus. You would have some students that would be extremely tall (like those on the basketball team) and others who were naturally shorter. However, the majority of students would be around the national average height.

Also, think of your own personal environments at school or work. If you recorded a decent number of adult heights, you would likely end up with a dataset with a normal distribution. While normal distribution happens often, it would be rare to see your data perfectly distributed like this simply due to natural variances. Instead, you will most likely see some skewness.

Understanding Skewness

Skewness is when you see the majority of the data consolidated toward one side of the curve. To give you a visual example, let's first analyze a left-skewed distribution, as shown in Figure 4-7.

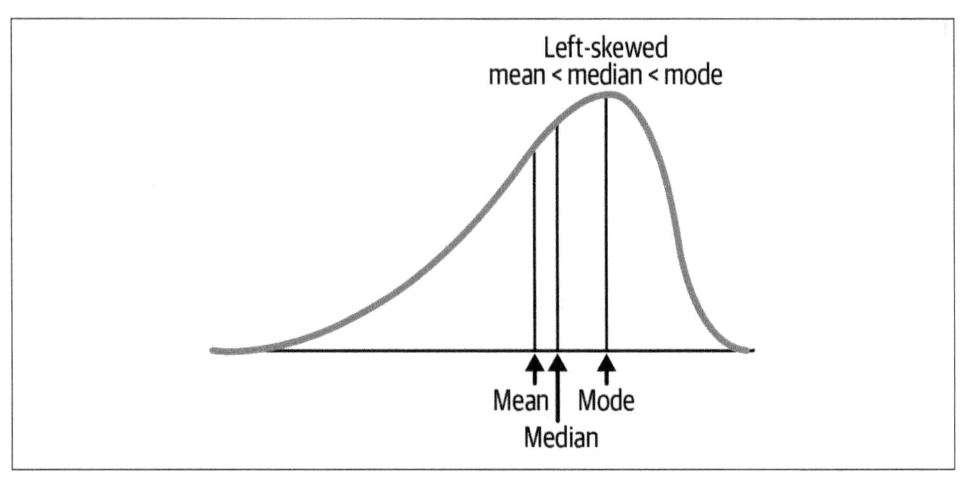

Figure 4-7. Left-skewed distribution

Mathematically, to be left skewed, the mean is less than the median, which is less than the mode. Naturally your eyes are probably drawn to the data on the right side of Figure 4-7. You may be asking yourself, Isn't this right skewed? When it comes to skewness, you should be more interested in what is causing the data to be skewed. In this case, there is some extreme value (or values) on the left side that are causing the data to be skewed. That is why there is a long "tail" to the left side and why you would refer to this type of skewness as left-skewed.

Now let's analyze the opposite of this, which is to be right skewed, as shown in Figure 4-8.

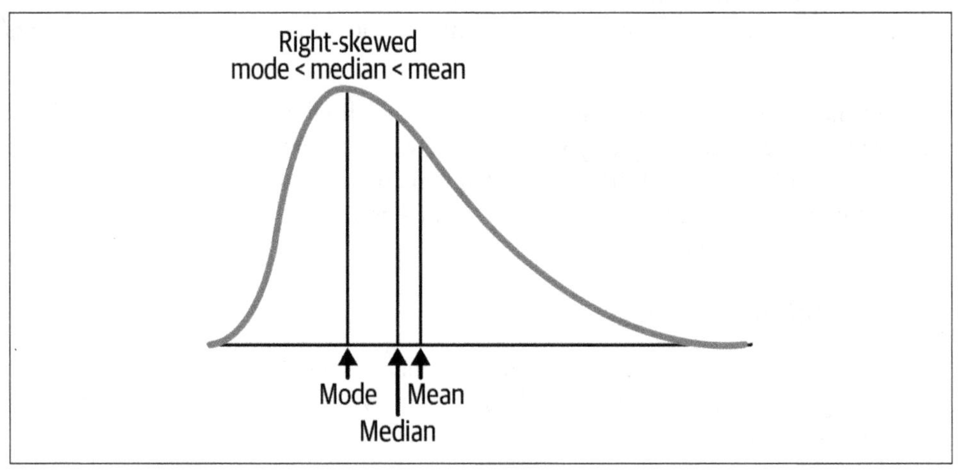

Figure 4-8. Right-skewed distribution

To be right skewed, the mode is less than the median, which is less than the mean. As you can see, the majority of the data is bunched to the left side with a long tail on the right side. To give you an example, imagine if the data you collected was the price of cars sold in the United States. You would probably observe some very cheap cars, but the majority of cars sold would be around the national average price ($30,000–$50,000). However, you would also have some cars sold at extremely high prices that would give you this long tail effect. Most likely, there wouldn't be a lot of vehicles sold at those high prices; however, the prices would drag our mean away from the median.

Remember that when you are thinking about skewness, you should be more concerned about what is causing the skewness. This will also help you remember the correct name for both left- and right-skewness.

Accounting for Skewness

Many models operate with the assumption that the data you have is normally distributed. With that said, skewed data can throw off the results of these models, giving statistically significant results when there aren't or suggesting there isn't a statistically significant result when indeed there is one. For this reason, you need to know how to account for skewness by transforming the data in some way.

Hopefully, you can understand how skewness can happen, and maybe you are already thinking of ways you could clean your data to make it more normally distributed. For instance, in the car example, you could simply exclude the more expensive cars from your analysis or segment them out into a separate analysis.

There are many ways you can account for skewness, and each of them has its own pros and cons. Here are a few techniques that you can apply to skewed data to make it normally distributed for your analysis:

Log transformation

A common transformation is to take the log of the data. This returns the log values for the data and essentially pushes the extreme values closer to each other. This will usually give you a normal distribution you can work with for modeling. In Tableau, you can use the LOG function and apply it to the measure you are analyzing.

Pros: You may not have to remove any of your data, as this technique will get the values close enough to each other to leave everything in most cases.

Cons: It is difficult to explain and interpret your data after the transformation.

Remove extreme values

This would entail filtering all the extreme values or outliers from your analysis. In Tableau, you can simply use a conditional filter to do this. Or you can use the

Explain Data feature in Tableau to find the exact observations that are causing the extreme values and remove them.

Pros: This is very easy to explain to your stakeholders and document for your colleagues.

Cons: You could exclude extremely valuable information from your analysis or create bias.

Other statistical transformations
: There are many more transformation techniques that are more methodical and scientific. Some examples include winsorization and Box–Cox transformations.

Pros: The assumptions you are making are clearly defined and expected in the industry.

Cons: Providing details of these transformation techniques can be difficult for some stakeholders to understand.

How to Visualize Distributions in Tableau Using Histograms

Now that you have an understanding of different types of distributions, let me show you how to visualize the distribution of your data in Tableau.

To get started, connect to the Sample - Superstore dataset. This is the default example data that comes with any version of Tableau and will be displayed at the bottom-left corner of the data source connection screen as a data source. Once connected, you need to create a *bin* in Tableau. Bins are primarily used to turn measures into discrete dimension members. This is key when trying to determine the distribution of data because you want each bar in the histogram to represent a group of values.

To give you an example, Table 4-1 is a table of data that has been binned by increments of 20.

Table 4-1. Example of bins

Order	Profit	Profit bin
US-0001	$5.00	0
US-0002	$10.00	0
US-0003	$15.00	0
US-0004	$25.00	20
US-0005	$35.00	20
US-0006	($5.00)	–20
US-0007	($25.00)	–40

In each order, Tableau is assigning a bin to the raw value in increments of 20. So the first three values fall between 0 and 20, meaning their bin is 0. Orders 4 and 5 have profit that falls between 21 and 40, so they are assigned to bin 20, and so on.

Tableau actually helps you along by providing a preconstructed bin to use in the Sample - Superstore dataset called Profit (bin). I am going to walk you through how to create one from scratch, but if you would rather use that bin and jump ahead, then feel free to do so.

To create a bin, right-click on Profit and hover over the Create option to choose Bins from the menu (see Figure 4-9).

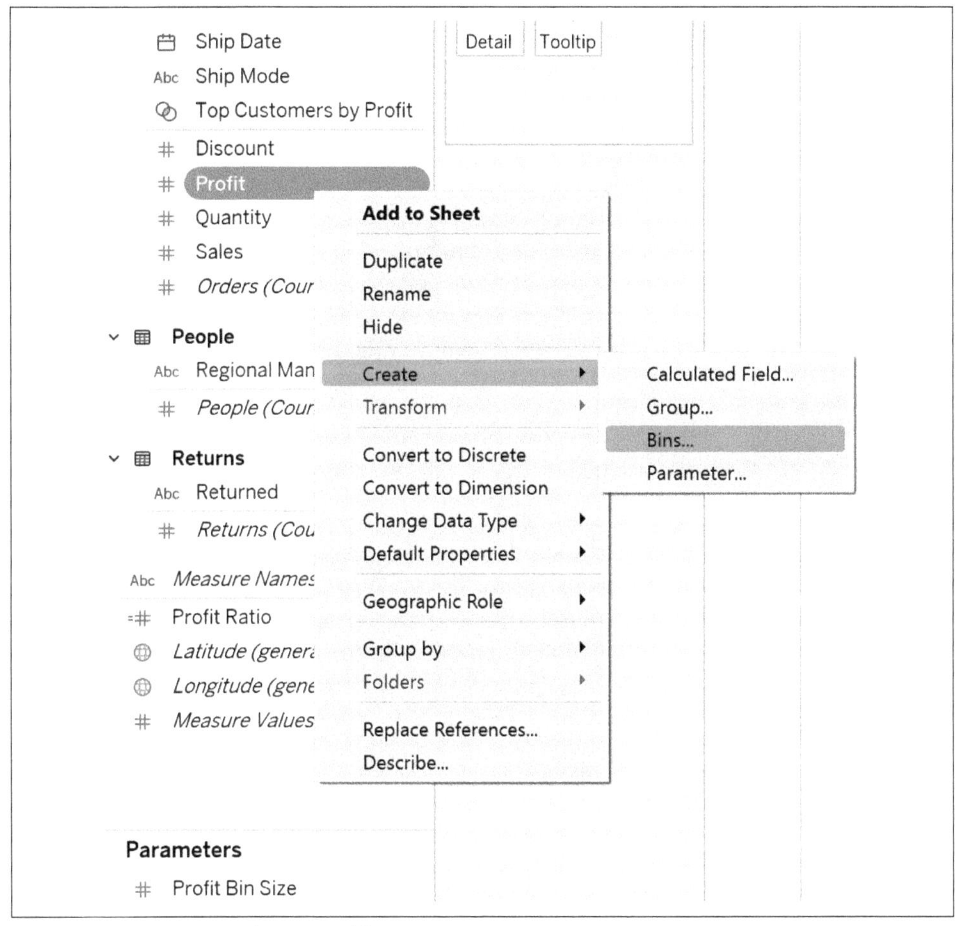

Figure 4-9. Creating a bin in Tableau

This will open a new dialog box with information about the profit measure, as shown in Figure 4-10.

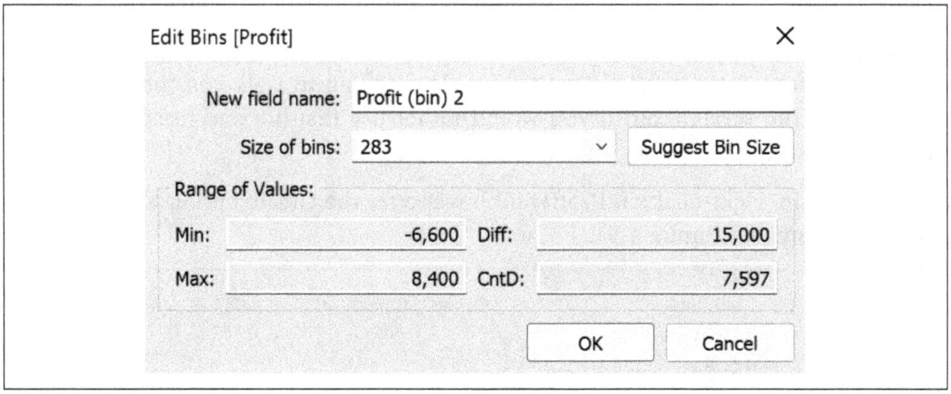

Edit Bins [Profit]

New field name: Profit (bin) 2

Size of bins: 283 Suggest Bin Size

Range of Values:

Min:	-6,600	Diff:	15,000
Max:	8,400	CntD:	7,597

OK Cancel

Figure 4-10. Edit Bins menu in Tableau

You can see that Tableau will assign a "Size of bins" value automatically. In this case, it is 283, which means that each bar in the histogram will "bin" the profit values in these increments. As a better example, if a product generated $200 in profit, then it would get binned together with values ranging from $0 to $283. If a product generated $400 in profit, it would get binned together with values ranging from $284 to $567.

As you can imagine, the larger the bin size, the wider and fewer the bars will be, whereas the smaller bin sizes give you thinner and more abundant bars. It may be useful to test various bin sizes to get a better view of your data's distribution.

For now, let's use the suggested bin amount and refine it later. Before I move on, I want to explain the other information displayed in this menu. You can see a Min and Max value; these are the minimum and maximum profit values of all the records in the dataset. You can also see a CntD value, which stands for *count distinct*. This is a distinct count of all the individual profit amounts in the data, so I have 7,545 distinct values in this dataset. The Diff value is just the difference from the min and max values.

When you're done, click OK. This is going to create a new dimension named "Profit (bin) 2" in the Data pane. Drag that dimension to the Columns shelf and then drag the *Orders (Count)* measure to the Rows shelf, which you can see in Figure 4-11.

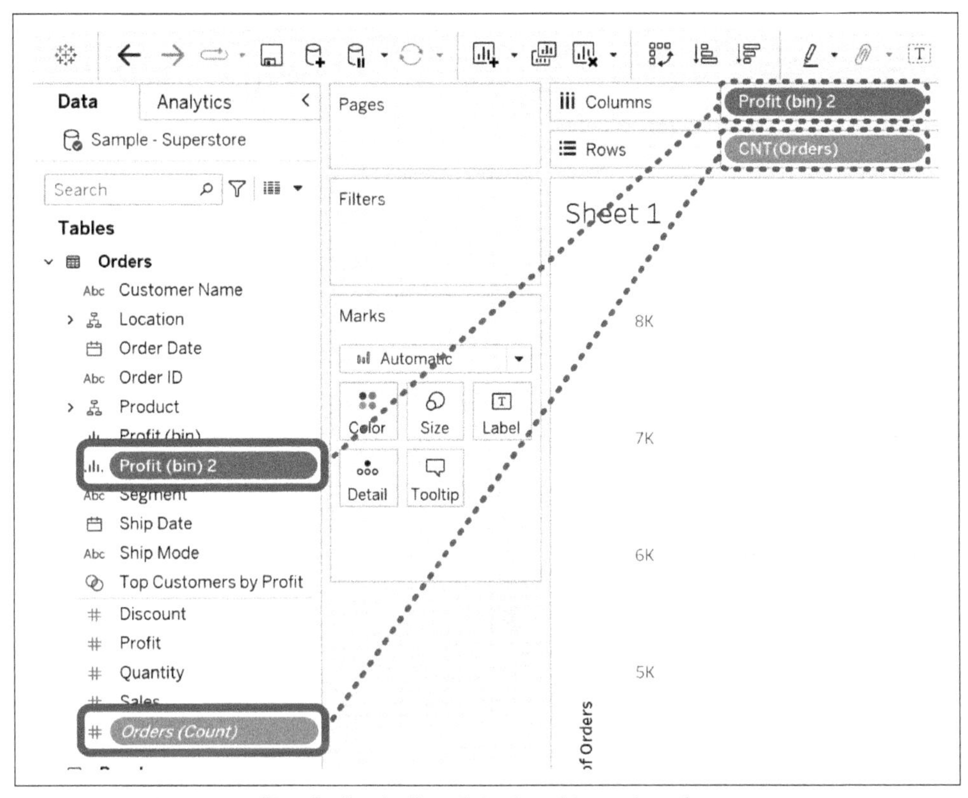

Figure 4-11. Dragging "Profit (bin) 2" and Orders (Count) to the canvas

This gives you a nice histogram that appears to have a normal distribution, which you can see in Figure 4-12.

You can see that the majority of the CNT(Orders) generate profits ranging from about –$849 to $849. At a high level, this looks very informative, but let's examine it a bit closer. Let's edit the bin size from 283 to 20. This will bin the data at a more granular level. To do this, right-click on the "Profit (bin) 2" dimension and change the "Size of bins" to 20, as shown in Figure 4-13, then click OK.

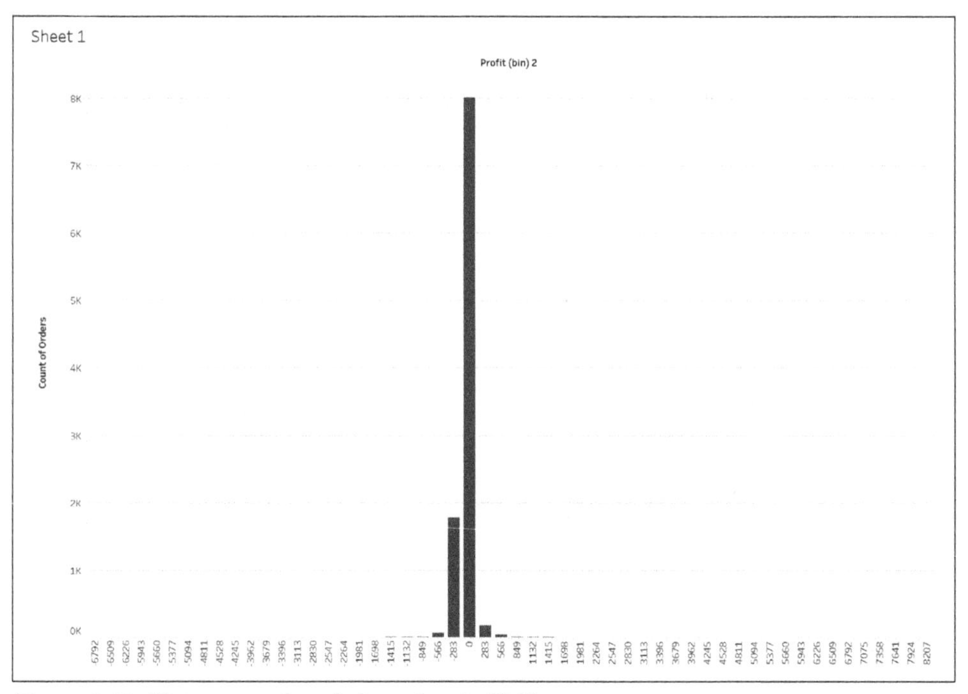

Figure 4-12. Histogram of profit by orders in Tableau

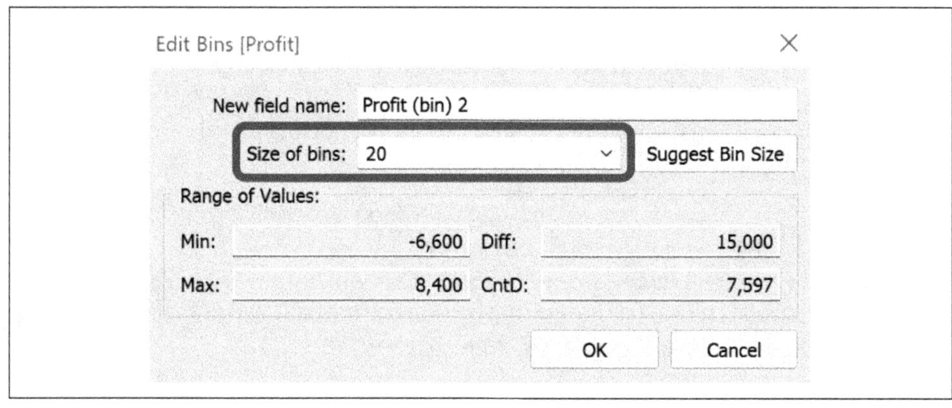

Figure 4-13. Editing "Size of bins" to 20

As I mentioned earlier, the smaller the bin size, the thinner and more abundant the bars will be. By editing the size of the profit bin, you can now compare Figure 4-12 to what is shown in Figure 4-14.

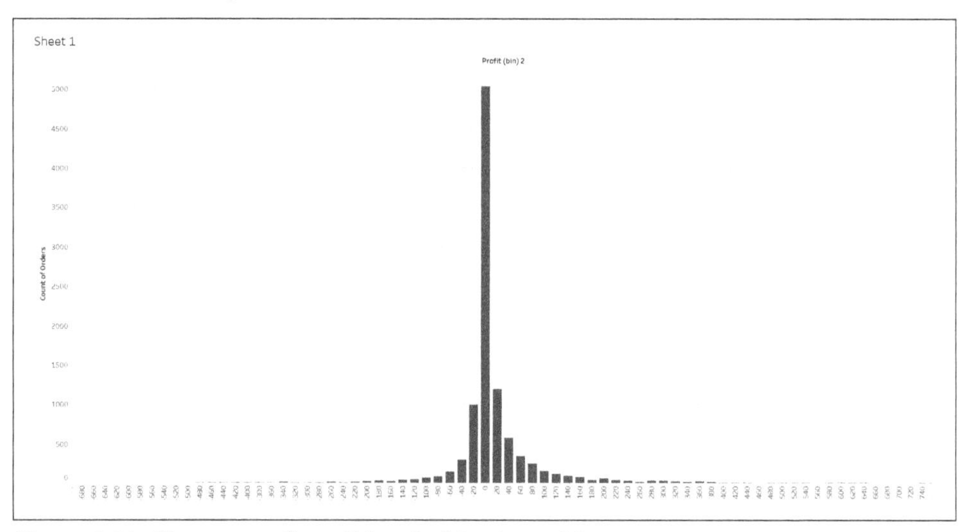

Figure 4-14. Histogram after editing the bin size to 20

You can see that some orders generated profit bins to either extreme on both the left and right sides. As an analyst, you may want to exclude these extreme values and check the bins in closer detail to ensure they are calculated correctly.

To demonstrate this, let's focus on the orders that are between –$300 and $300. Filter to these bins by clicking on the –$300 bin in the x-axis; then, while holding Shift, click on the $300 bin, which will highlight that bin and everything in between. With those bins selected, click on Keep Only from the tooltip command buttons, as shown in Figure 4-15.

Figure 4-15. Selecting a smaller range of bins

There are multiple other ways to apply this filter. As an example, you could also drag "Profit (bin) 2" to the Filters shelf and apply a dimension filter. That should give you a view that looks like Figure 4-16.

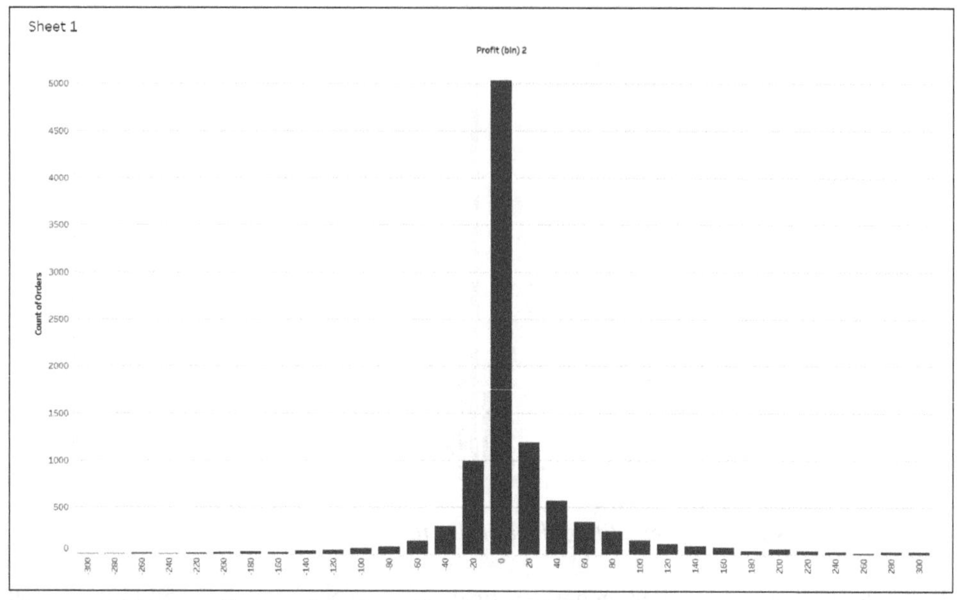

Figure 4-16. Normally distributed profit values in Tableau

You can see that the data follows a normal distribution here. In this dataset, there are some orders that generated negative profit and some that generated positive profit. What would the data look like if it were skewed, though? To show you a good example of this, let's create a bin of sales.

Start by right-clicking on Sales, navigate to Create, and then select Bins, as shown in Figure 4-17).

This will open the Edit Bins menu, as shown in Figure 4-18.

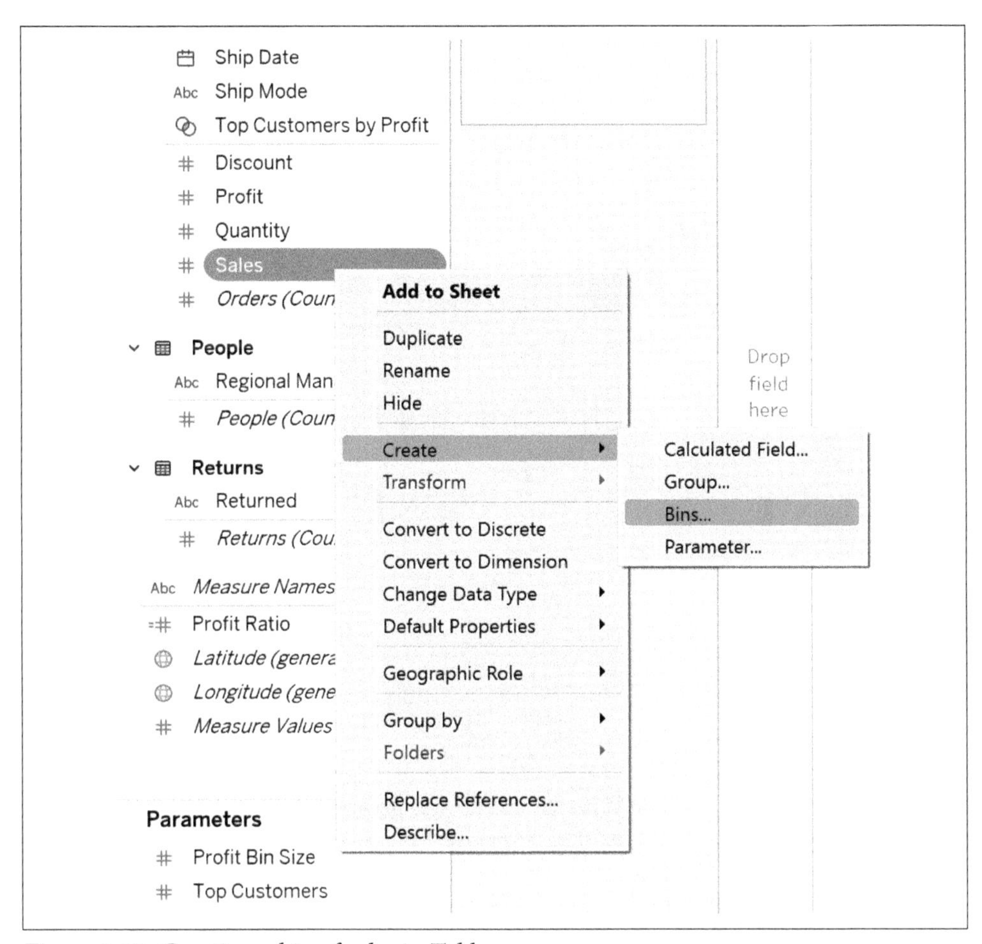

Figure 4-17. Creating a bin of sales in Tableau

Edit Bins [Sales]

New field name: Sales (bin)

Size of bins: 100 | Suggest Bin Size

Range of Values:

Min:	0	Diff:	22,638
Max:	22,638	CntD:	6,161

OK | Cancel

Figure 4-18. Edit Bins menu for Sales (bin)

For this example, change the "Size of bins" to 100 and then click OK to close the menu. Now create a new sheet and drag the new "Sales (bin)" field to the Columns shelf and *Orders (Count)* to the Rows shelf, as shown in Figure 4-19.

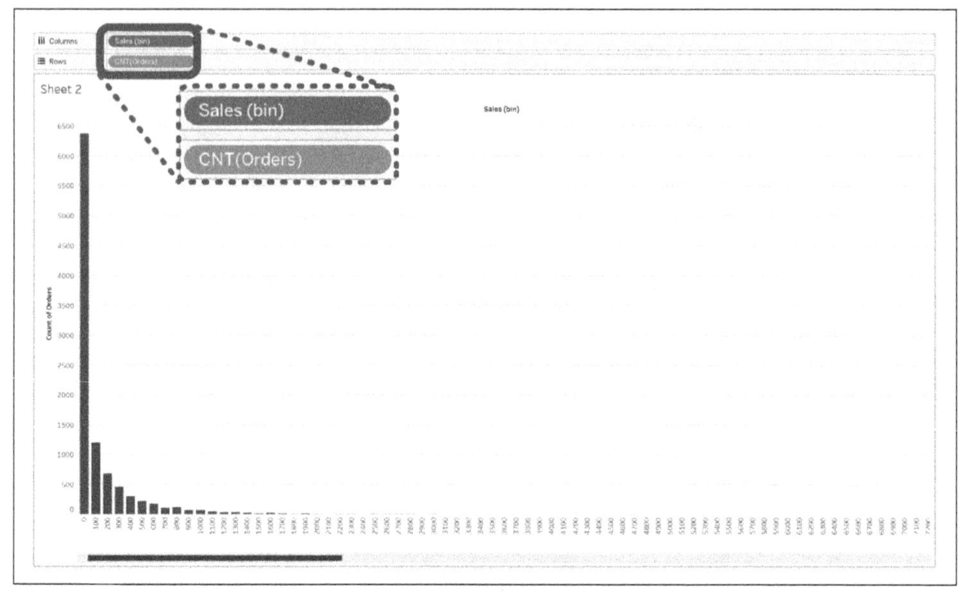

Figure 4-19. Histogram of sales by order count

You can see that changing the bin size to 100 has given you a right-skewed distribution. If you think about the sales data logically, this makes sense. There are few cases when a business would have negative sales from an order. That would essentially mean the business paid the customer for the product. With that said, it wouldn't make sense to try to apply a transformation to this data either. You would need to apply a nonparametric model or a model that does not assume a normal distribution to get the best results when working with this measure.

As I've mentioned previously, there are some models that require the data to be normally distributed to function properly. However, there are also other models that function properly even if your data is not normally distributed. I will cover some of those models in Chapter 7. To introduce the idea to you, let's briefly discuss parametric models and nonparametric models.

Parametric Models

Parametric models are statistical models that make assumptions about the distribution of the data. In parametric modeling, the goal is to estimate the parameters of the chosen distribution based on the available data. Once the parameters are estimated, the model can be used to make inferences and predictions or generate new data. The term *parametric* refers to the estimation of those parameters.

Some examples of parametric models include:

- Linear regression
- Exponential regression
- Poisson regression
- Logistic regression

The benefit of these models is that they are easy to interpret and explain to stakeholders. This makes them ideal models that a business can easily take action on and incorporate into their operations.

Nonparametric Models

Nonparametric models are statistical models that do not make strong assumptions about the underlying population distribution or its parameters. Unlike parametric models, which specify a fixed form for the distribution and estimate its parameters, nonparametric models aim to estimate the underlying data distribution directly from the data itself.

Nonparametric models are flexible and can handle a wide range of data distributions without assuming a specific functional form. They are particularly useful when the data does not adhere to the assumptions of parametric models or when there is limited prior knowledge about the data distribution. These models often focus on estimating the patterns, relationships, or rankings in the data rather than estimating specific parameters.

Some examples of nonparametric models include:

- *K*-nearest neighbors (*k*-NN)
- Decision tree
- Random forest
- Support vector machine (SVM)

The benefit of these models is their flexibility, but they are more challenging to interpret and communicate to stakeholders.

Throughout the next chapters, I will show you examples of these different types of models.

Summary

In this chapter, you learned about different types of distributions, what skewness is, and how to visualize histograms in Tableau. Being able to apply these techniques is a foundational skill you need to know before you can begin modeling.

This chapter also introduced the idea of parametric and nonparametric models. Knowing this information will help you decide which model to apply to certain data, depending on the distribution of your data.

CHAPTER 5
Understanding Confidence Intervals

Understanding *confidence intervals* (CIs) is a fundamental aspect of statistical analysis that provides valuable insights into the reliability and precision of data. A confidence interval is a range of values within which we can reasonably expect the true population parameter to lie, based on a sample from that population. It serves as a statistical tool to quantify the uncertainty inherent in inferential statistics, allowing researchers and decision makers to make informed conclusions about population parameters. A solid grasp of confidence intervals empowers individuals to interpret study results more effectively, make informed decisions, and communicate the degree of certainty or uncertainty associated with their findings.

In this chapter, you will learn about confidence intervals, what they mean, how to interpret them, and how to implement them into your work. There are different types of confidence intervals, but this chapter will cover only average and median two-tailed confidence intervals.

What Is a Confidence Interval?

A confidence interval is a range of values that you expect your estimate to fall between a certain percentage of the time if you were to run your experiment again or resample the population. Confidence intervals normally contain the average or median of the estimate as a central point and a plus and minus variation from that center point. This plus or minus variation is your confidence interval range.

A confidence level refers to the level of certainty or assurance we have in the results of a study or analysis. It's a way of expressing how confident we are that our findings are accurate and representative of the larger population or phenomenon we're studying. To set your interval range, you first have to decide on what your confidence level is. The standard level you will see is 95% confidence. However, you can increase it, which would widen your interval range. Or you could decrease your level of confidence, which would shorten your interval range.

Your desired confidence level is usually 1 minus the alpha (α) value.

$$\text{Confidence level} = 1 - \alpha$$

If you use an alpha value of $p < 0.05$, then your confidence level would be $1 - 0.05 = 0.95$, or 95%. If you want to be more confident, you would use an alpha value of $p < 0.01$, $1 - 0.01 = 0.99$, or 99%. If you wanted to be less confident you would use an alpha level of $p < 0.1$, $1 - 0.1 = 0.90$, or 90%.

To be clear, the alpha value is determined by you before conducting a statistical test or hypothesis test. It represents the maximum level of risk you are willing to accept for making a type I error (false positive), which is rejecting the null hypothesis when it is actually true. For example, if you are working with healthcare data where your results could be the difference between life and death, use a higher level of confidence. If you are estimating the height of the next person to walk into a classroom of college students, you have room to be less confident. Remember, when someone says they are 95% confident in an estimate they are basically saying that 95 out of 100 times, the average of the estimate would fall between the upper and lower values of the confidence interval.

Recall Chapter 4 on normal distribution. If you were to visualize a 95% confidence interval on a normal distribution, it would look like Figure 5-1.

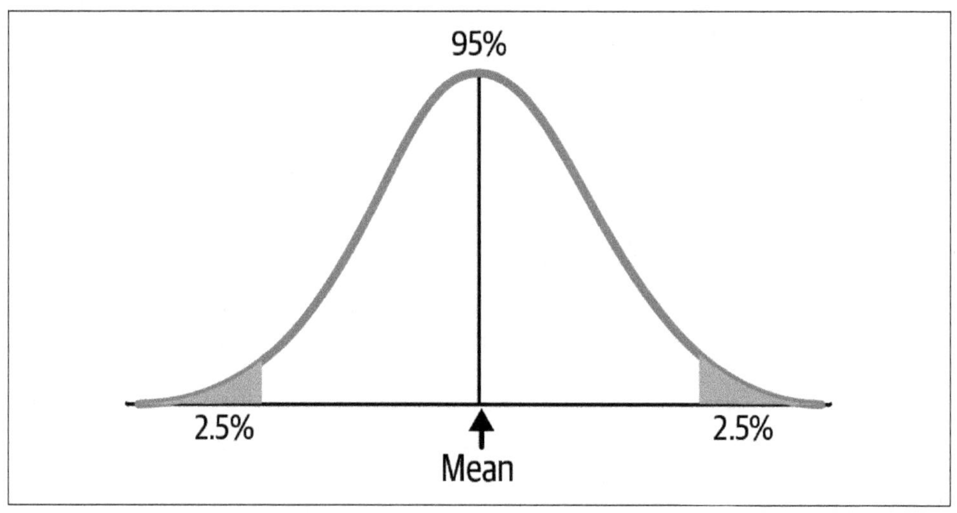

Figure 5-1. Confidence interval bell curve

You can see that there are two "tails" on this curve that represent the confidence interval. These tails make up 2.5% on either side of the distribution of data. Based on this level of confidence, you can infer that the true parameter of the sampled population would fall within that 95% zone. However, let's say I resampled the population and ended up with new results. The new average would fall in that 95% range 95% of the time, as shown in Figure 5-2.

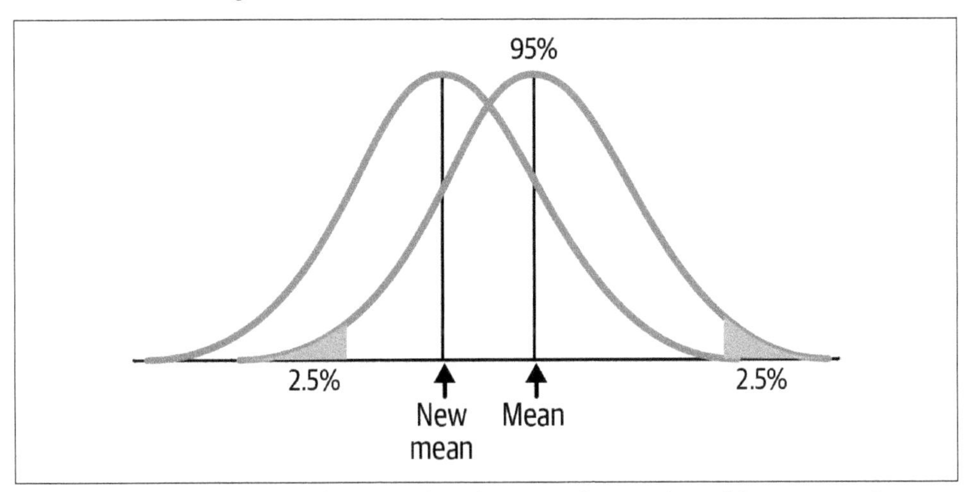

Figure 5-2. Resampling and falling within the original created confidence interval

On the other hand, you know that 5% of the time the average could fall outside the confidence interval range (see Figure 5-3).

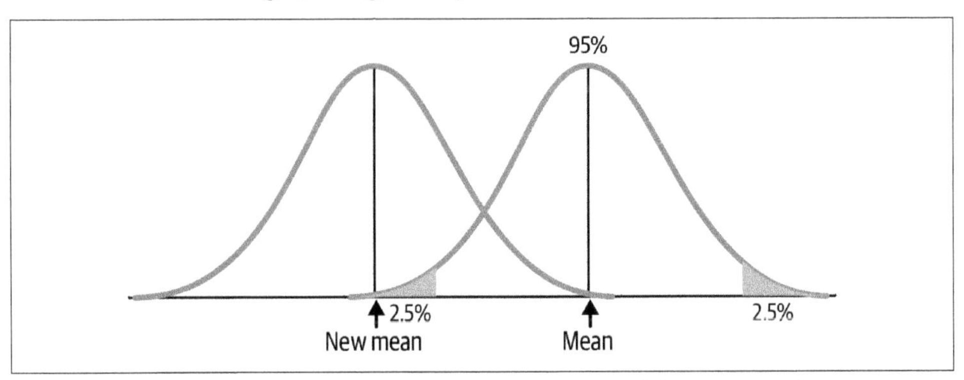

Figure 5-3. Resampling and falling outside the original confidence interval

This is why it is important to know how confidence intervals work because you can make some very powerful inferences about a population based on a sample of that population. This has many applications in business, healthcare, manufacturing, and more.

How to Calculate Confidence Intervals

Understanding the underlying formula used to calculate confidence intervals can be extremely beneficial to you. There are several different formulas used to calculate confidence intervals. Tableau always assumes that you are working with a sample population versus the total population, so that is the formula I will show you.

The formula to calculate the confidence interval is expressed as:

$$CI = \bar{x} \pm t\left(\frac{s}{\sqrt{n}}\right)$$

where

CI = confidence interval

$\bar{x}$ = sample average

t = t-score for the desired level of confidence and degrees of freedom

s = sample standard deviation

n = sample size

In this example, I will be calculating a 95% confidence interval using 12 test scores: 80, 95, 80, 80, 85, 85, 90, 85, 75, 95, 90, and 80. To begin, you need to find the sample average. You can do this by adding up all the scores and dividing them by the total number of tests:

$$80 + 95 + 80 + 80 + 85 + 85 + 90 + 85 + 75 + 95 + 90 + 80 = 1020 \div 12 = 85$$

$$\bar{x} = 85$$

Next, you need to calculate the standard deviation, which I covered in detail in Chapter 1. Start by subtracting the average from each test score and square each result:

$$(80 - 85)^2 + (95 - 85)^2 + (80 - 85)^2 + (80 - 85)^2 + (85 - 85)^2 + (85 - 85)^2 + (90 - 85)^2 + (85 - 85)^2 + (75 - 85)^2 + (95 - 85)^2 + (90 - 85)^2 + (80 - 85)^2 =$$

$$(-5)^2 + (10)^2 + (-5)^2 + (-5)^2 + (0)^2 + (0)^2 + (5)^2 + (0)^2 + (-10)^2 + (10)^2 + (5)^2 + (-5)^2 =$$

Now add each square and divide the result by the sample size (n) – 1. Since there are 12 test scores, you will divide by 11. This will give you what is called the *sum of squares*:

$$25 + 100 + 25 + 25 + 0 + 0 + 25 + 0 + 100 + 100 + 25 + 25 = 450 \div 11 = 40.91$$

Now find the standard deviation, which is the square root of the sum of squares:

Looking back at the formula, you only have one more number to find, which is t. You can find this value by looking it up in a t-table. However, when you have more than 1,000 rows of data, this value starts to normalize to a standard value as it moves toward infinity. That standard value is different for each level of confidence, but for 95% confidence, it would normalize to 1.96.

When using a t-table, you need to calculate the degrees of freedom (DF), which equals the sample size (n) minus 1. Since there are 12 rows of data, the DF = 12 – 1 = 11. In the t-table, you will go to DF 11 and move across to the 0.95 column, which is the confidence level. You should have ended up with a t-value of 2.201. You now have all the data points needed to work through the equation:

$$CI = \bar{x} \pm t\left(\frac{s}{\sqrt{n}}\right)$$

$$CI = 85 \pm 2.201\left(\frac{6.396}{\sqrt{12}}\right)$$

$$CI = 85 \pm 2.201\left(\frac{6.396}{3.464}\right)$$

$$CI = 85 \pm 2.201(1.864)$$

$$CI = 85 \pm 4.064$$

After completing this equation, you will have an upper confidence interval of 85 + 4.064 = 89.064 and a lower confidence interval of 85 – 4.064 = 80.936:

Upper confidence interval = 85 + 4.064 = 89.064

Lower confidence interval = 85 – 4.064 = 80.936

Interpreting the Results

With these results, you can say that the average test score will center around 85% and fall between 80.94 and 89.06 for the total population of students 95% of the time.

You can visualize the individual test scores as a distribution, as shown in Figure 5-4.

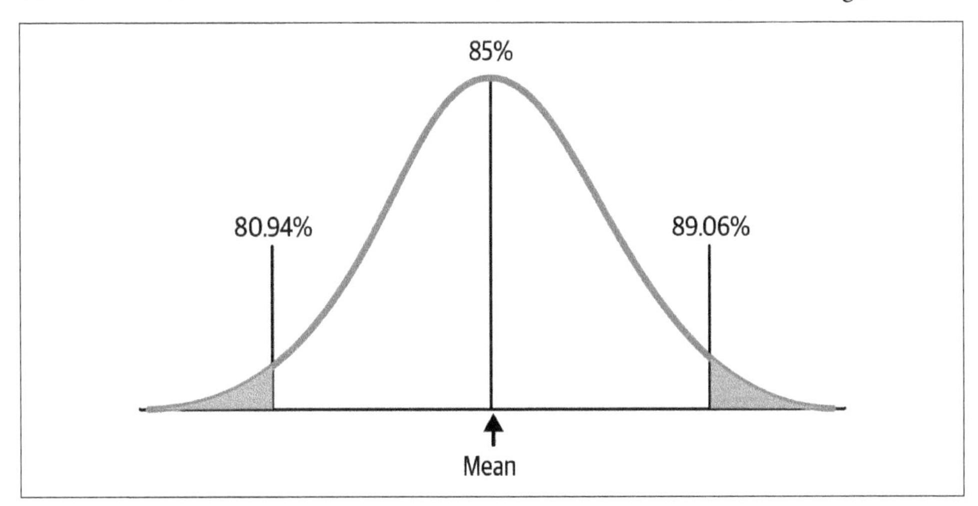

Figure 5-4. Confidence interval from the example

While this alone is powerful insight into the data, there is another quick statistic you can calculate that will give you further insight: the standard error. The *standard error* is calculated by dividing the standard deviation (s) by the square root of the sample population (n). Pulling those values from our formula, I get the following results:

$$6.396 \div \sqrt{12} = 1.846$$

The standard error tells you how accurately the sample data will reflect the total population. The higher the standard error, the more volatility you can expect from the total population; the lower the standard error, the less volatility. Since you have a relatively low standard error you can say that the test scores in future rounds will be close to the sample. However, 12 data points is a small amount. In statistics, a rule of thumb is that any data with over 30 observations is a decent sample size to make inferences on. Be careful of the assumptions you make with anything less than 30 rows of data.

How to Calculate Confidence Intervals in Tableau

You now have a solid understanding of confidence intervals and how to calculate them. You can also see that anything beyond a couple dozen data points would be too cumbersome to calculate by hand. That said, open up Tableau and check your work from the preceding example. Then you'll implement confidence intervals on the Sample - Superstore dataset.

Check the Confidence Interval You Solved by Hand

To start, I re-created the test score dataset in Excel. To follow along, you can do the same by copying Table 5-1 into a Microsoft Excel sheet and following the rest of the instructions.

Table 5-1. Test score dataset

Test scores	Student ID
80	A
95	B
80	C
80	D
85	E
85	F
90	G
85	H
75	I
95	J
90	K
80	L

Next, open Tableau and from the Start Page, click Microsoft Excel from the list of data connectors to the left, as shown in Figure 5-5.

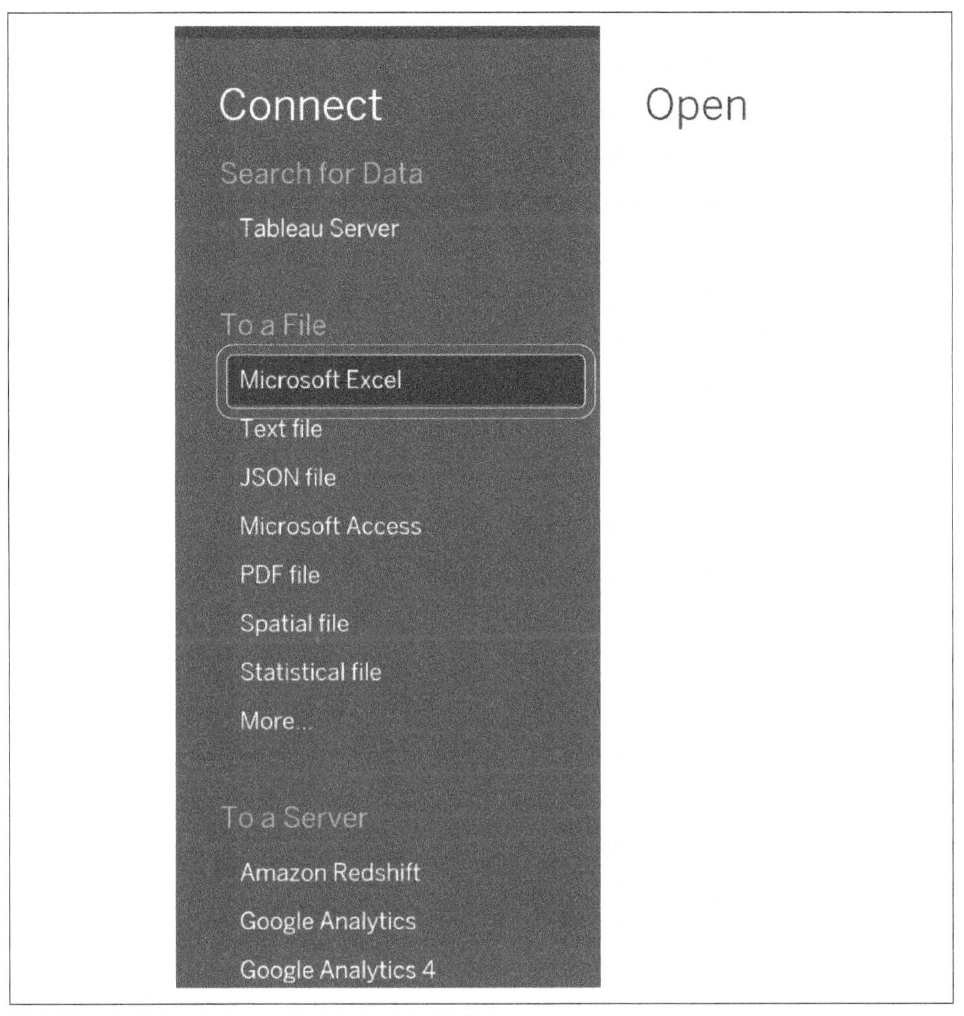

Figure 5-5. Connecting to Excel data in Tableau

A window will appear and ask you to choose the file you want to connect to. Navigate to the file you just created, select it, and click Connect. From the Data Source page, click Sheet 1 at the bottom left of the page, as shown in Figure 5-6.

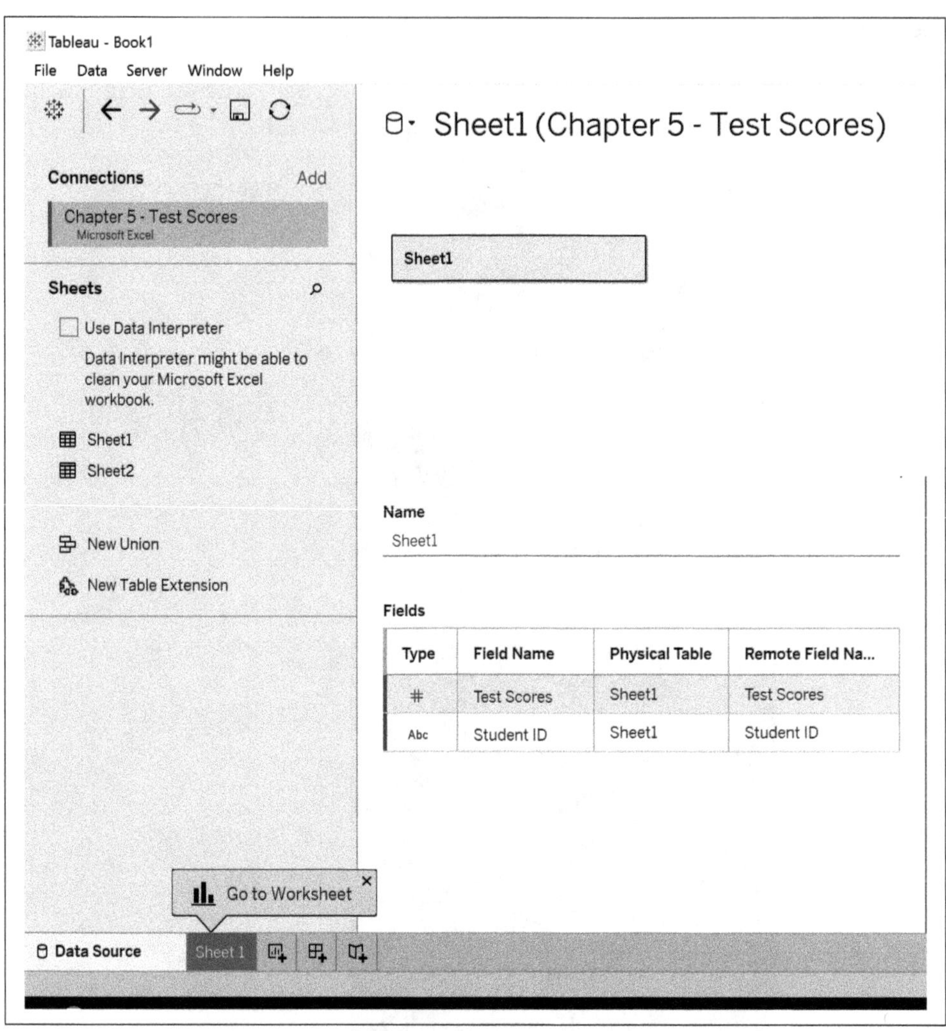

Figure 5-6. Data Source page in Tableau after connecting to the Test Scores dataset

Drag Test Scores onto the Rows shelf and Student ID to the Columns shelf. This will give you a vertical bar chart in the view, as shown in Figure 5-7.

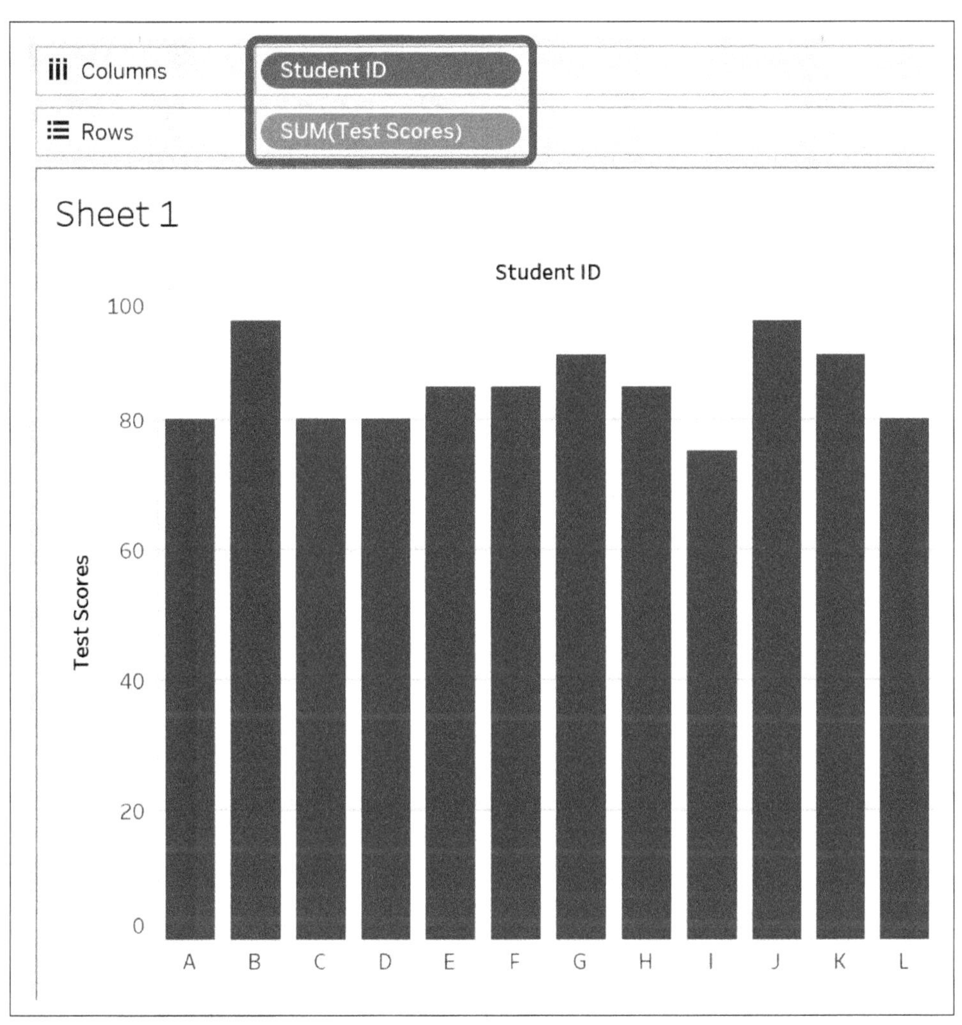

Figure 5-7. Bar chart of test scores

From here, tab to the Analytics pane, drag "Average with 95% CI" onto the view, and drop it on Table, as shown in Figure 5-8.

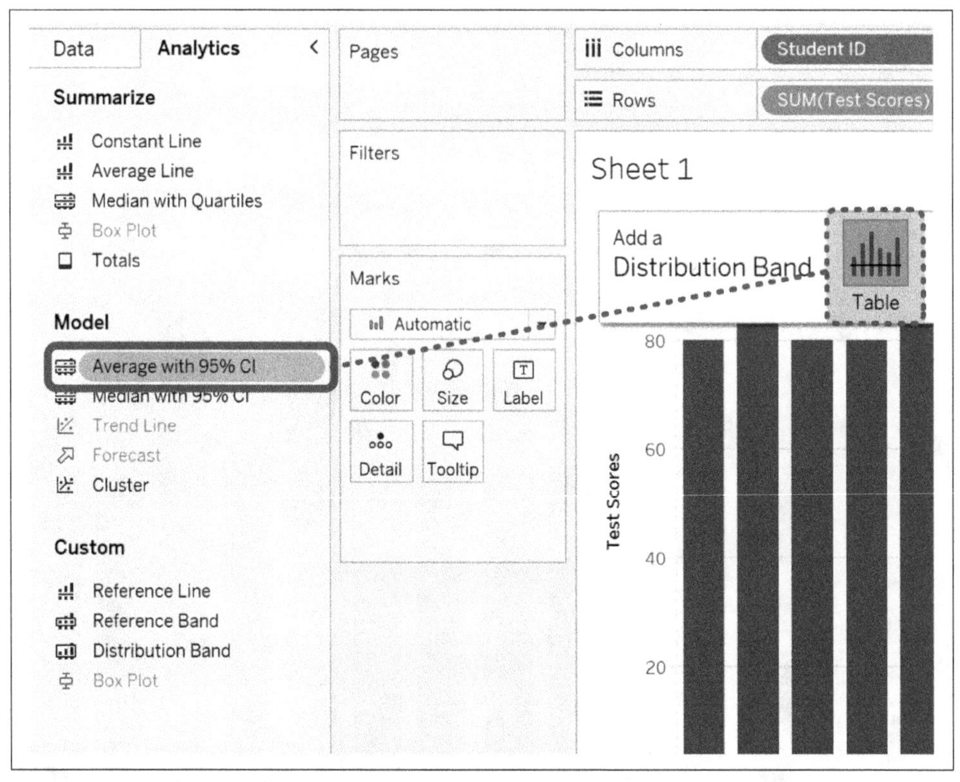

Figure 5-8. Adding confidence intervals to the view from the Analytics pane

This is going to add the average with a 95% confidence interval to the view, as shown in Figure 5-9.

If you hover over the upper and lower bounds of the confidence interval, a tooltip will activate. This shows the upper and lower range values as well as the sample average when you hover over each line. By doing this, you will see that the confidence intervals match exactly what you got when you calculated them by hand earlier in the chapter.

Getting the confidence intervals in Tableau took 2 minutes total versus the 15–20 minutes it took me to calculate the results by hand. Not only that, but Tableau is designed to compute these results quickly, even when you begin to get more data.

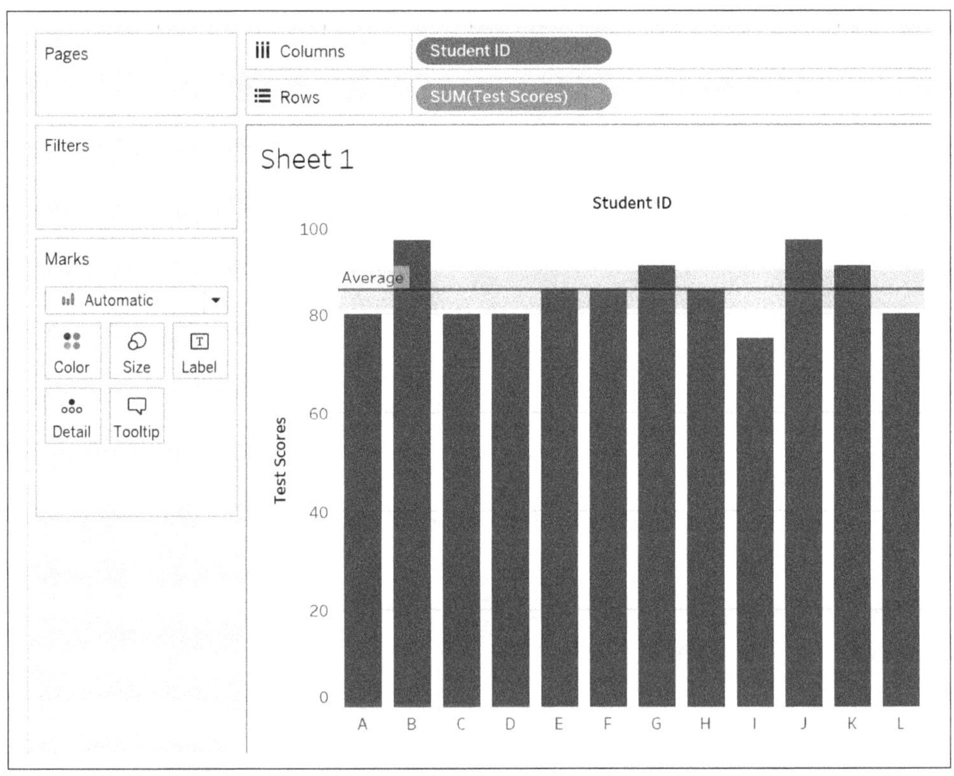

Figure 5-9. Confidence interval in Tableau Desktop

Implement Confidence Intervals on a Sample Dataset

As you can see from the previous example, Tableau can be extremely flexible for these types of analysis. This is one of the main advantages of using a visual analytics tool like Tableau versus doing this type of analysis using a coding language or by hand. You also have the advantage of filtering your data or changing what measure you are interested in on the fly. To demonstrate the flexibility of Tableau, connect to the Sample - Superstore dataset and implement a confidence interval.

To start, click Data from the top navigation and select New Data Source from the menu, as shown in Figure 5-10.

From the Connect menu, choose Sample - Superstore, which is near the bottom of the left side of the menu, as shown in Figure 5-11.

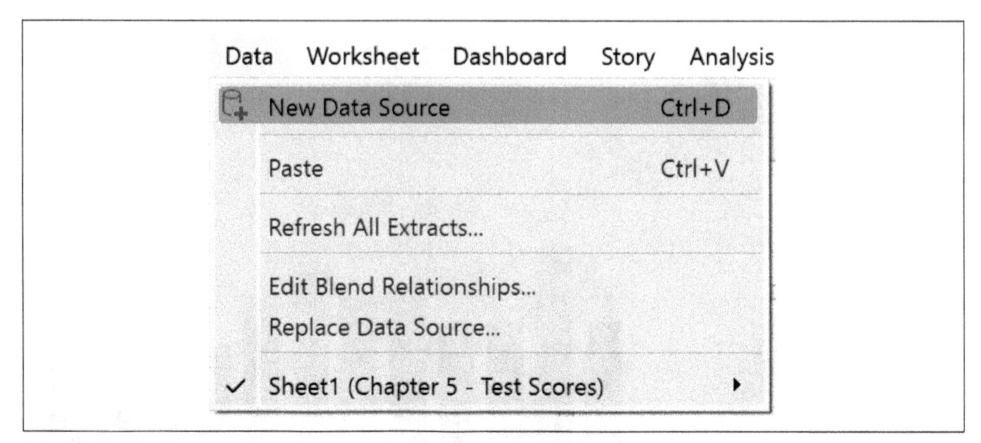

Figure 5-10. Connecting to a new data source in Tableau

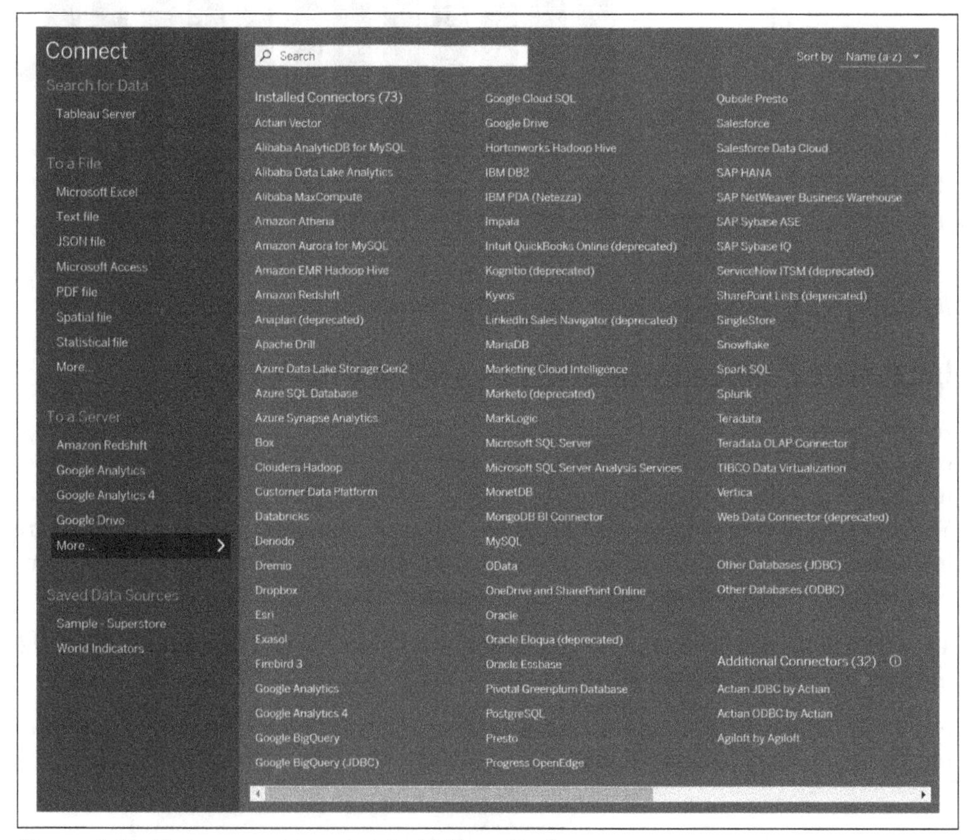

Figure 5-11. Data connection page in Tableau Desktop

Tableau will add a new data source into the top of the Data pane, and you will see a list of the measures and dimensions loaded in the Tables section of the Data pane, as shown in Figure 5-12.

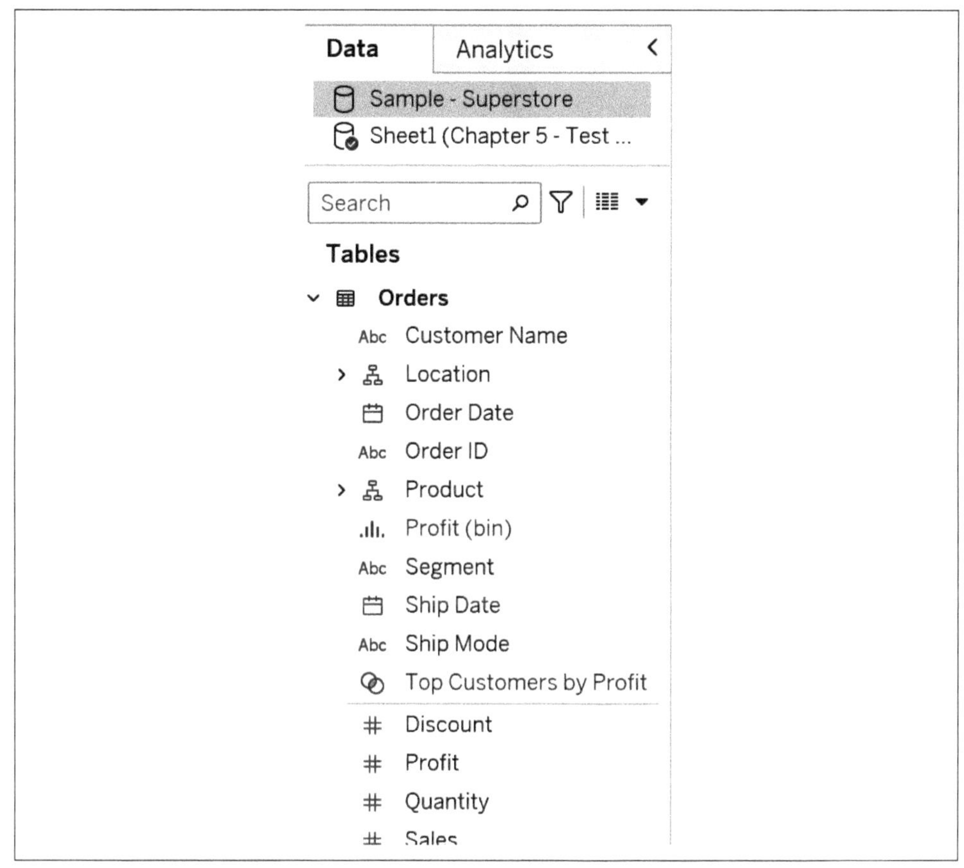

Figure 5-12. Data pane in the authoring interface of Tableau Desktop

Create a new sheet and drag Sales onto the Rows shelf and Sub-Category onto the Columns shelf, as shown in Figure 5-13. This will create a vertical bar chart that displays the sum of sales by the subcategory dimension.

Toggle to the Analytics pane and drag "Average with 95% CI" onto the view. As you can see in Figure 5-14, Tableau automatically calculates this figure for you in a matter of seconds.

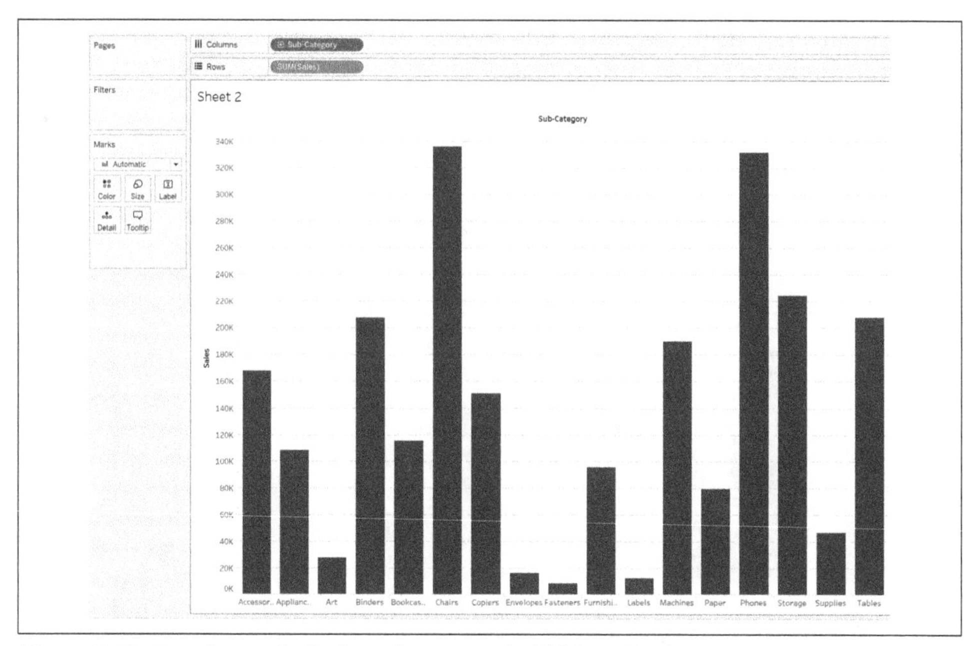

Figure 5-13. Bar chart of sales by subcategory in Tableau Desktop

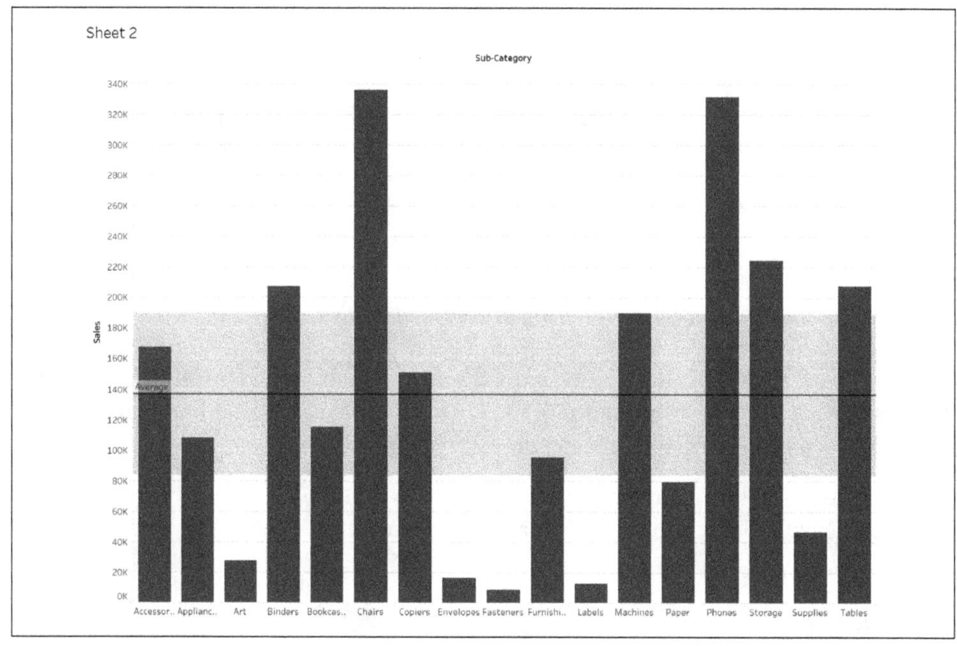

Figure 5-14. Implementing a confidence interval on bar chart of sales by subcategory in Tableau Desktop

Now that the confidence intervals are incorporated into the view, you can also change the level of detail, and Tableau will recalculate the results for you on the fly. For example, replace Sub-Category with State/Province by dragging and dropping the State/Province dimension on top of Sub-Category in the Columns shelf. Tableau will calculate the confidence interval for you immediately at this new level of detail, as shown in Figure 5-15.

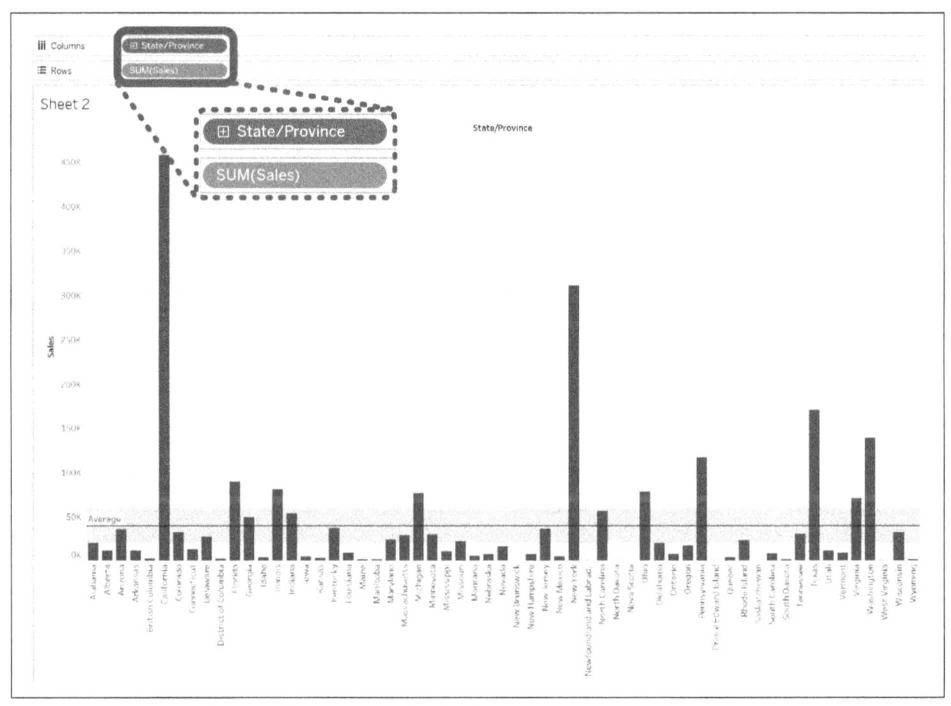

Figure 5-15. Bar chart with 95% CI of sales by state

Summary

In this chapter, you learned how to calculate confidence intervals by formula. This will allow you to better understand and communicate the results to your stakeholders. You also learned how to implement this model within Tableau using the built-in feature from the Analytics pane. This allows you to apply this method to large amounts of data. Finally, you saw the flexibility Tableau has when changing the level of detail or switching to different measures and dimensions to analyze. This allows you to move very quickly when business requirements change.

CHAPTER 6

Anomaly Detection on Normally Distributed Data

There are many ways to detect outliers in your data. You have already been introduced to one way in Chapter 4 on histograms. However, visualizing those outliers in a histogram only gets you so far. What if you want to quantify that anomaly and communicate those findings to your stakeholders? In this chapter, you will learn three different techniques you can use to quantify and visualize outliers using Tableau.

By the end of this chapter, you will be able to use standard deviations, median with quartiles, and z-scores to flag outliers in your data and present them visually to your stakeholders. It is also important to note that these methods should be used on data that fit a normal distribution, which you have learned about in previous chapters.

Understanding Standard Deviations

Standard deviation is a statistical measure that quantifies the amount of variability or dispersion within a set of data points. It measures how spread out the values are in a dataset from the mean, or average, of the data.

Mathematically, the standard deviation is calculated as the square root of the variance. The variance is obtained by taking the average of the squared differences between each data point and the mean. Recall the empirical rule (68–95–99.7) shown in Figure 6-1 (previously shown in Chapter 4).

Since standard deviations are a function of variability, they are a perfect measure to detect outliers within a dataset. To put it simply, data points that fall far from the mean are considered outliers. Standard deviation helps identify these outliers by quantifying their deviation from the average value.

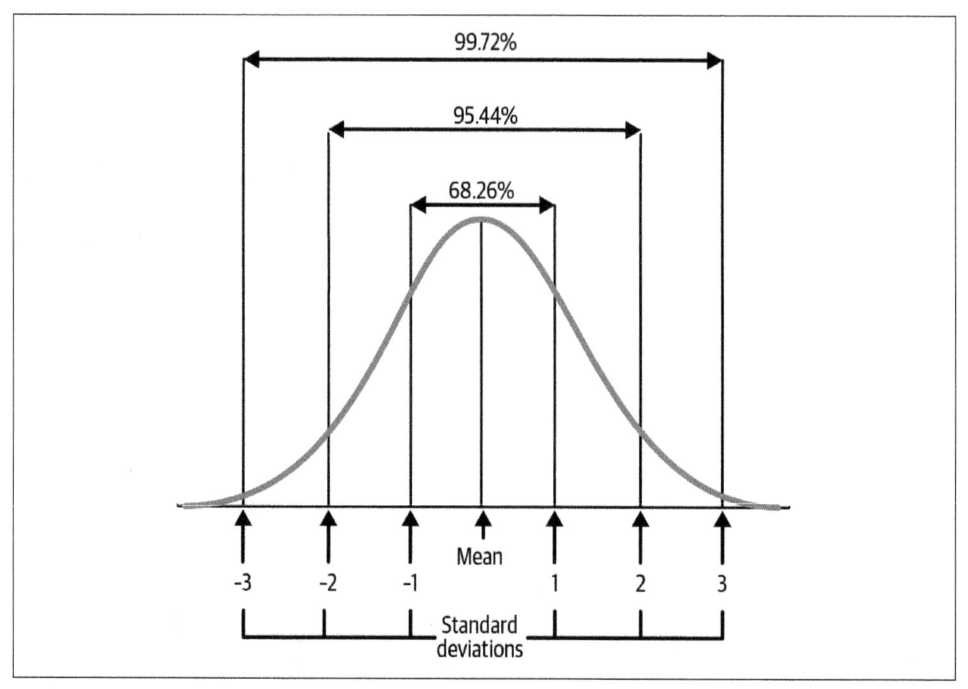

Figure 6-1. Standard deviations on a normal distribution

There are two different formulas to calculate the standard deviation:

- Sample standard deviation =

$$s = \sqrt{\frac{\Sigma_{i=1}^{n}\left(x_i - \bar{x}\right)^2}{(n-1)}}$$

- Population standard deviation =

$$\sigma = \sqrt{\frac{\Sigma_{i=1}^{n}\left(x_i - \mu\right)^2}{(n)}}$$

In Tableau, you have access to both these formulas in the form of functions. If you create a calculated field, you will find two aggregate functions and two table calculations for standard deviations (see Figure 6-2).

The functions denoted with a "P" at the end are the population standard deviation; the others are used for the sample standard deviation. The difference between the two is truly in the name. You use the sample standard deviation if you have a sample of the total population, and you use the population standard deviation if your dataset contains the total population.

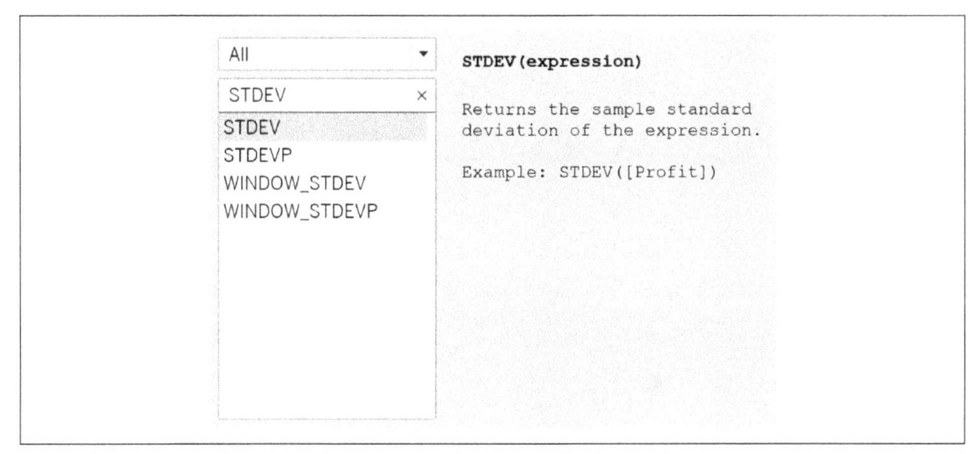

Figure 6-2. Standard deviation functions in Tableau

To give you an example, let's say you surveyed half of the students at a university. Since you only included 50% of the total population of that college campus, you would use the sample standard deviation. If you surveyed the entire campus and got 100% responses, then you would use the population standard deviation.

Seems straightforward, right?

The real answer is: it depends. Just like everything, you have to use your best judgment. Let's say the survey you sent out was asking students what political party they preferred. Even if you did survey the entire student population, if the analysis was to predict the next president in an upcoming election, the analysis would not encompass the entire voting population.

How to Implement Standard Deviations in Tableau to Find Anomalies

For this first method, I am going to use standard deviations to identify outliers of different sales months. You can plot a dual-axis line chart of the sum of profit by month of order date, as shown in Figure 6-3.

The intention here is to flag everything above 1 standard deviation from the average, as well as to add a visual cue for anything above 2 standard deviations. In Tableau, this chart type is perfect for setting this up. We will use the line to visualize the trend and the circles to highlight the outliers above 2 standard deviations. Recalling the empirical rule, 95% of the data is within 2 standard deviations from the average. This means that everything beyond that threshold could be considered extreme values in your data.

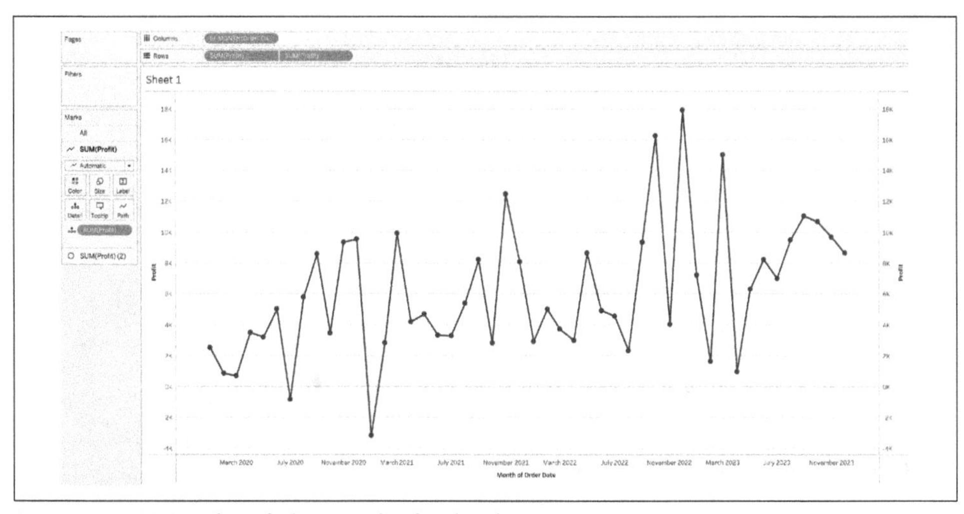

Figure 6-3. SUM of profit by month of order date

To begin, you need to add a distribution band that shows +/– 2 standard deviations. To add that band, toggle to the Analytics pane by clicking on that pane in the top left of the authoring interface, as shown in Figure 6-4.

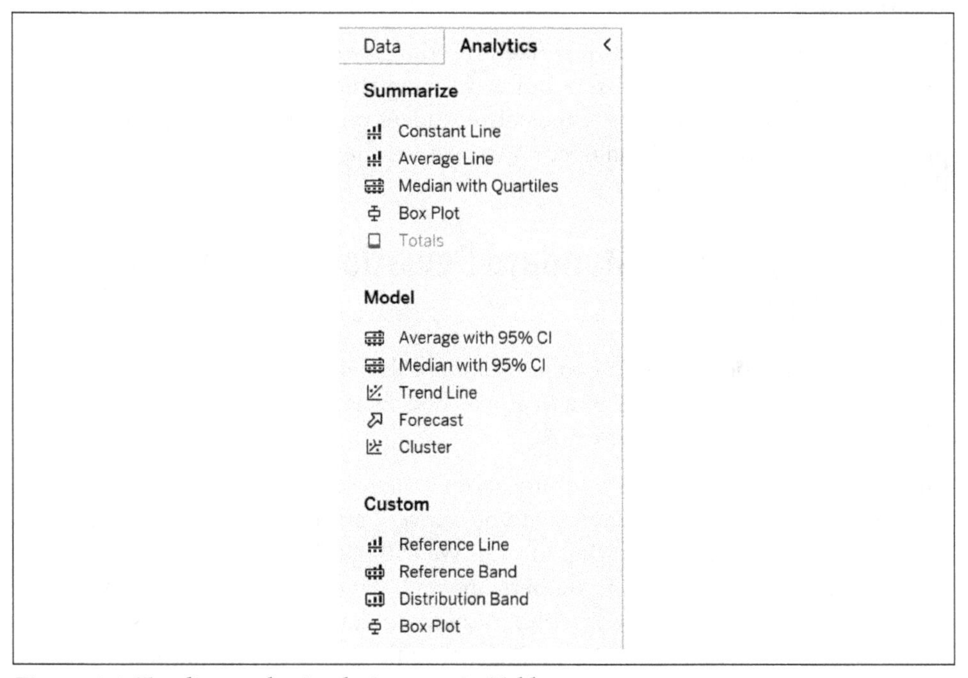

Figure 6-4. Toggling to the Analytics pane in Tableau

Click and drag the Distribution Band option onto the view and select Table to apply the band to the entire table, as shown in Figure 6-5.

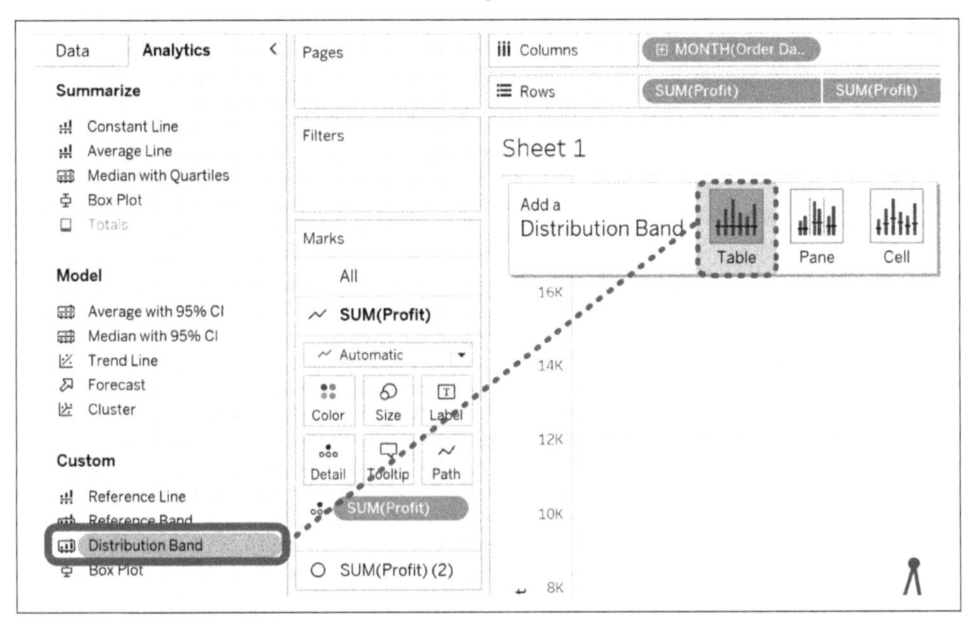

Figure 6-5. Dragging Distribution Band onto the canvas in Tableau

After adding the band to the view, you will be presented with the editing menu. Click the Value drop-down and select the Standard Deviation button. In the Factors box, type "-2,2" (see Figure 6-6). This signifies plus or minus 2 standard deviations. Click OK to close the menu.

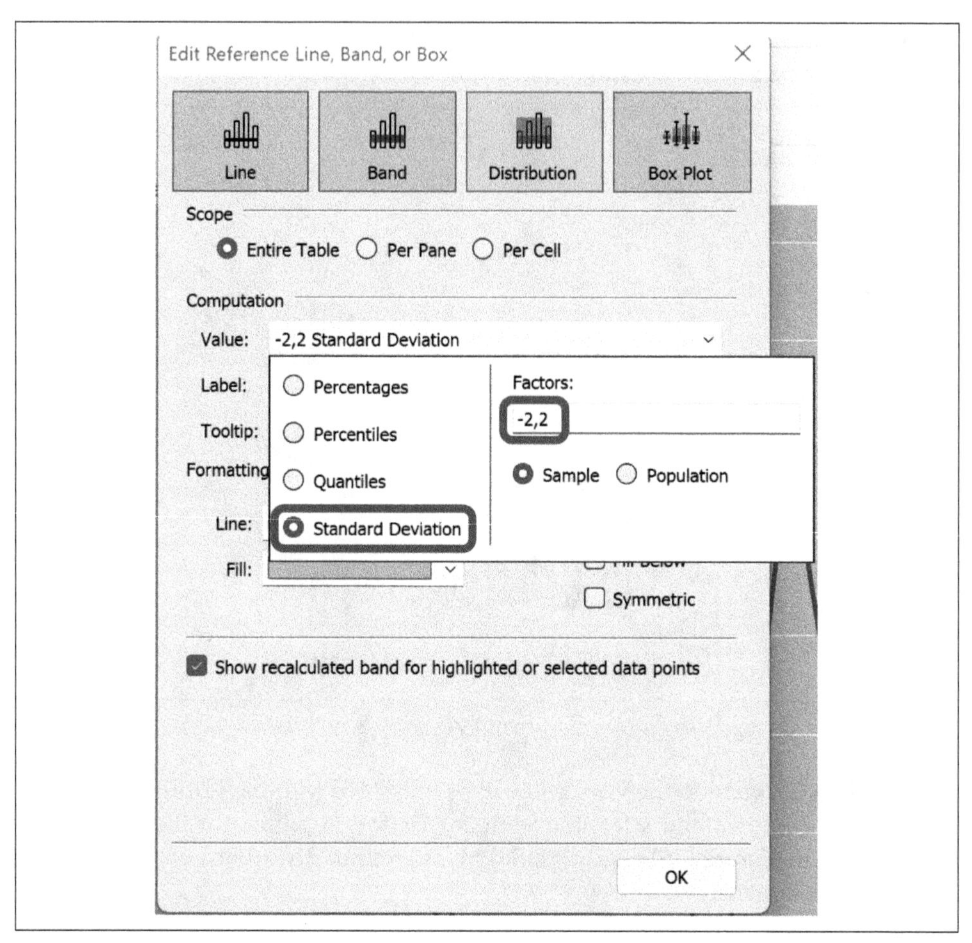

Figure 6-6. Distribution Band editor menu

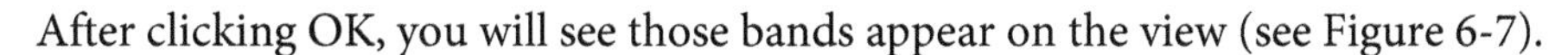

After clicking OK, you will see those bands appear on the view (see Figure 6-7).

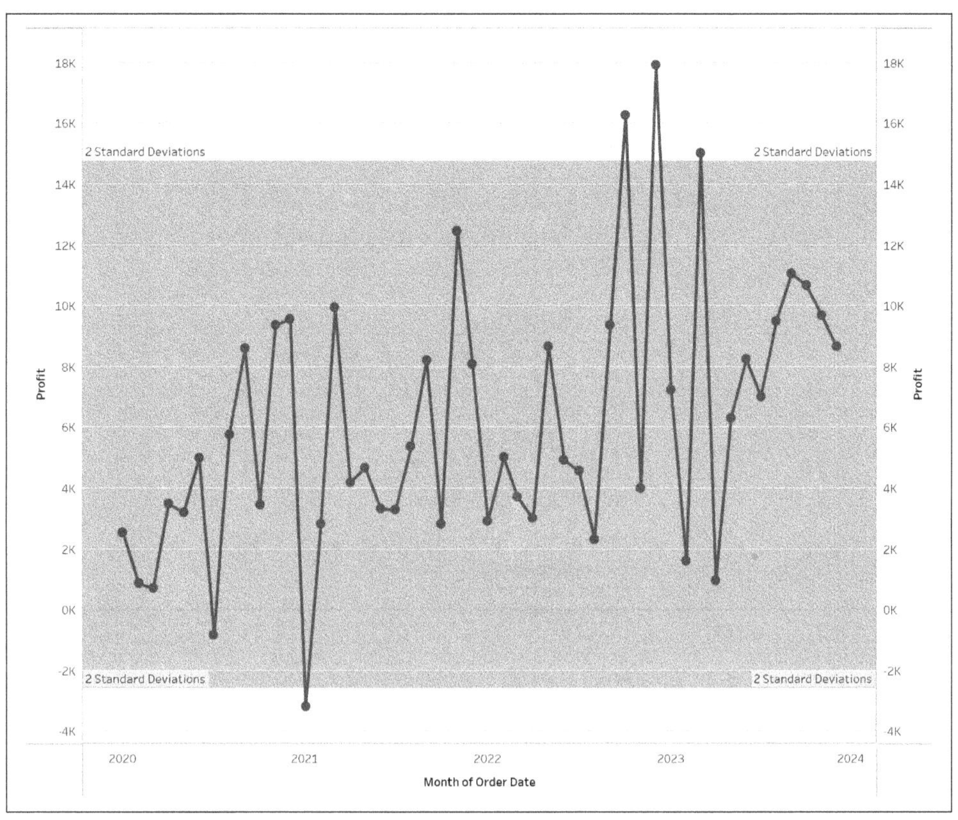

Figure 6-7. Adding plus or minus 2 standard deviations to the view in Tableau

Again, anything above or below these bands should be considered an outlier in your data. Roughly 95% of your data should fall within 2 standard deviations.

Next, you want to add another set of reference lines to signify plus or minus 1 standard deviation. To do this, repeat the steps you just completed, but leave "-1,1" in the Factors section of the menu, as shown in Figure 6-8.

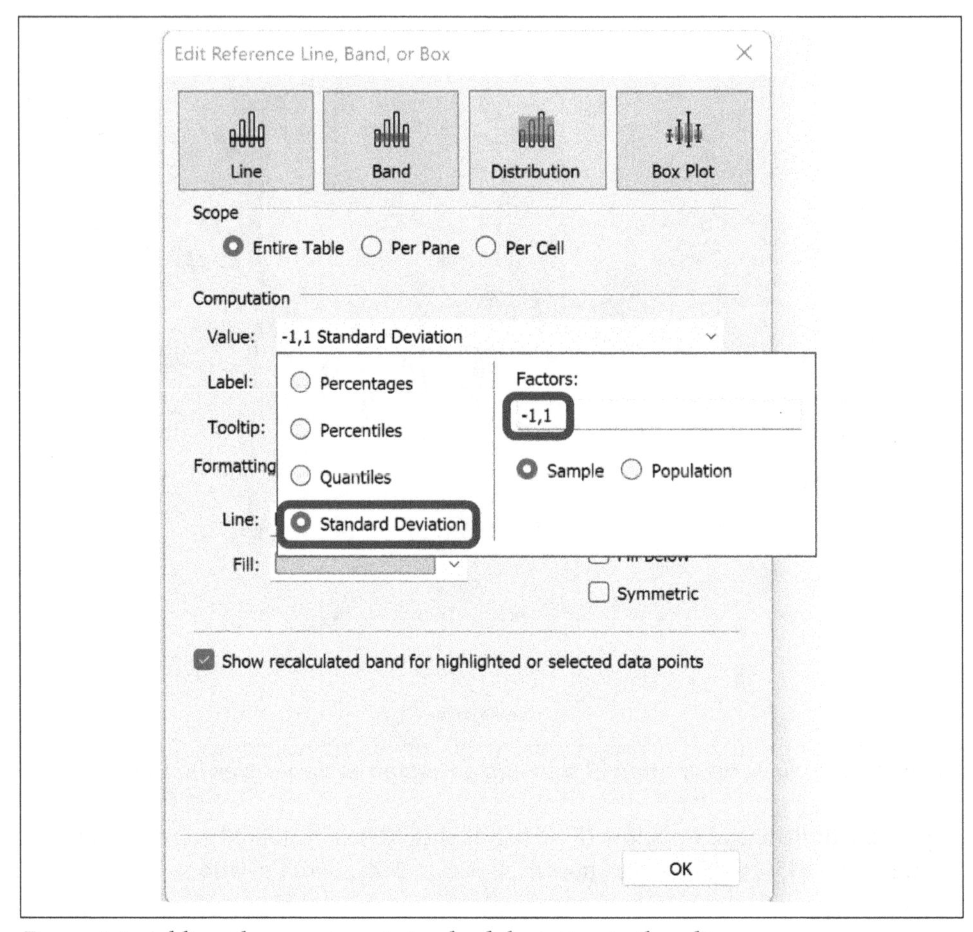

Figure 6-8. Adding plus or minus 1 standard deviation in the editor menu

Last, clean up the view a little and add conditional formatting on the marks. You can use the calculation in Figure 6-9 for the formatting:

```
IF SUM([Profit]) < (WINDOW_AVG(SUM([Profit])) -
    WINDOW_STDEV(SUM([Profit]))) THEN 'Bad Anomaly'
ELSEIF SUM([Profit]) > (WINDOW_AVG(SUM([Profit])) +
    WINDOW_STDEV(SUM([Profit]))) THEN 'Good Anomaly'
ELSE 'Expected'
END
```

In this calculation, you are flagging anything +/– 1 standard deviation and highlighting the circle in the dual axis combination chart. Add that dimension to the Color property of the Marks card for the circle. After adding the conditional formatting and editing the lines and labels of the distribution bands, the view now appears as shown in Figure 6-10.

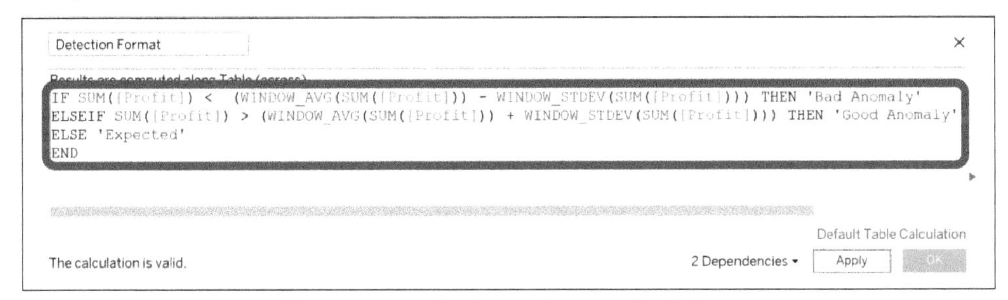

Figure 6-9. Conditional formatting calculation for standard deviation

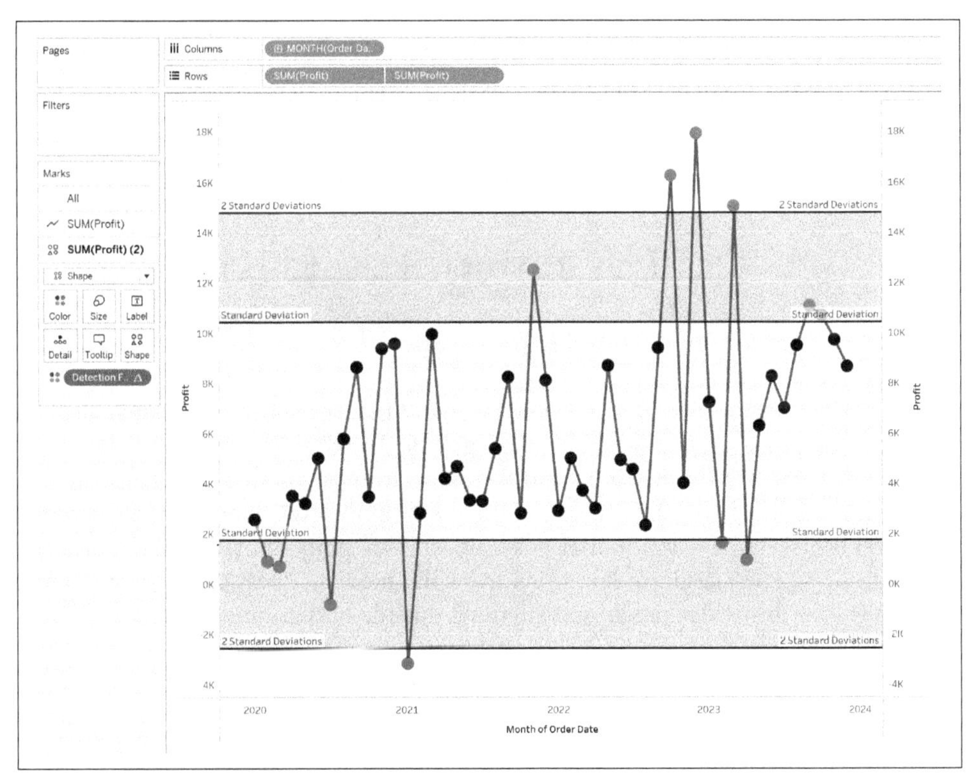

Figure 6-10. Final view after conditional formatting

In this view, you can see the outliers that fall outside the 2 standard deviation ranges as well as the data that has less variance in the middle.

Understanding Median with Quartiles

To understand the Median with Quartiles, see Figure 6-11.

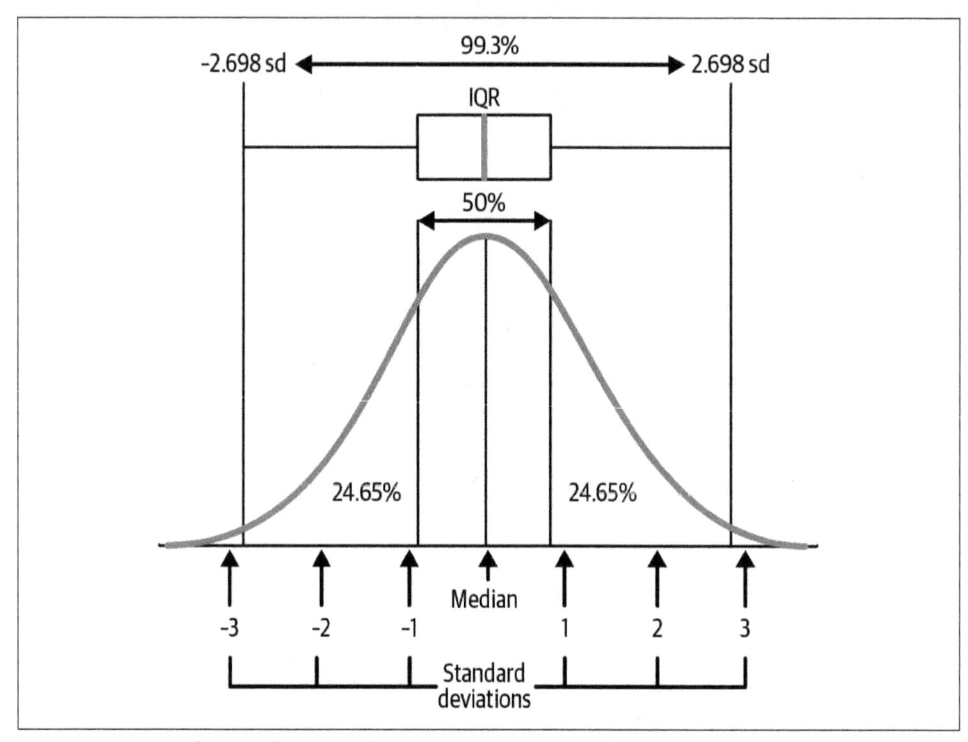

Figure 6-11. Median with Quartiles example image

Figure 6-11 is very similar to the empirical rule. First, there is a 50% band called an interquartile range (IQR), which makes up the "box" of a traditional box-and-whisker plot. The IQR captures 50% of the data within it, and the middle line is represented by the median. Then there are the upper and lower quartiles that represent the outer whiskers of a traditional box-and-whisker plot. Those whiskers measure anything +/− 2.698 standard deviations from the median. Most of the data (99.3%) will fall within this wider range, and anything outside of that range would be considered an outlier.

How to Use Median with Quartiles in Tableau to Find Anomalies

To start, use the same dual-axis line chart as before, which you can reference in Figure 6-12.

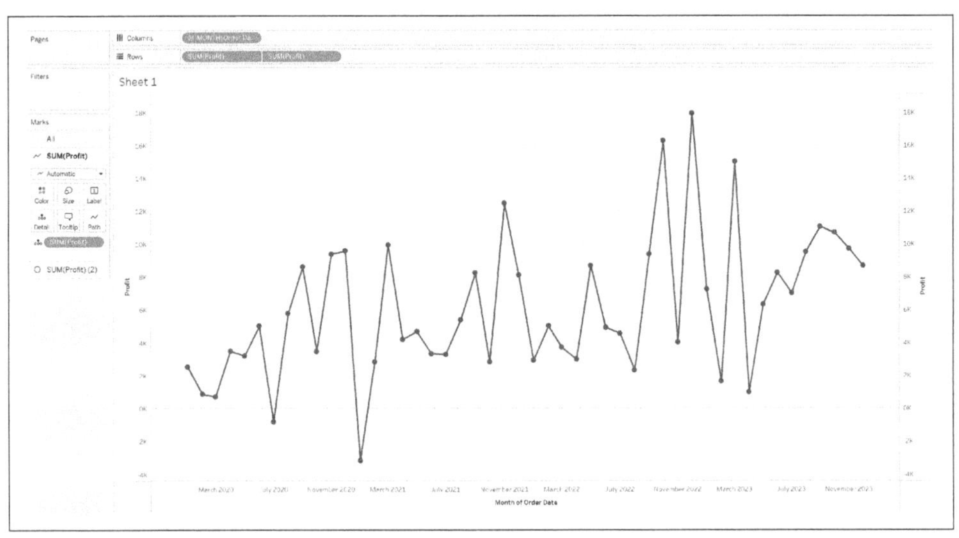

Figure 6-12. Sum of profit by month of order date

Toggle to the Analytics pane again and drag Median with Quartiles onto the view, dropping it on the Table option, as shown in Figure 6-13.

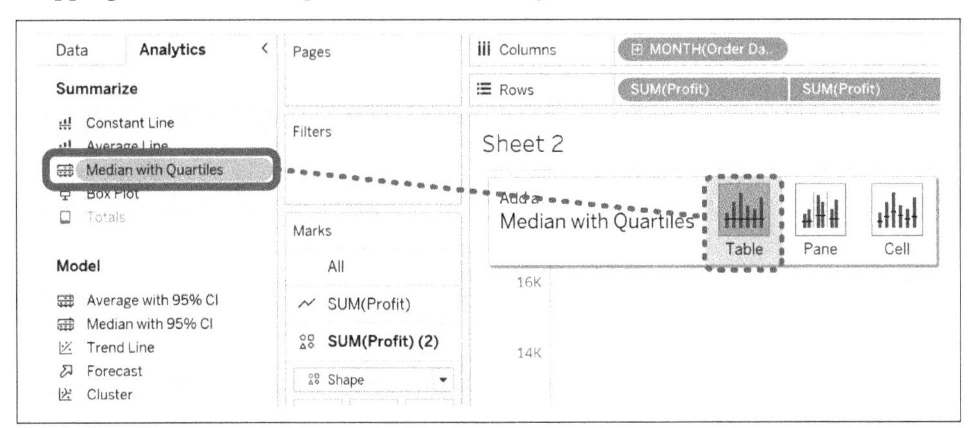

Figure 6-13. Adding Median with Quartiles to the canvas

This will display the median and the IQR on the view, as shown in Figure 6-14.

In a traditional box-and-whisker plot, this would be the "box" or IQR, and 50% of your data falls within this range. You are missing the "whisker," which would represent your upper and lower bounds and would visualize outliers.

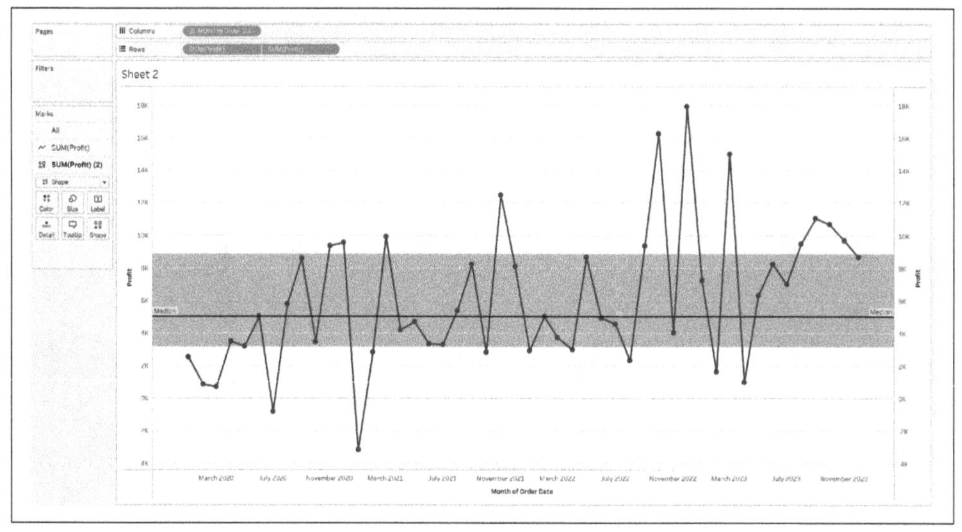

Figure 6-14. Adding the Median with Quartiles banding in Tableau

To add the "whisker," drag Distribution Band from the Analytics pane onto your view, as shown in Figure 6-15.

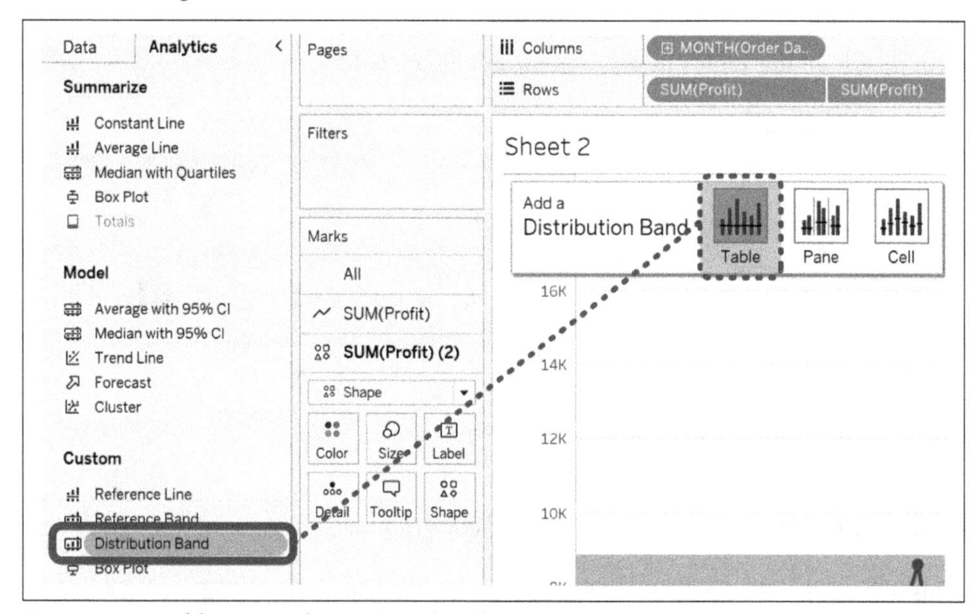

Figure 6-15. Adding Distribution Band to the view

In the menu that pops up, click on the Value drop-down, select Standard Deviation, and enter "-2.698,2.698" into the Factors text box (see Figure 6-16).

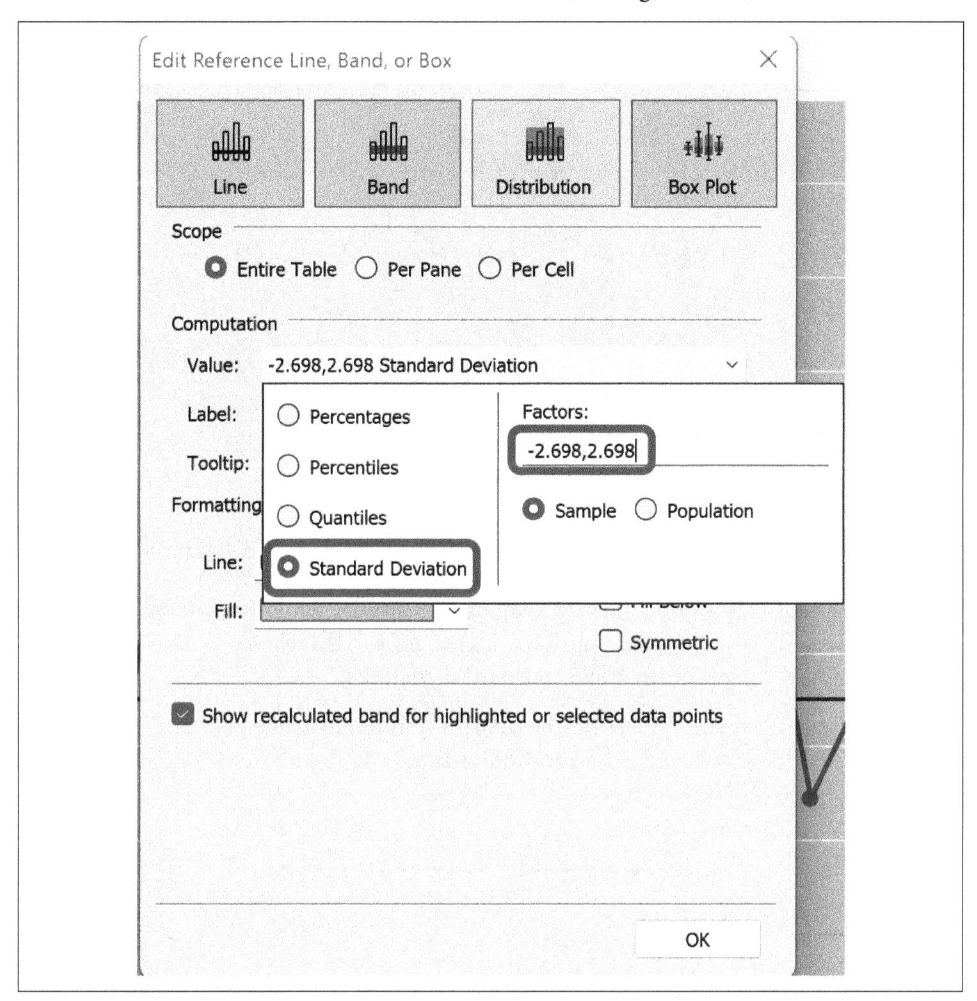

Figure 6-16. Adding the upper and lower whiskers using banding

Then click OK to close the menu and format it similarly to the last section. You can see the final view displayed in Figure 6-17.

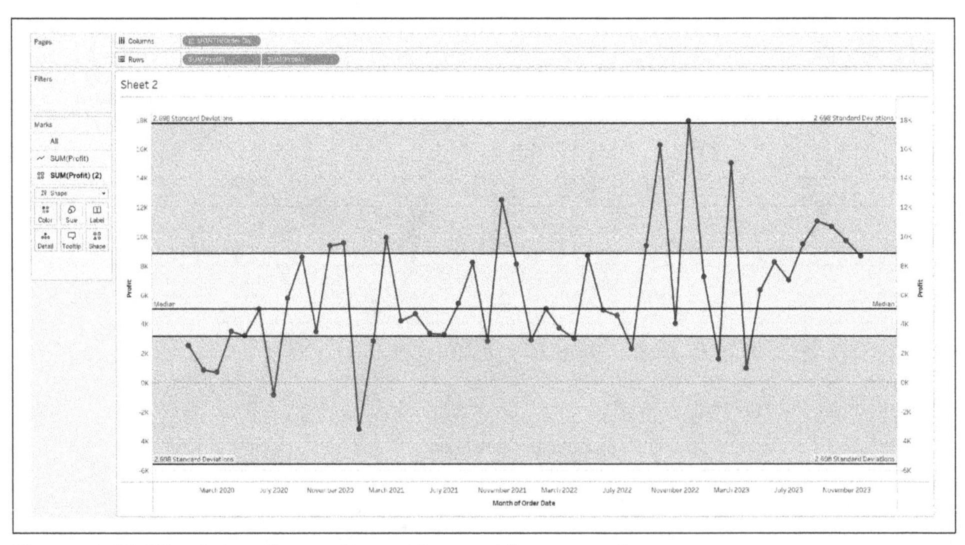

Figure 6-17. Final view for median with quartile approach

You can see only one mark on the view that falls above the upper bounds. Compare that to the final view from the standard deviation approach we saw in Figure 6-10. You can see that this approach only flagged one outlier, while the last approach had several others. This is because you have increased the threshold of allowable values from 2 standard deviations to 2.698 standard deviations.

Also, compare this technique using the distribution bands from the Analytics pane to just implementing a box-and-whisker plot, as shown in Figure 6-18.

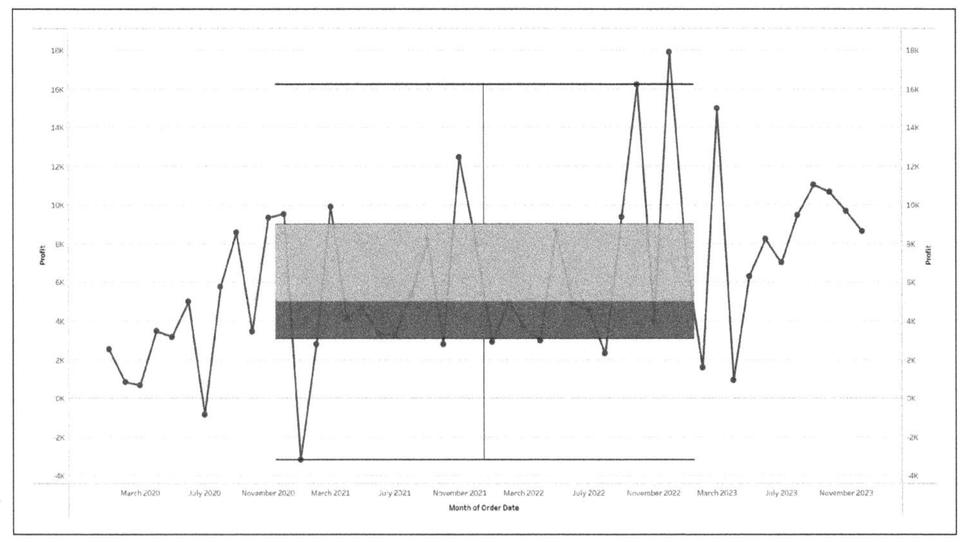

Figure 6-18. Implementing a box-and-whisker plot

While you do get the same data points, the view is poorly laid out and could raise more questions than give answers. Ultimately, you have to remember that statistics is already a complex discipline, and you don't want to add to the cognitive load that will already be on your stakeholders. Making the view easier to understand will help drive adoption of your analysis and insights.

Understanding Z-Score Tests

For the final method in this chapter, I will show you how to implement z-scores. Calculating out a z-score is a great way to index the marks in your view, and it gives you a simple way to explain outliers to your stakeholders. A *z-score* (or standard score) is a statistical measure that quantifies how far away a particular data point is from the mean of a group of data. It is expressed in terms of standard deviations from the mean. The formula to calculate a z-score takes the value of the data point, subtracts the average, and then divides by the standard deviation. The mathematical formula is:

$$z = (x - \mu) \div \sigma$$

z-score = observed value − mean of sample ÷ sample standard deviation

This formula will return how many standard deviations above or below that mark is from the mean. For instance, if the z-score of a mark is 2, that means that the data point is 2 standard deviations from the average. If a z-score is zero, that means the data point equals the average, and so on.

There are a few final things to note about z-scores. You need to have a decent sample size to get accurate results using this method. My rule of thumb is to have at least 30 data points or observations in the view. However, this depends on your data. The best practice is to see if your data follows a normal distribution by plotting a histogram (see Chapter 4). This will help you decide if this is a good method to use.

Here's the last caveat: it's standard to consider any z-score +/− 3 as an outlier, but, as I mentioned in the other sections, this is up to interpretation depending on how accurate you want to be. For this demonstration, I will flag anything +/− 2 as an outlier.

How to Use Z-Scores in Tableau to Find Anomalies

To calculate a z-score in Tableau, let's use the following formula (see Figure 6-19):

```
(SUM([Profit]) - WINDOW_AVG(SUM([Profit]))) / WINDOW_STDEV(SUM([Profit]))
```

To see this method implemented, add it to the Text property of your Marks card using the same dual-axis line chart from the other examples, as you can see in Figure 6-20.

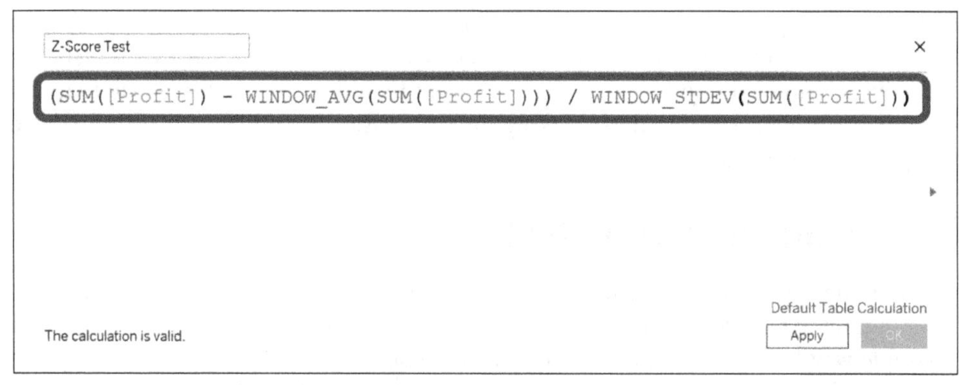

Figure 6-19. Z-score formula in Tableau

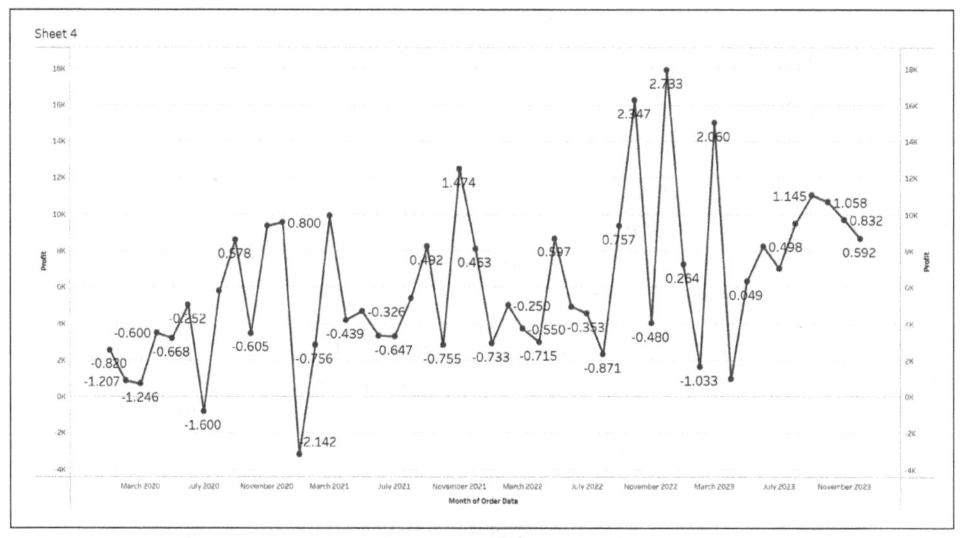

Figure 6-20. Adding z-scores to each mark in the view

Adding this formula to the Marks property is a great way to check the values to see if they make sense and if the formula is working properly. You can see the values diverge, moving down negatively and up positively from the average. To make this clearer, add an average line to the view by dragging Average Line from the Analytics pane and dropping it on Table, as shown in Figure 6-21.

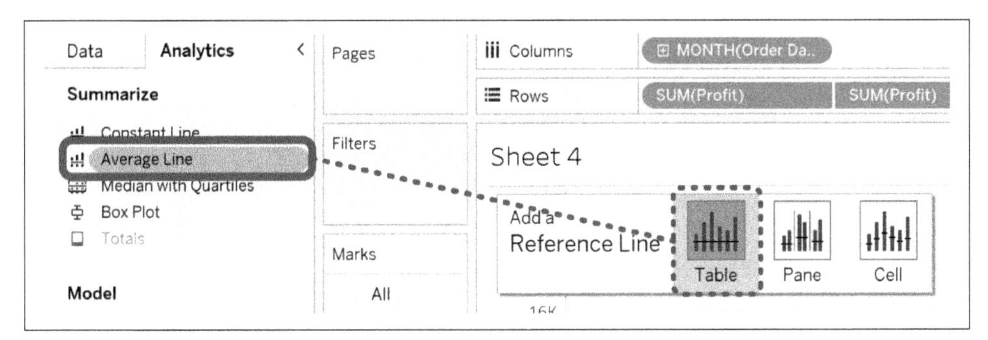

Figure 6-21. Adding an Average Line to the view

You can see that the positive and negative z-scores oscillate around that line, as shown in Figure 6-22.

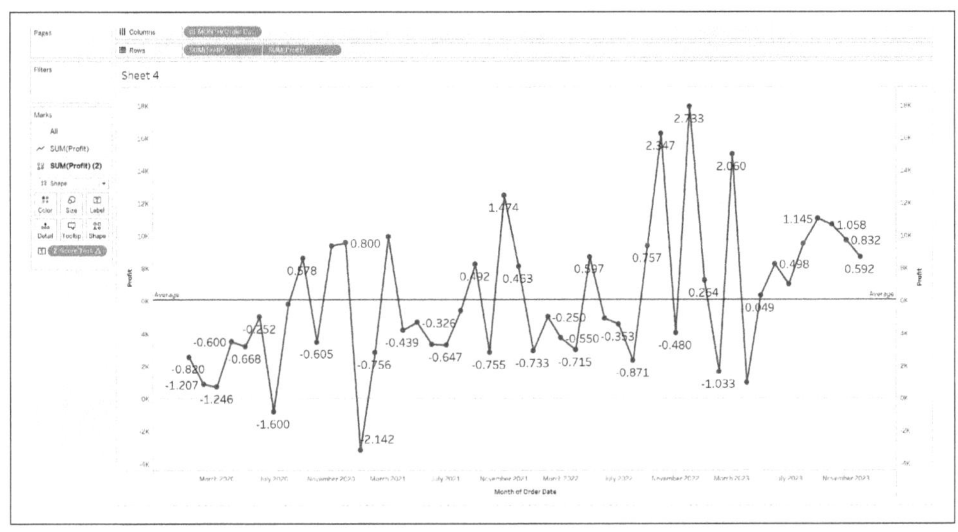

Figure 6-22. Checking the z-scores using an average line

Next, add some more reference lines to indicate what is outside of the +/– 2 range. To do this, drag Distribution Band from the Analytics pane onto the view and drop it on Table. In the menu that appears, click the Value drop-down, select Standard Deviation, and change the Factors to -2,2 (see Figure 6-23).

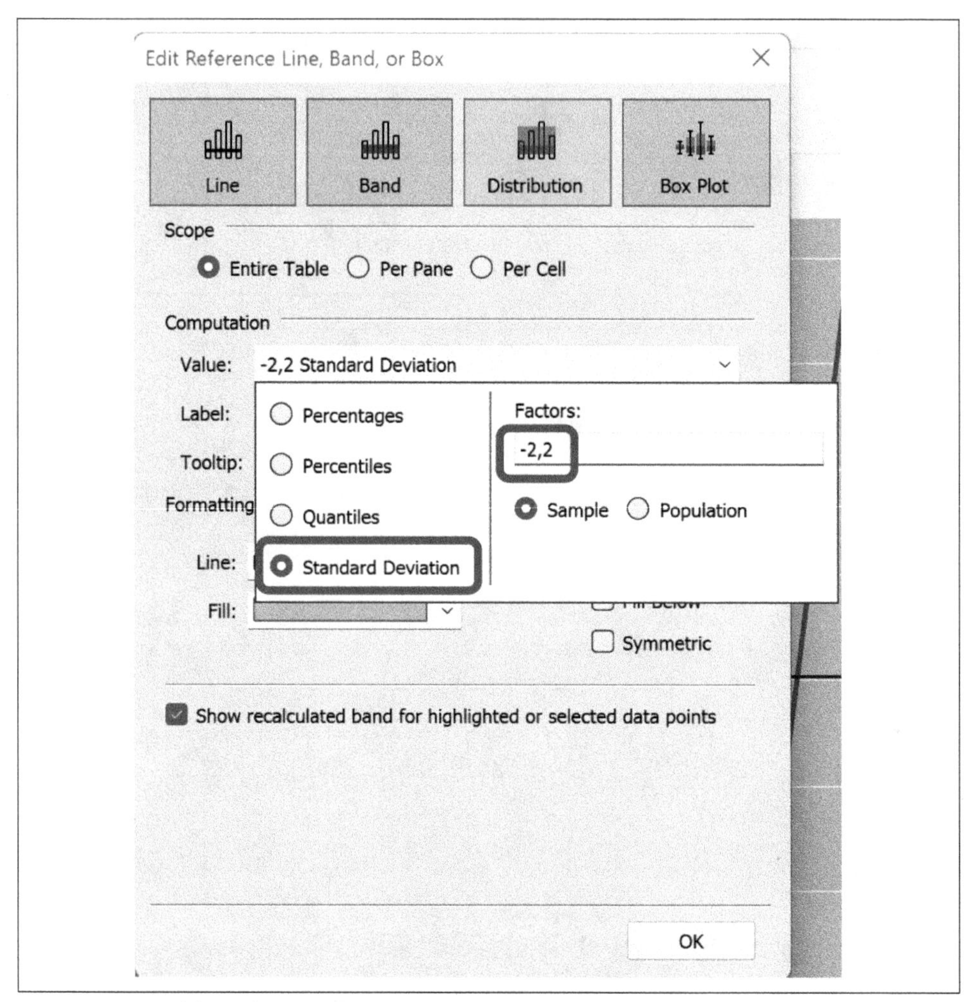

Figure 6-23. Adding plus or minus 2 standard deviations to the view

Next, format the band fill, remove the text labels, and add some conditional formatting that will highlight the outliers. Here is the formula for the conditional formatting (see Figure 6-24):

```
IF (SUM([Profit]) - WINDOW_AVG(SUM([Profit]))) /
    WINDOW_STDEV(SUM([Profit])) > 2 THEN "Good Anomaly"
ELSEIF (SUM([Profit]) - WINDOW_AVG(SUM([Profit]))) /
    WINDOW_STDEV(SUM([Profit])) < -2 THEN "Bad Anomaly"
ELSE "Expected"
END
```

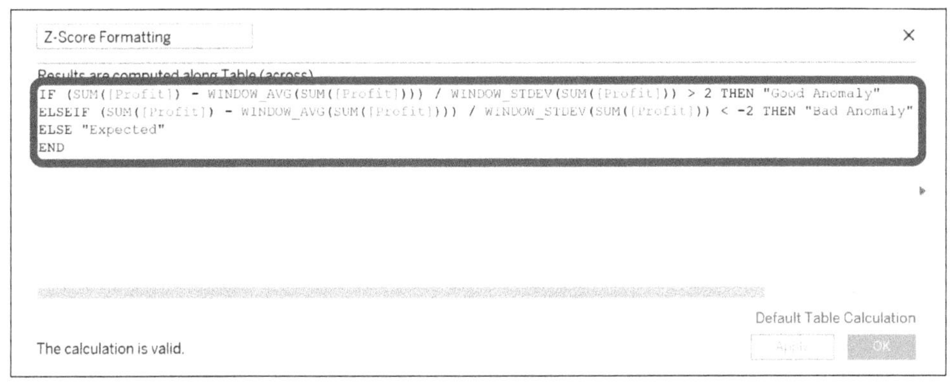

Figure 6-24. Conditional formatting formula for z-scores approach

The final view after adding the conditional formatting and formatting the lines properly can be seen in Figure 6-25.

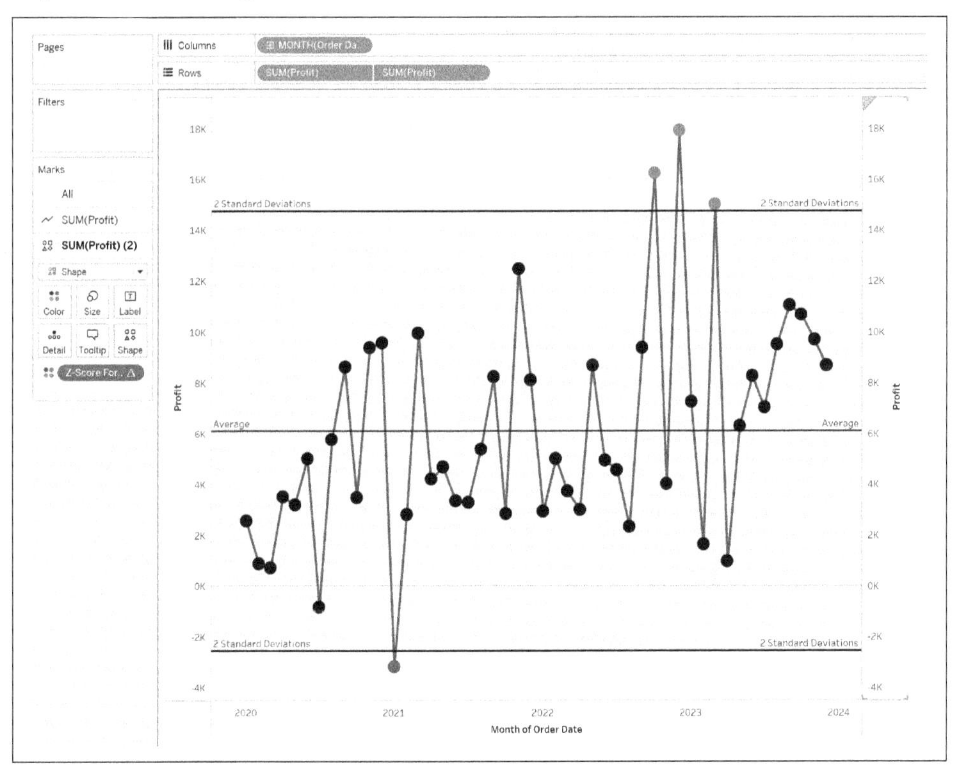

Figure 6-25. Final view using z-score approach

Summary

In closing, there are many methods you can use to visualize outliers in your data. They all have their own pros and cons. My recommendation is to know your audience and have an understanding of their level of comfort and what model would make the most sense to them. My rule of thumb with these types of decisions has always been: the simpler the better. Don't try to overcomplicate things for the additional accuracy percentile unless your use case calls for it.

In this chapter, I showed you three techniques you can implement in Tableau to visualize anomalies in your data. I covered an approach using standard deviations from the average and median with quartiles, as well as z-scores. Each of these techniques assumes a normal distribution of your data. In Chapter 7, I will cover three more techniques you can use when your data is not normally distributed.

CHAPTER 7

Anomaly Detection on Nonnormalized Data

In Chapter 6, I showed you three ways to visualize outliers when your data is normally distributed. However, oftentimes you will come across data that isn't normally distributed. Using methods that assume a normal distribution could lead to false conclusions or misguided decisions by you and your stakeholders. That is why the exploratory tactics covered in Chapter 4 are so important.

In this chapter, I will show you three methods you can implement to visualize outliers when you are working with nonnormalized data. The methods are mean absolute deviation, Tukey's fences, and modified z-score test.

Understanding Median Absolute Deviation

The *median absolute deviation* (MAD) is a statistical measure that quantifies the dispersion or variability of a dataset. It is calculated by finding the absolute deviation of each data point by subtracting the median from each value and taking the absolute value of the result. Then you find the median of the absolute deviations, which gives you the MAD. The mathematical formula to calculate the MAD is as follows:

$$\text{MAD} = \text{Median}(|X_i - \text{Median}|)$$

where

MAD = median absolute deviation

X_i = each value

Median = median value

The steps to find the MAD are very simple when you break this formula down. Consider this dataset as an example: 5, 10, 12, 15, 18. Here are the steps to find the MAD from this sample dataset:

1. Find the median. In this dataset you can see that the median value is 12.
2. Next, you will calculate the absolute deviations from the median by subtracting each value by the median and then taking the absolute value of the result:

   $|5 - 12| = 7$

   $|10 - 12| = 2$

   $|12 - 12| = 0$

   $|15 - 12| = 3$

   $|18 - 12| = 6$

3. Now calculate the median of the absolute deviations. The absolute deviations are 7, 2, 0, 3, 6. When arranged in ascending order you can see that the median is 3 (0, 2, **3**, 6, 7).

Using the MAD statistic, you can then calculate an upper and lower bound. Anything that falls outside of this range could be considered an outlier. The formulas for those calculations are:

$$\text{Lower MAD} = \text{median} - k \times \text{MAD}$$

$$\text{Upper MAD} = \text{median} + k \times \text{MAD}$$

where k is a constant value that is typically set to 3. You can think of this k value as a multiplier that sets the threshold of the analysis. Similar to looking at 2 or 3 standard deviations from the average, the k constant ensures your threshold is far enough away from the median that the value would be considered an outlier.

While the execution of this approach may seem simple, this technique is very efficient at detecting outliers in a dataset due to its unique properties. Since the primary measure used is the median, this technique is less affected by extreme values in the dataset. This technique is also an example of a nonparametric method, which was introduced in Chapter 4. Recall that nonparametric models are flexible and can handle a wide range of data distributions without assuming a specific functional form.

To sum it up, this method is simple to implement and is a powerful method that has a wide range of applications.

How to Implement Median Absolute Deviations in Tableau

To implement this method in Tableau, connect to the Sample - Superstore dataset and create a line chart of the sum of sales by continuous month of order date. I also like to add a dual axis of the same measure and encode those marks as circles, as shown in Figure 7-1. This will come in handy later when implementing conditional formatting to visualize the outliers.

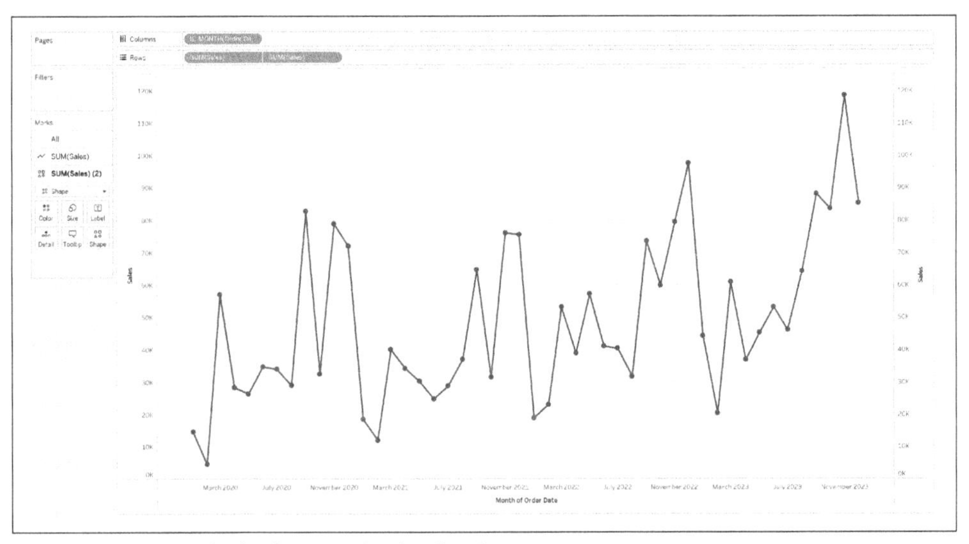

Figure 7-1. Sum of sales by month of order date

Next, add a reference line of the median sum of sales to the view. Just like everything in Tableau, there are multiple ways to do this; I've discussed several in previous chapters. Here is another way: right-click on the y-axis and select Add Reference Line from the options that appear, as shown in Figure 7-2.

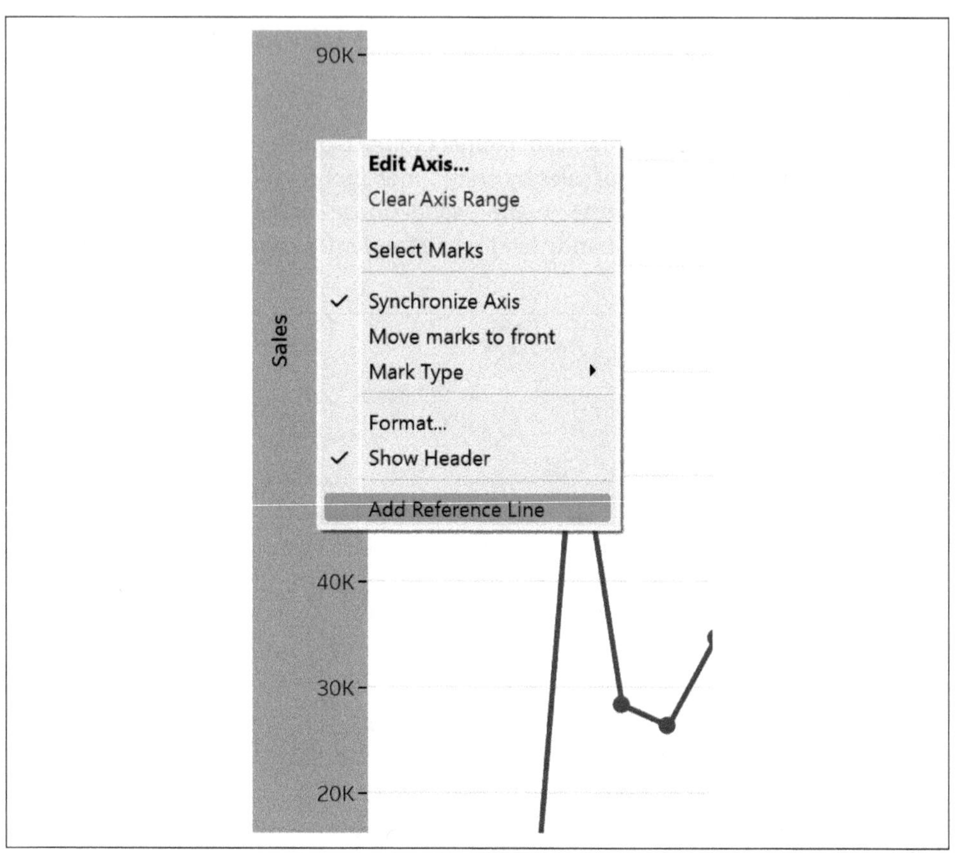

Figure 7-2. Add a reference line of the median sum of sales.

After you make this selection, you should see the "Add Reference Line, Band, or Box" menu appear. Choose Median as the aggregation of the reference line. You will see the line appear on the view behind the menu after making the selection, as shown in Figure 7-3. When you are finished, click OK to close the menu.

This reference line will help you visualize the outliers later, and it gives you the median value so you can validate our calculations in the following steps.

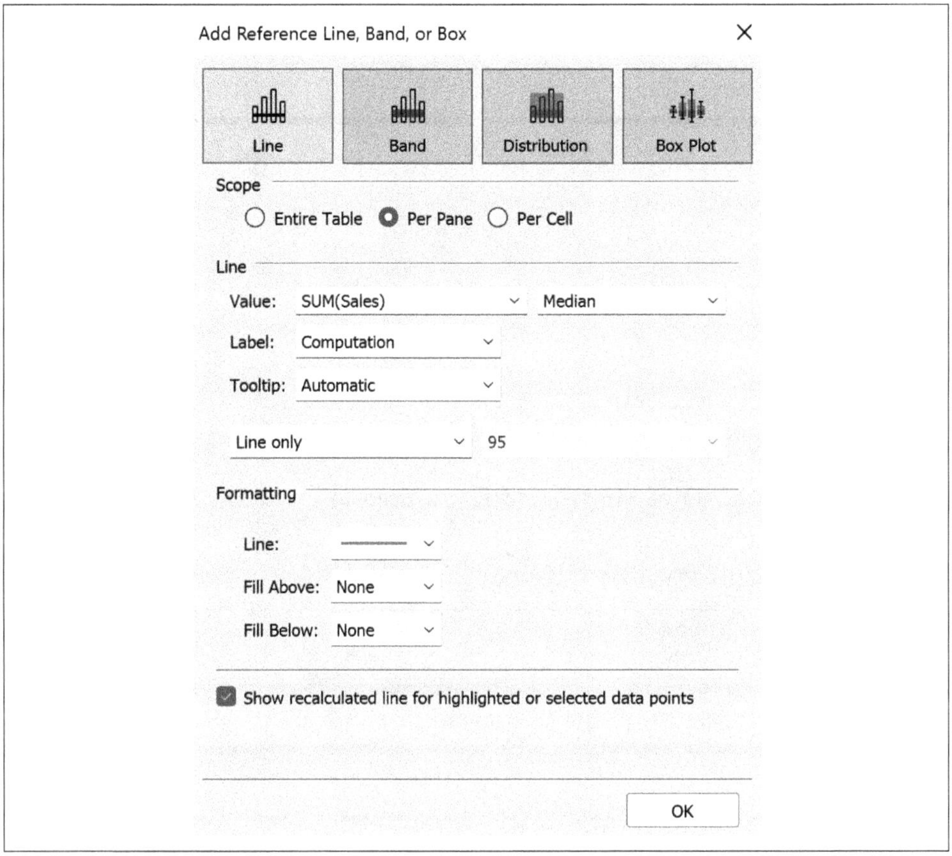

Figure 7-3. "Add Reference Line, Band, or Box" menu and selections

Now you can begin creating the calculated fields needed to execute this method. Recall from the example at the beginning of the chapter (see "Understanding Median Absolute Deviation" on page 117) that you have three calculations you need to create: the MAD, a lower bound, and an upper bound. I will start by creating the MAD statistic (see Figure 7-4):

```
WINDOW_MEDIAN(ABS(SUM([Sales])-WINDOW_MEDIAN(SUM([Sales]))))
```

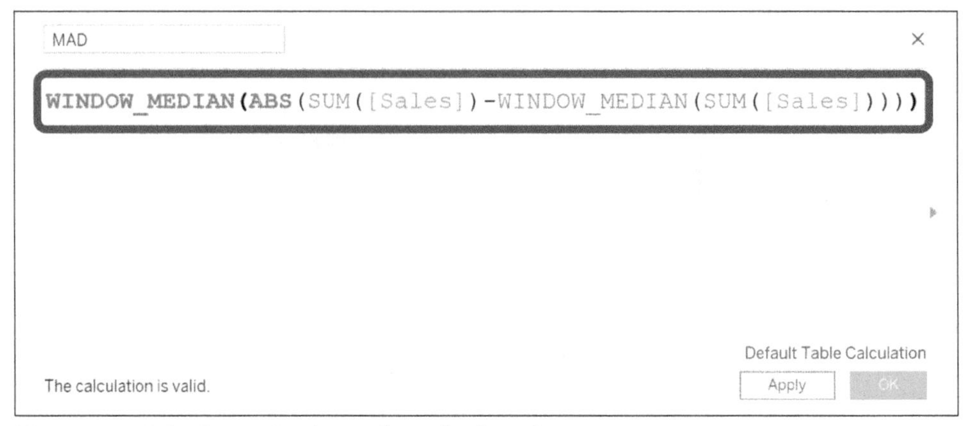

Figure 7-4. Calculation for the median absolute deviation

This gives you a MAD of $16,370. Next, you need to create the lower bound (see Figure 7-5):

```
WINDOW_MEDIAN(SUM([Sales])) - 3 * [MAD]
```

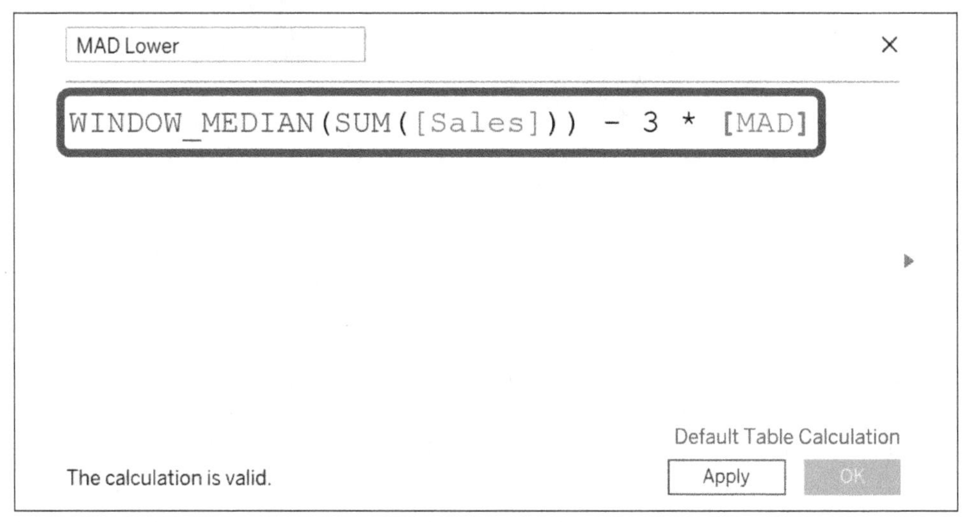

Figure 7-5. Median absolute deviation lower bound

Next, create the upper bound calculation (see Figure 7-6):

```
WINDOW_MEDIAN(SUM([Sales])) + 3 * [MAD]
```

You can see that this calculation is basically the same as the lower bound, but it adds the *k* constant of 3 instead of subtracting it.

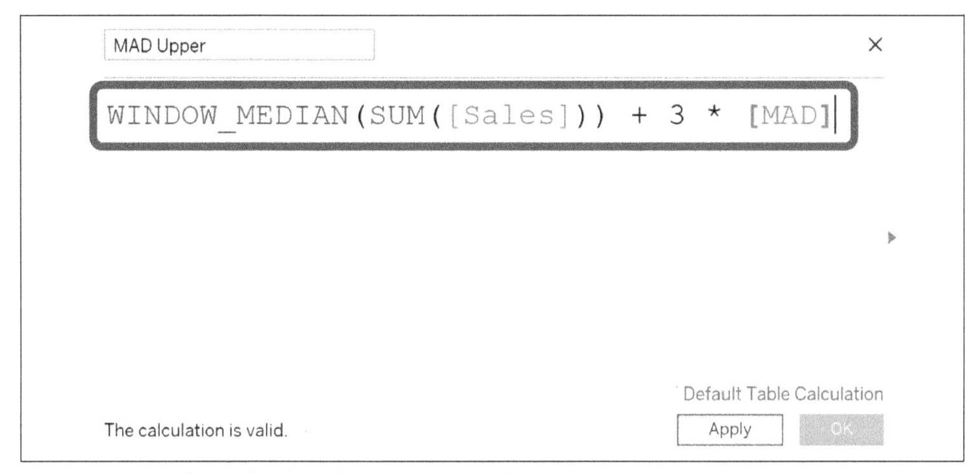

Figure 7-6. Median absolute deviation upper bound

Now you are ready to visualize this technique by adding your upper and lower bound calculations to the view using reference lines. To begin, you will add both calculated fields to the Detail property of the Marks card, as shown in Figure 7-7.

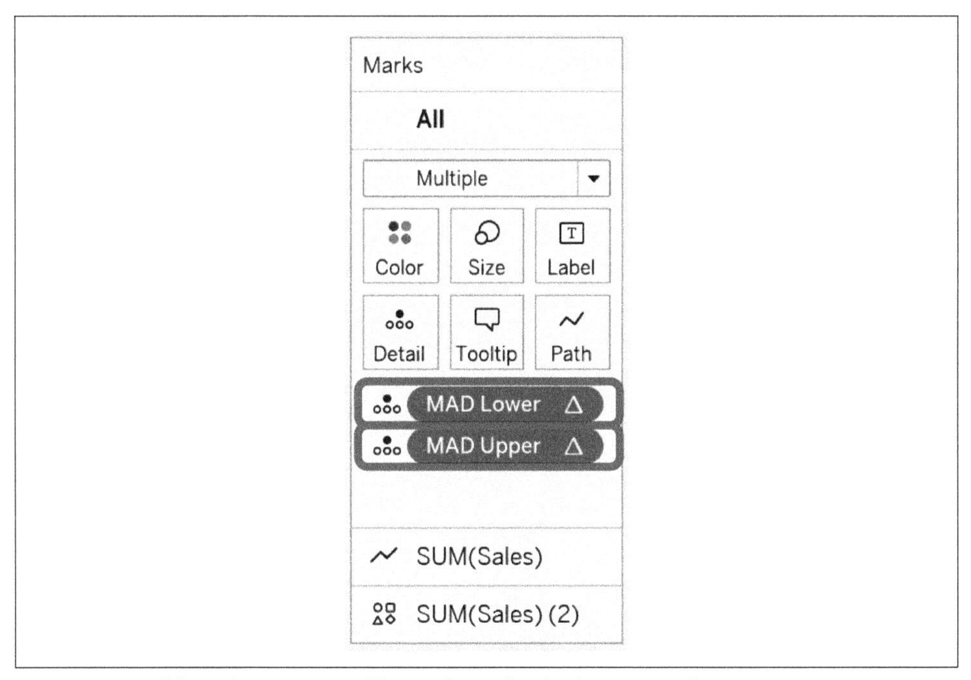

Figure 7-7. Adding the upper and lower bound calculations to the view

Right-click on the y-axis and select Add Reference Line. With those calculations in the Detail property of the Marks card, you will be able to select them as options from this menu. In the Line section, click on the drop-down next to Value and select MAD Lower, as shown in Figure 7-8.

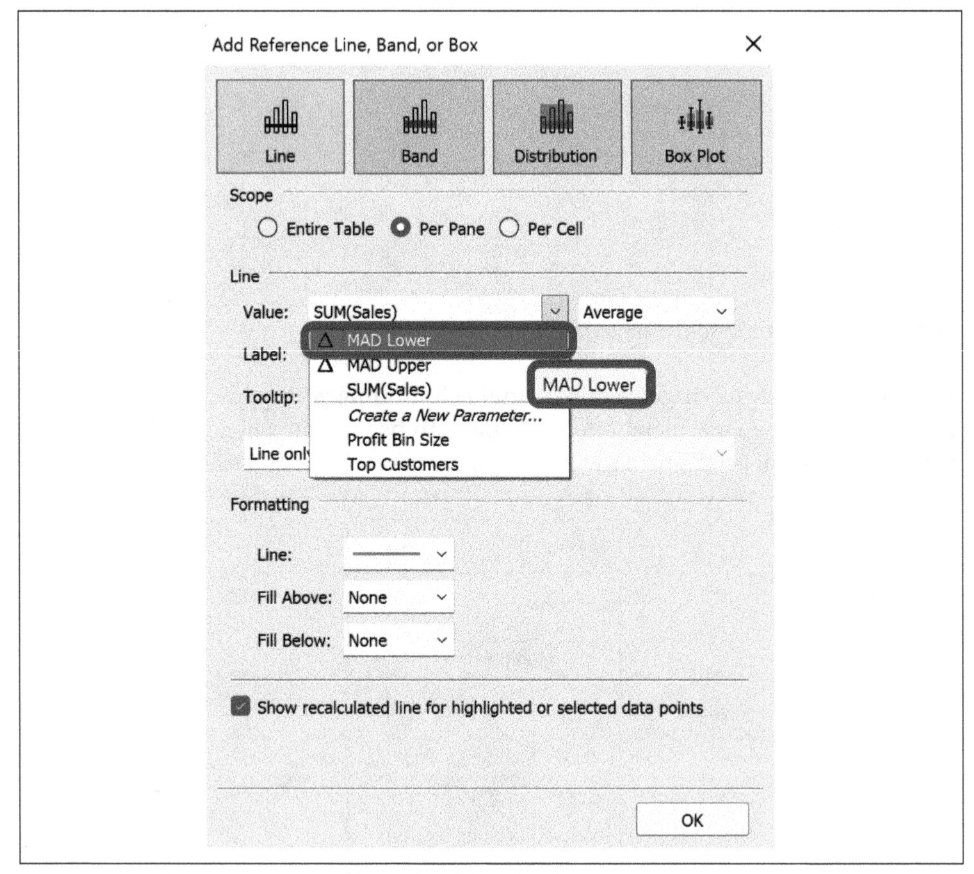

Figure 7-8. Select MAD Lower from the drop-down menu

You will see the lower bound reference line appear on the view. Now follow the same steps for the upper bound line. You can see both implemented on the view in Figure 7-9.

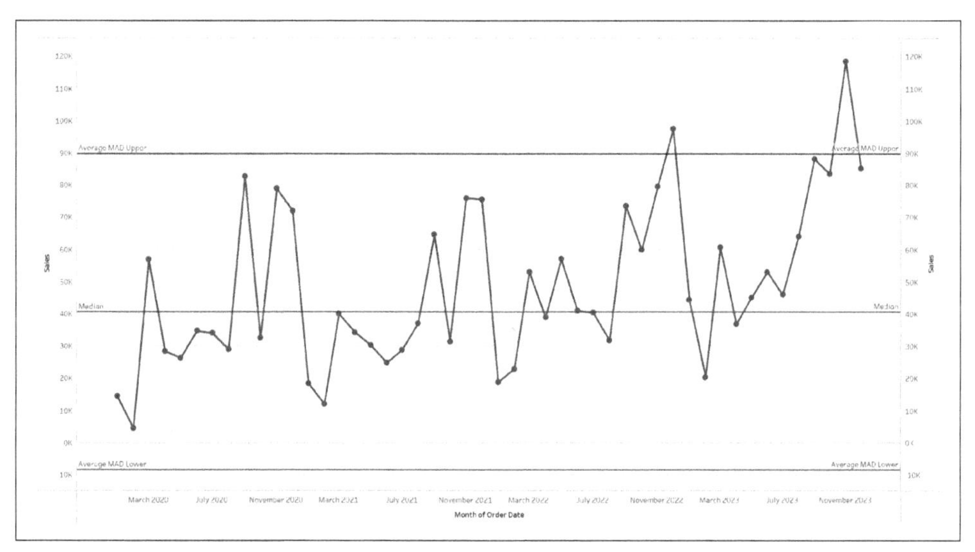

Figure 7-9. Median absolute deviation upper and lower bounds on the view

You can see that there are two values that exceed the upper bound. You could consider those outliers or deviations from the norm with this data and conduct a more in-depth analysis on those months to find out why.

Understanding Modified Z-Score

In Chapter 6, I introduced a method that used z-scores to detect outliers. The original formula took the value, subtracted the average, and then divided by the standard deviation. Mathematically this formula is as follows:

$$z = (x - \mu) \div \sigma$$

The standard z-score is dependent on a normal distribution due to its reliance on averages and the standard deviation. You can modify this formula, however, to make it more robust by using the median and incorporating the MAD. Here is the new mathematical formula after the modification:

$$\text{Modified z-score} = (x - \text{Median}) \div \text{MAD}$$

Since you are using the median, your calculation is less impacted by extreme outliers. Using this new modified z-score, all of the scores will oscillate around the median. Typically you would call anything that has a modified z-score of a +/– 3 an outlier.

How to Implement Modified Z-Scores in Tableau

To begin, start with the same line chart you created in the previous section, as shown in Figure 7-10.

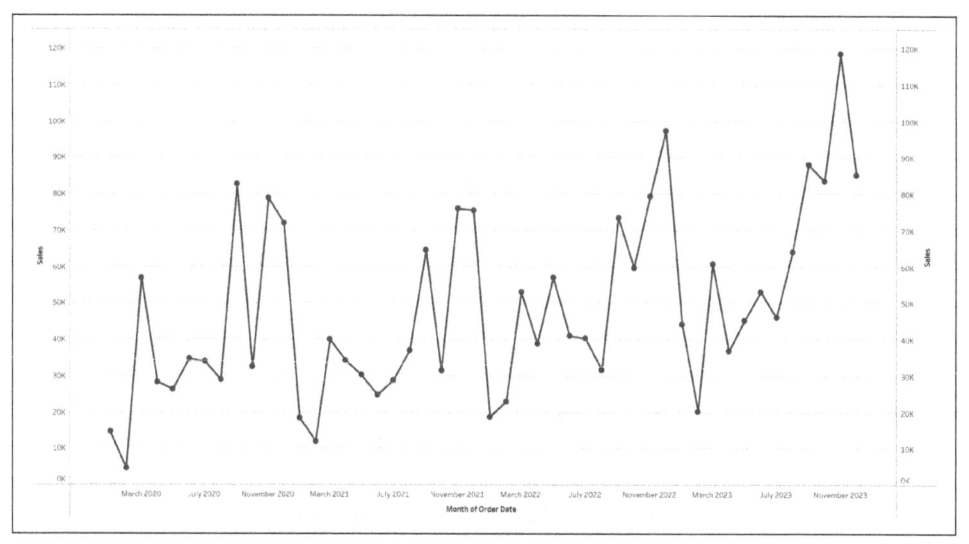

Figure 7-10. Sum of sales by month of order date line chart

Since you are using the median of the data, add the median as a reference line, as shown in Figure 7-11. Right-click the y-axis, choose Add Reference Line, and select Median from the aggregations to the right of the Value drop-down.

Now that the view is set up, it is time to create the calculations. For this method, you need a calculation for the MAD and one for the modified z-scores. Refer back to Figure 7-4 for this MAD calculation:

```
WINDOW_MEDIAN(ABS(SUM([Sales])-WINDOW_MEDIAN(SUM([Sales]))))
```

Using the MAD calculation, you can now create the modified z-score calculation (see Figure 7-12):

```
(SUM([Sales]) - WINDOW_MEDIAN(SUM([Sales]))) / ([MAD])
```

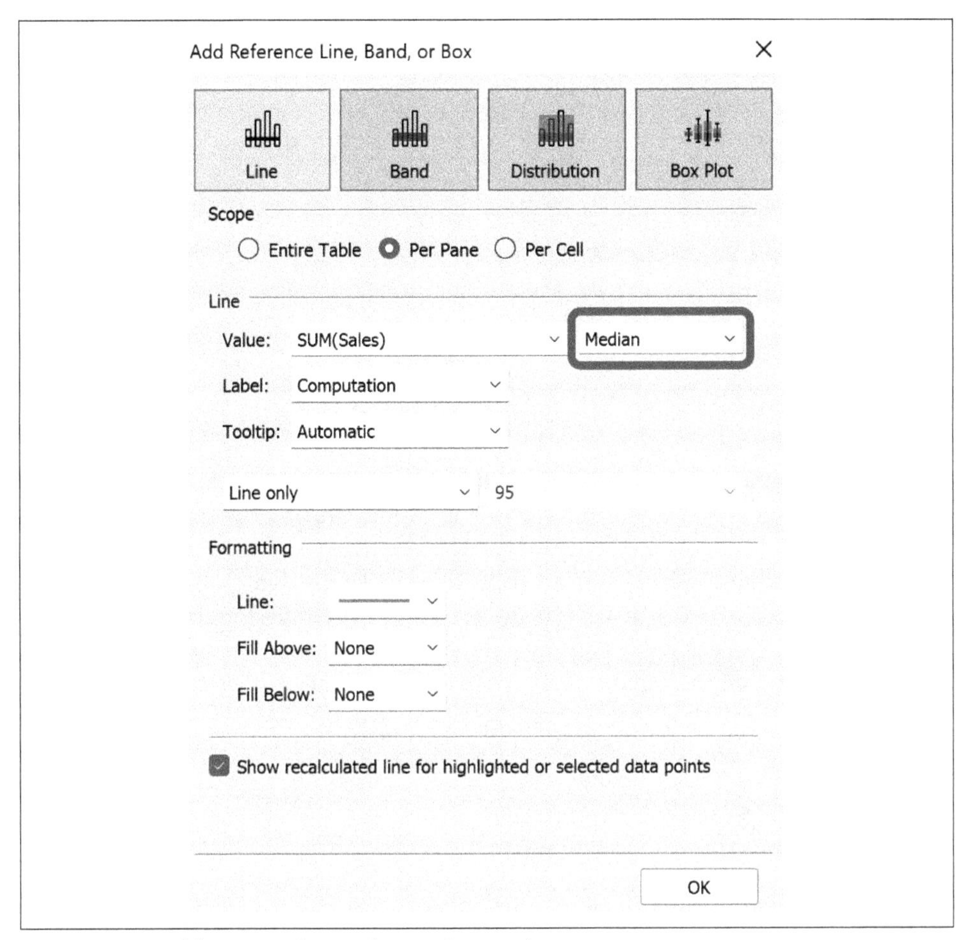

Figure 7-11. Adding a median reference line to the view

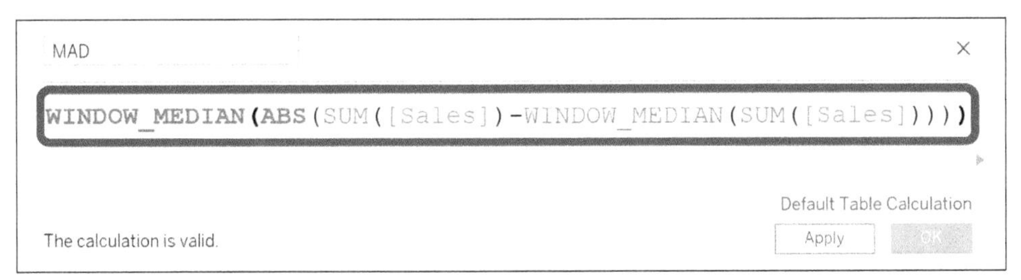

Figure 7-12. Calculation for the modified z-scores

The next step is to apply the modified z-score calculation to the view. To incorporate this new measure, drag the calculation to the Text property of the Marks card, as shown in Figure 7-13.

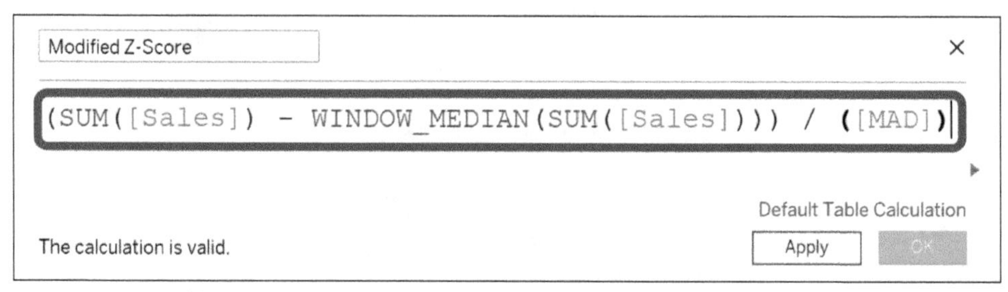

Figure 7-13. Adding the modified z-score to the Text property of the Marks card

You can see that as the sum of sales in any given month approaches the median, it gets closer to zero. If the sum of sales for a month equaled the median exactly, it would be zero. As it gets further from the median, though, it gets more positive or negative depending on the direction from the median. Again, you would typically consider anything +/– 3 as an outlier. To get a really nice effect and make the visual pop even more, you can apply conditional formatting on this view to make the outliers change colors, depending on if it is above or below that +/– threshold.

To do this, create one more calculated field (see Figure 7-14):

```
IF [z-score] >= 3 THEN "Above"
ELSEIF [z-score] <= -3 THEN "Below"
ELSE "Expected"
END
```

To apply this conditional formatting, select the second Marks card that you encoded as circles and drag this new formula to the Colors property. As you can see, this technique highlights two marks in the view as above the threshold and colors them differently, as shown in Figure 7-15.

This technique works great and is a good way to assign scores to each individual mark, which sets it apart from some of the other methods. This allows you to get very precise with what you consider an outlier.

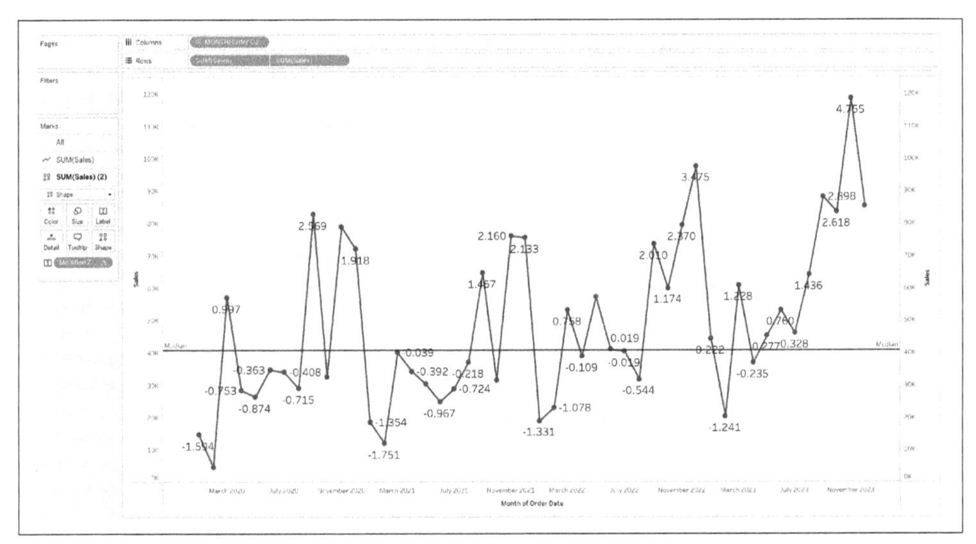

Figure 7-14. Conditional formatting for modified z-scores

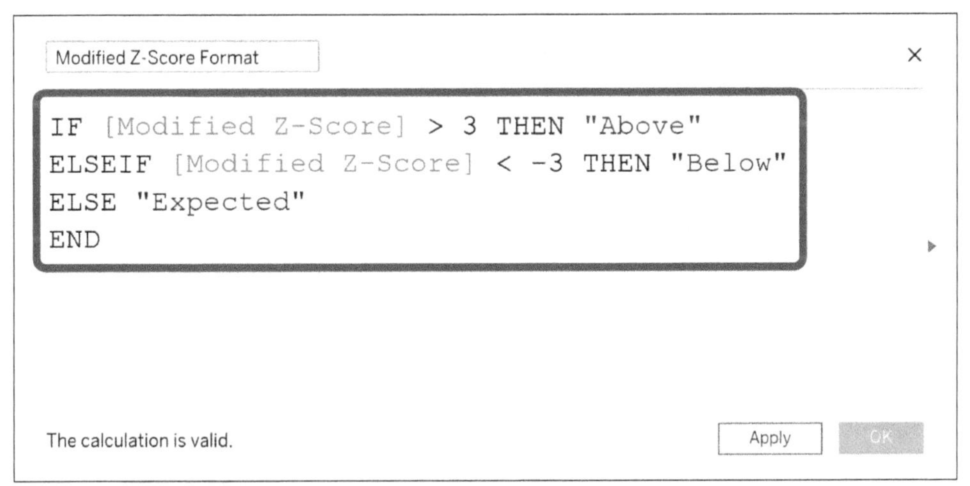

Figure 7-15. Conditional format applied to the view

Understanding Tukey's Fences

This method has similar properties as box plots and uses some terms that I introduced in Chapter 6. Recall this image from that chapter (Figure 6-11), shown here as Figure 7-16.

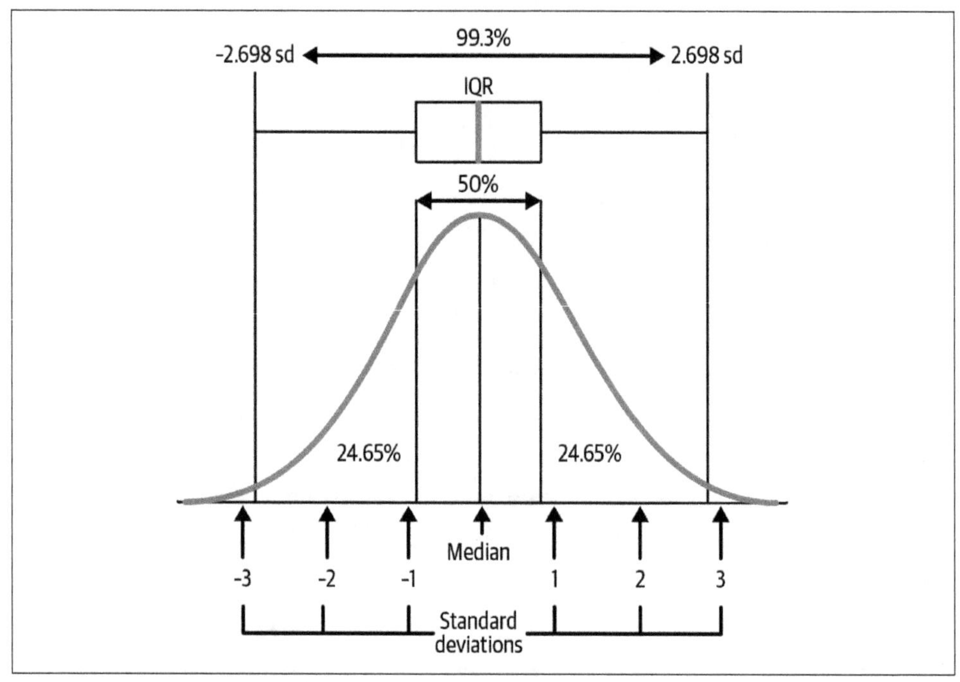

Figure 7-16. Explanation of median with quartiles and box plots

Remember that zone in the middle, referred to as the IQR. This range has an upper and lower bound, referred to as the 75th and 25th percentile of the data, and consists of 50% of the data. Tukey's fences method uses these percentiles of the IQR to calculate new upper and lower bounds, referred to as *fences*. Anything beyond the fences can be considered an outlier.

There are two calculations for Tukey's fences: an upper fence and a lower fence. These calculations are as follows:

$$\text{Upper fence} = Q3 + 1.5 \times (Q3 - Q1)$$
$$\text{Lower fence} = Q1 - 1.5 \times (Q3 - Q1)$$

where

$$Q1 = \text{25th Percentile}$$
$$Q3 = \text{75th Percentile}$$

Figure 7-17 shows you a modified version of Figure 7-16 after you incorporate Tukey's fences method into the view.

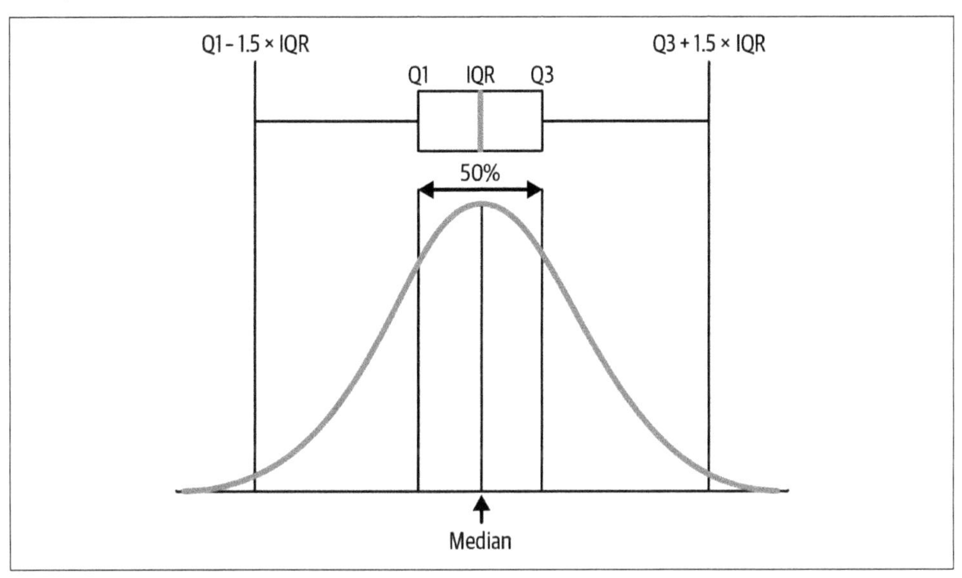

Figure 7-17. Visualizing Tukey's fences

How to Implement Tukey's Fences in Tableau

To incorporate this technique in Tableau, start by creating the same line chart we've been using, as shown in Figure 7-18.

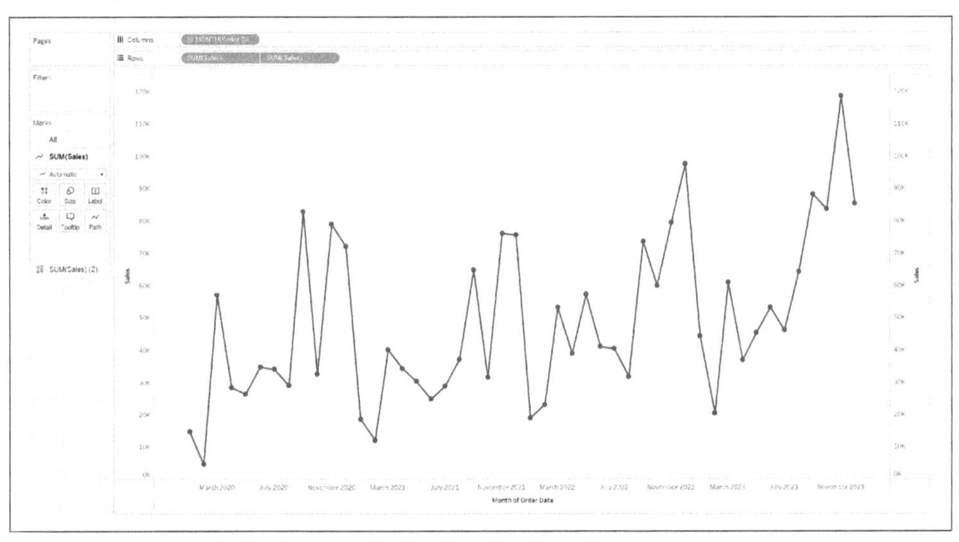

Figure 7-18. Sum of sales by month of order date line chart

Since this technique is derived from the "Median with Quartiles," you should first toggle to the Analytics pane and drag "Median with Quartiles" onto the view, as shown in Figure 7-19.

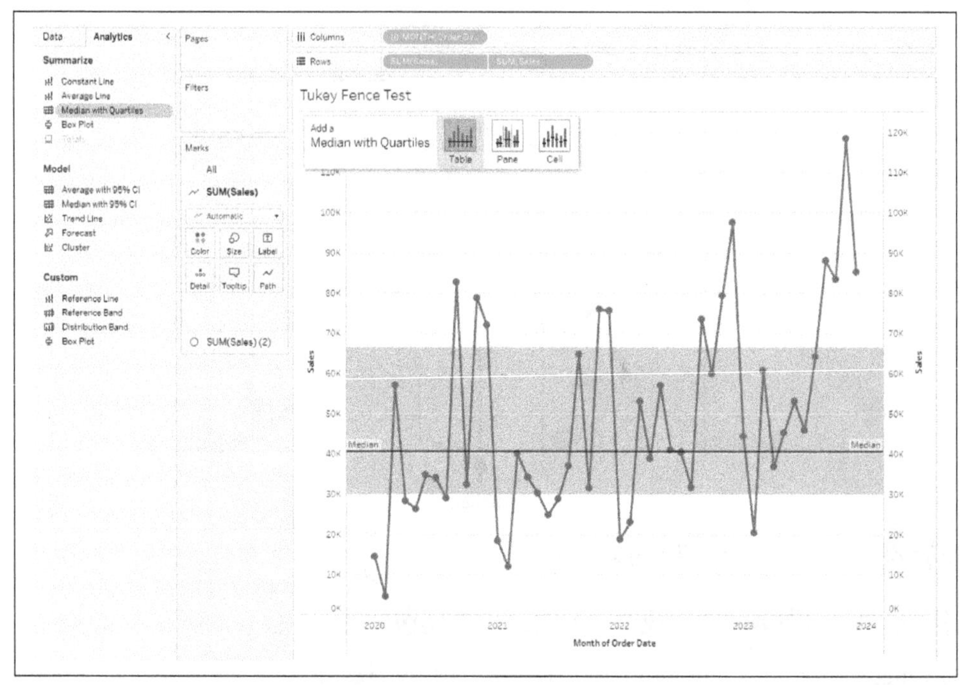

Figure 7-19. Dragging "Median with Quartiles" into the view

With the view set up, you now create the calculated fields of the upper and lower fences. Start by calculating the lower fence (see Figure 7-20):

```
WINDOW_PERCENTILE(SUM([Sales]),0.25) - MIN(1.5) *
(WINDOW_PERCENTILE(SUM([Sales]),0.75)-WINDOW_PERCENTILE(SUM([Sales]),0.25))
```

You can see I am using a new function called `WINDOW_PERCENTILE`. This function can be used to isolate the value of a certain percentile and is an easy way to incorporate the 25th and 75th percentiles dynamically within this view.

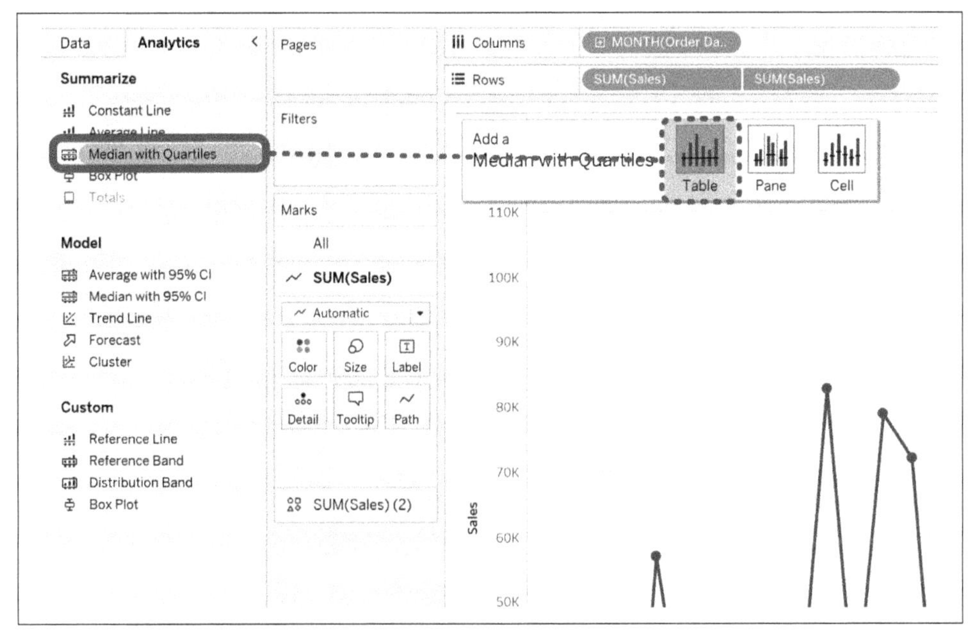

Figure 7-20. Tukey's lower fence calculated field

Next, you will use a similar approach to calculate the upper fence (see Figure 7-21):

```
WINDOW_PERCENTILE(SUM([Sales]),0.75) + MIN(1.5) *
(WINDOW_PERCENTILE(SUM([Sales]),0.75)-WINDOW_PERCENTILE(SUM([Sales]),0.25))
```

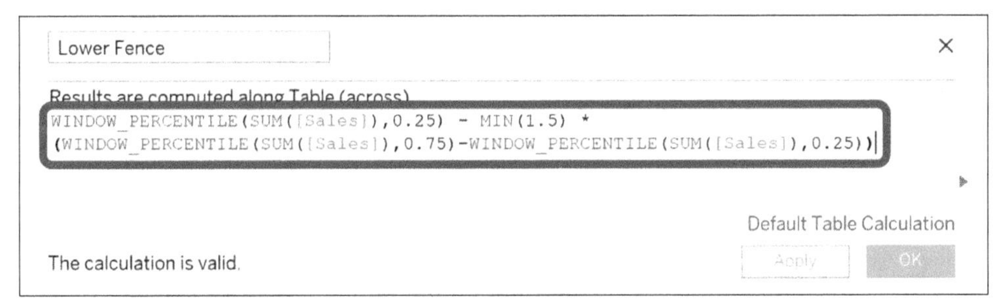

Figure 7-21. Tukey's upper fence calculated field

This is everything you need to begin incorporating this model visually within your view. Start by dragging the Upper Fence and Lower Fence calculations to the Details property of the Marks card, as shown in Figure 7-22.

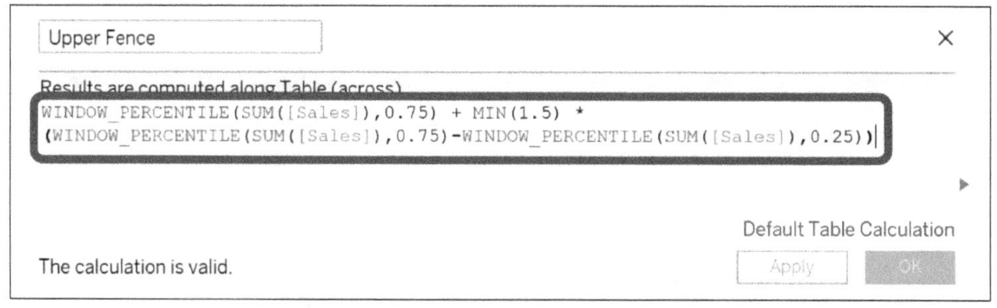

Figure 7-22. Adding the upper and lower fence calculations to the Details property

Now right-click on the y-axis and select Add Reference Line from the list of options. In the menu, change the Value to Lower Fence, as shown in Figure 7-23.

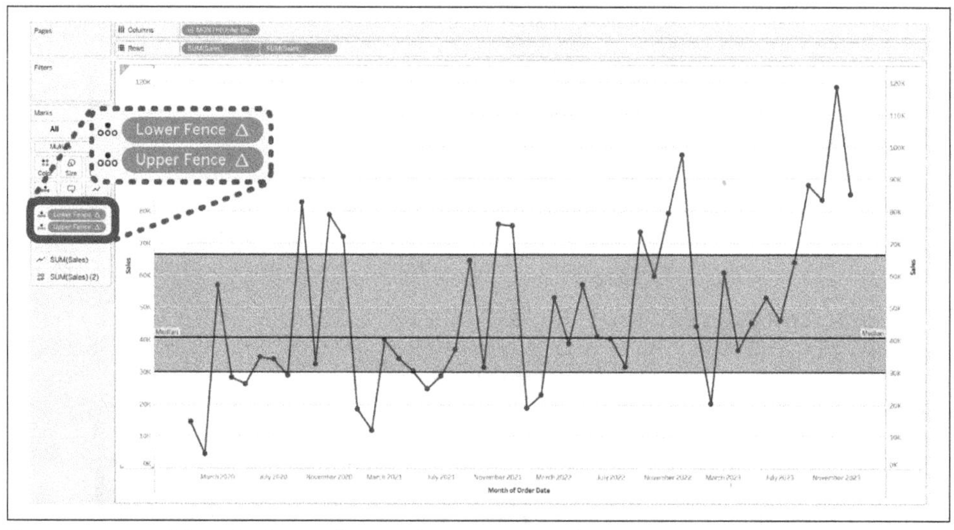

Figure 7-23. Adding the lower fence reference line to the view

Once you have it configured appropriately, click OK and follow the same steps to add the upper fence. You should end up with something similar to Figure 7-24.

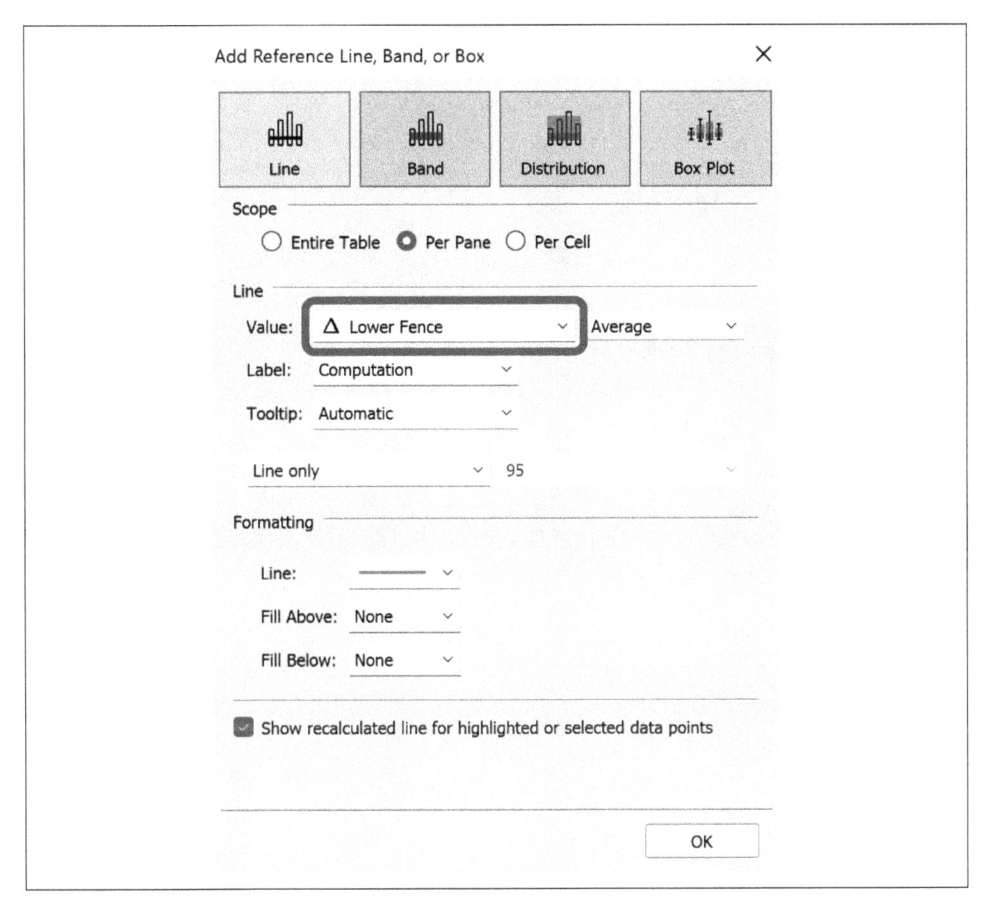

Figure 7-24. Adding the upper and lower fences to the view

If you analyze the view after incorporating this technique, you can see that there wasn't an outlier identified by this technique.

Summary

In this chapter, you learned three models that you can implement to detect outliers in data that have a nonnormal distribution: median absolute deviation, modified z-scores, and Tukey's fences. In the next chapters, I will begin showing you predictive models you can incorporate into your work.

CHAPTER 8

Linear Regression in Tableau

In the previous chapters, you learned different methods you can use to understand your data better. Confidence intervals allow you to make inferences about the data, anomaly detection can help you find outliers, and knowing the distribution of your data is a foundational skill every analyst should have. Starting in this chapter, you will learn how to incorporate predictive models within your visualizations, starting with simple *linear regression*. This is a great model that allows you to make inferences about your data and predict future outcomes, and it is very easy to explain to your stakeholders.

By the end of this chapter, you will have a basic understanding of simple linear regression, know how to implement it in Tableau, and be able to interpret the results of the statistical summary.

Linear Regression Model

Linear regression has many applications and is used across many different industries. A few examples of these applications are:

Predicting housing prices
: In real estate, linear regression can be used to predict housing prices based on various factors such as the size of the house, number of bedrooms, location, and other relevant features. By fitting a linear regression model to historical data of house sales, one can estimate the price of a new property on the market.

Sales forecasting
: Linear regression can be employed in sales and marketing to forecast product sales based on factors such as advertising spending, seasonality, pricing, and competitor data. Companies can use this information to make better inventory management decisions and plan their marketing strategies more effectively.

Climate change analysis

Linear regression can be applied to analyze climate change data, such as temperature variations over time. By fitting a linear regression line to historical temperature data, scientists can extrapolate future temperature trends and better understand the impact of climate change.

Financial analysis

In finance, linear regression can be used to model the relationship between a company's stock price and various financial indicators such as earnings, revenue, and interest rates. Analysts can utilize this information to assess the factors influencing stock performance and make investment decisions.

Medical research

Linear regression can be employed in medical research to study the correlation between variables such as dosage and treatment response, age and health outcomes, or genetic factors and disease susceptibility. By fitting a linear regression model to medical data, researchers can gain insights into these relationships and make informed decisions for patient care.

Linear regression is a statistical method used for modeling the relationship between a dependent variable and an independent variable. It is one of the simplest and most widely used techniques in machine learning and statistics for predictive analysis and understanding the correlation between variables.

Linear Regression Expressed Mathematically

In its most basic form, linear regression assumes a linear relationship between the *dependent variable* (also known as the *response variable* or *target variable*) and the *independent variable* (also called *predictor variable*). The goal of linear regression is to find the best-fitting straight line through the data points that minimizes the overall error or difference between the predicted values and the actual values of the dependent variable.

Simple linear regression can be expressed mathematically with the following formula:

$$Y_i = \beta_0 + \beta_1 X_i + \epsilon_i$$

where

Y_i = the value of the dependent variable

β_0 = the intercept

β_1 = the coefficient calculated by the model

X_i = a known value from the dataset at the ith point

ϵ_i = a random error that occurs

Simple Linear Regression Example

To explain this formula a little more, consider this example. Let's say you surveyed 10 people for their age and their income. Then you took the data and visualized it using a scatterplot, as shown in Figure 8-1.

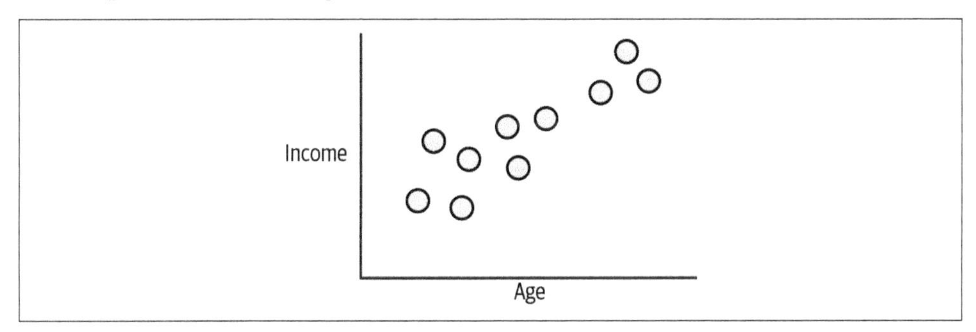

Figure 8-1. Scatterplot of age by income

You can begin to make some assumptions about the data by just looking at the visual. For instance, as age goes up, so does income. This is often referred to as a *positive correlation*. You could also assume that the audience surveyed was younger because, after a certain age around retirement, you would think income would begin to decrease.

While these assumptions alone are informative, applying a regression model could help you confirm these assumptions and give you a mechanism to predict this information moving forward. Applying a regression model to this data, you would get similar results to what's shown in Figure 8-2.

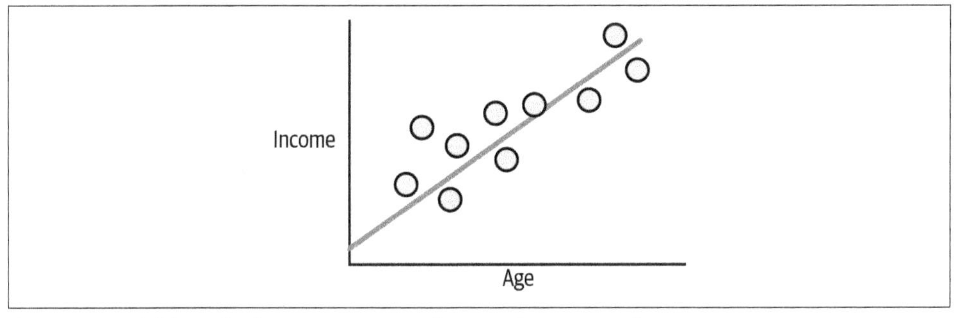

Figure 8-2. Adding the linear regression line to the scatterplot

You can see the positive trend of this data very clearly in Figure 8-2, but the benefits of this line go beyond just visualizing the trend.

As mentioned previously, the goal of the regression line is to find the best-fitting line that minimizes the distance between the predicted values and the actual values of the dependent variable, as shown in Figure 8-3.

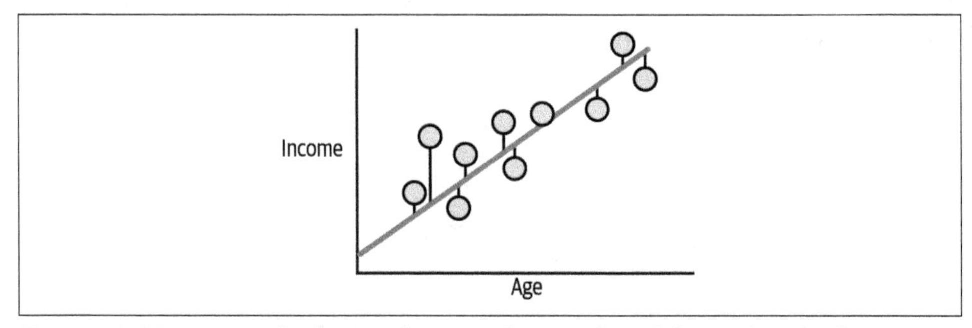

Figure 8-3. Minimizing the distance between the actuals and the predicted value

As you can see in Figure 8-3, I have drawn lines from the actuals (the dots) to the regression line. The point on the regression line that intercepts at the actual value would be the predicted value for that given income and age. In this case, it is predicting someone's age given their income. Figure 8-4 shows you how the equation maps out to the visualization, so that you can begin putting the pieces together.

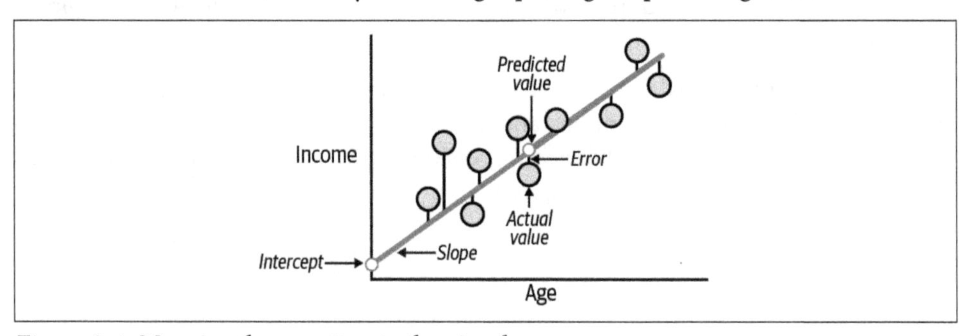

Figure 8-4. Mapping the equation to the visual

Recall the simple linear regression equation from earlier in this section:

$$Y_i = \beta_0 + \beta_1 X_i + \epsilon_i$$

where

Y_i = the predicted value

β_0 = the intercept

β_1 = the slope

X_i = the actual value

ϵ_i = the error between the predicted value and the actual value

Later in the chapter, you will learn how to extract these predictions, interpret the results, and implement this model in Tableau.

Assumptions of Linear Regression Models

There are five major assumptions when working with simple linear regression modeling.

Linearity
: The relationship between the dependent variable and the independent variable is assumed to be linear. This means that a change in the independent variable is associated with a constant change in the dependent variable.

Independence
: The residuals (the differences between the observed and predicted values) should be independent of each other. In other words, the value of the dependent variable for one observation should not be influenced by the value of the dependent variable for any other observation.

Homoscedasticity
: The variance of the residuals should be constant across all levels of the independent variable. In simpler terms, the spread of the residuals should be roughly the same for all values of the independent variable.

Normality of residuals
: The residuals should follow a normal distribution. This assumption is more critical with smaller sample sizes. If the residuals are approximately normally distributed, it suggests that the statistical inferences made using the model are more reliable.

No perfect multicollinearity
: In the context of simple linear regression, multicollinearity is not a concern as there is only one independent variable. However, in multiple regression (with more than one independent variable), it is assumed that there is no perfect linear relationship among the independent variables.

Another thing to consider is that you need a decent sample size to get a less volatile prediction model. To prove this, let's go back to the example of predicting age by income. Let's say you surveyed five more people and visualized the results shown in Figure 8-5. The new five are represented by the open circles.

Adding these new data inputs is going to change our regression model and therefore change your predictions. If you plot a new regression model on the visual using a dashed line, you will see the difference (see Figure 8-6).

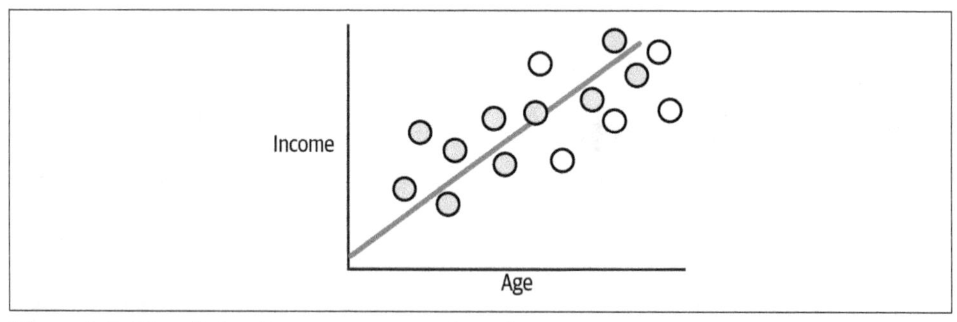

Figure 8-5. Visualizing five new surveys of age by income

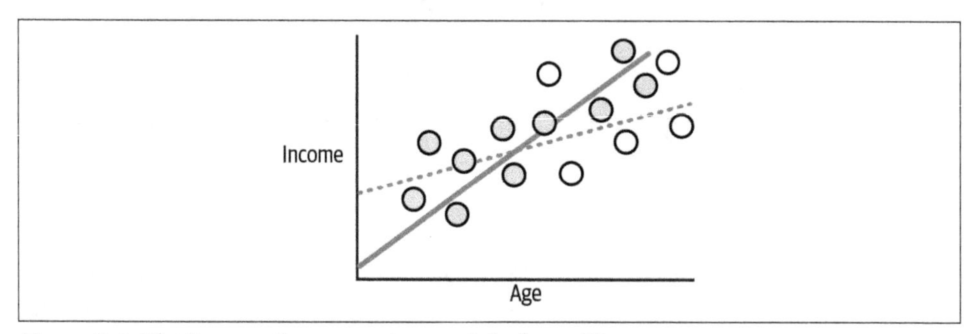

Figure 8-6. Plotting another regression model after adding more surveys

You can see that the new line fits differently than the original. Thinking through the visual, you can see that the slope of the line has begun to flatten out. There is still a positive relationship between age and income, but it has less magnitude than the prior model. It is important to understand that every time you add new data your model will change. However, if you have a larger sample size, the change will be less drastic. Moreover, there could be times when external factors change and impact your model.

Think about it; in this example, each of these data points represents a person. There are many external factors that could affect people's income on micro and macro levels. There could have been an economic recession from the first survey to the next; you could have surveyed one group in a large city and another group in a rural area where income levels would change; and the list goes on.

It is up to you as the analyst to understand the data and all the factors that could affect it, even external factors outside your control. Thinking through these external factors can make or break your analysis and will set you apart from your peers. It's also just the ethical thing to do.

Implementing Linear Regression in Tableau

You have learned some very broad conceptual information about the simple linear regression model, but how do you actually apply this in Tableau and begin making predictions? Tableau makes it so easy for you to implement this model. For this next example, you will be analyzing the relationship that discounts have on profit by manufacturers. You will start by connecting to the Sample - Superstore dataset. Then you will add the sum of profit to the Rows shelf and the discrete month of order date to the Columns shelf. You should end up with a view similar to what is shown in Figure 8-7.

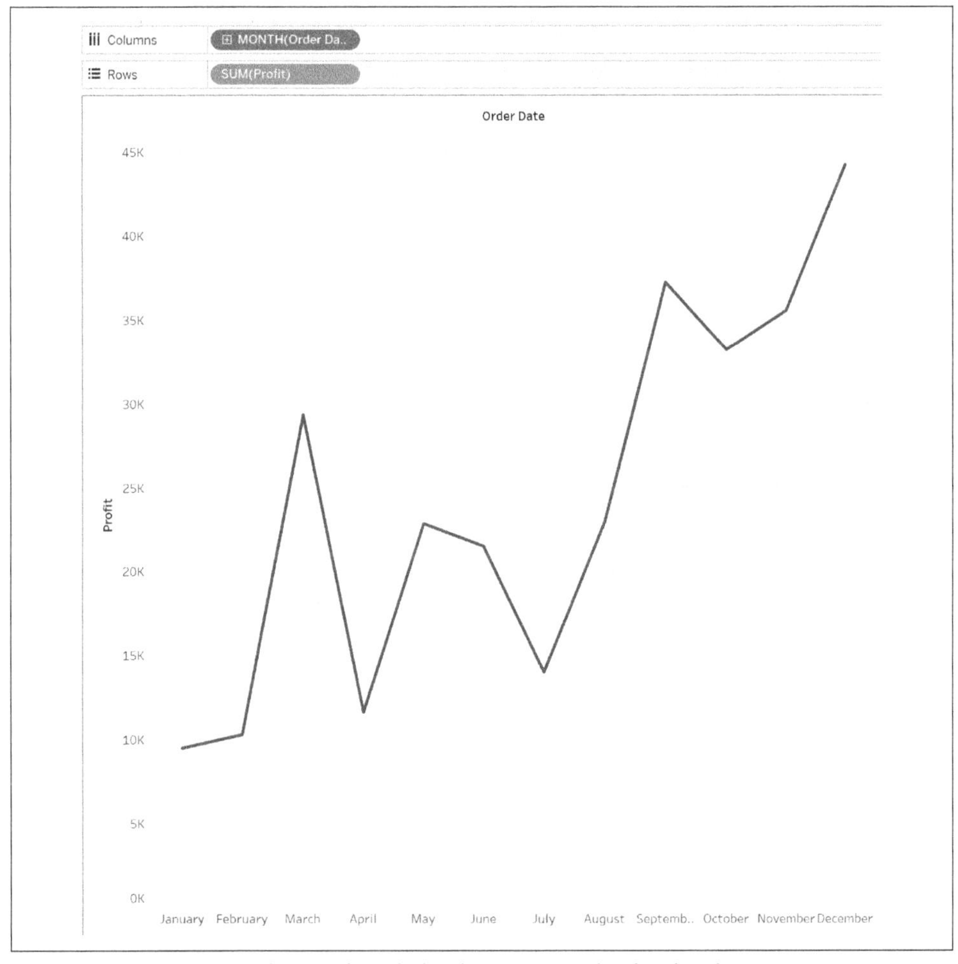

Figure 8-7. Line chart of sum of profit by discrete month of order date

To apply the model, toggle to the Analytics pane and drag Trend Line to Linear, as shown in Figure 8-8.

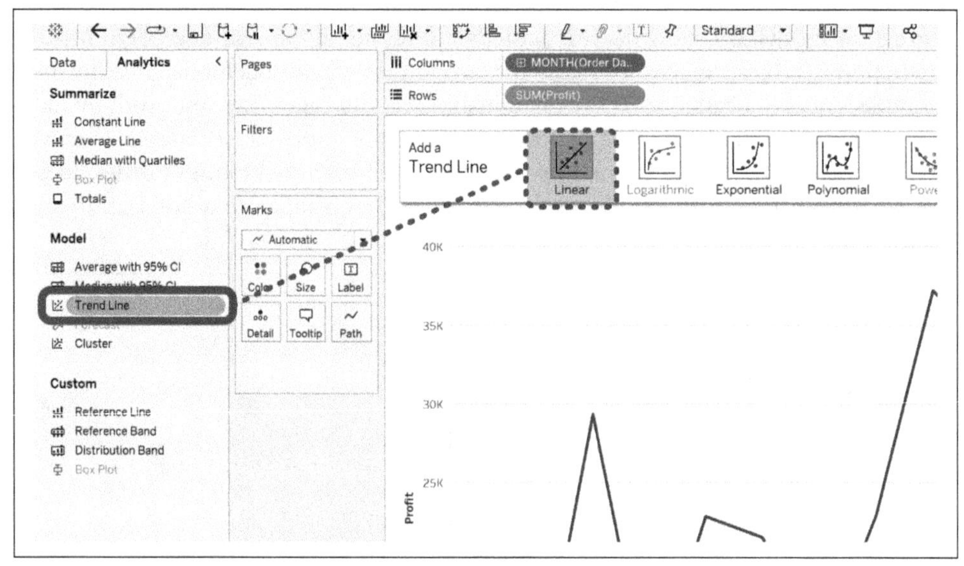

Figure 8-8. Applying a linear regression model to the view from the Analytics pane

The Linear Trend Line option from the Analytics pane is actually a simple linear regression model. By dragging it into the view you should have something similar to Figure 8-9.

If you hover over the regression line, you will see an equation in the tooltip, as shown in Figure 8-10.

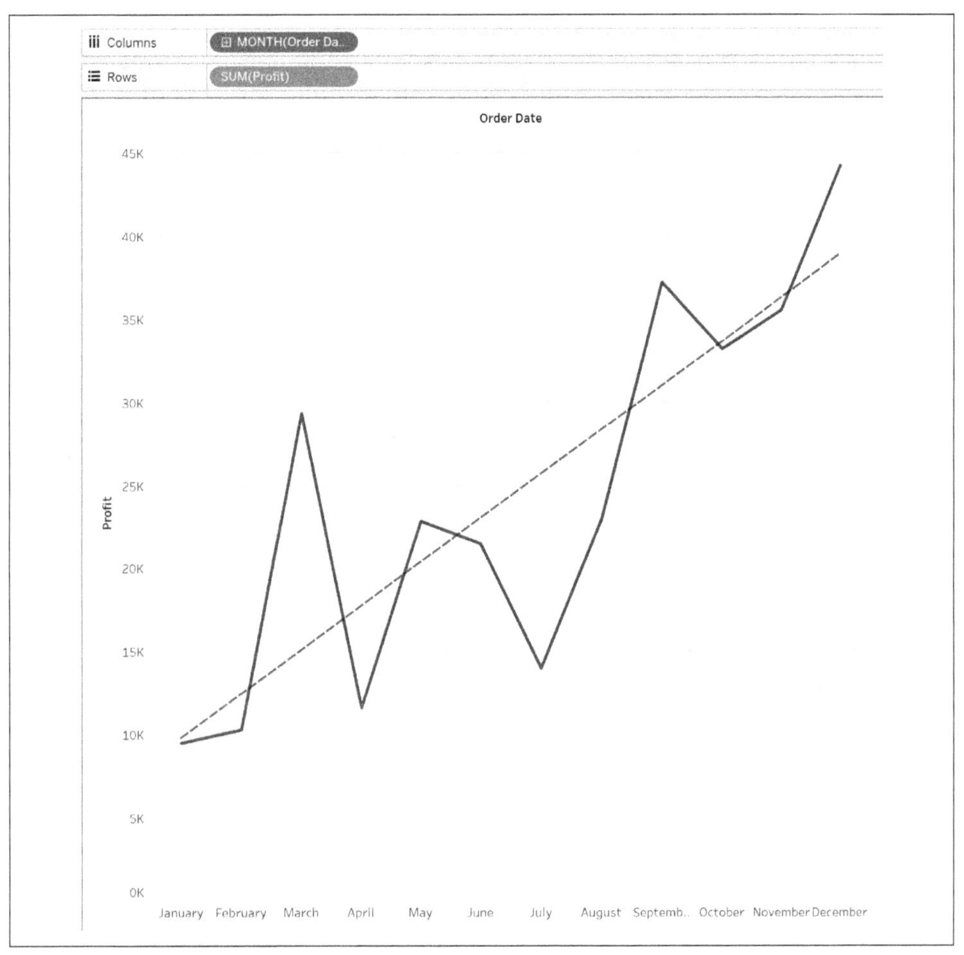

Figure 8-9. Linear regression model added to the view

Profit = 2647.53*Month of Order Date + 7149.15
R-Squared: 0.673069
P-value: 0.0010791

Figure 8-10. Regression model summary statistics within the tooltip

This is showing you the linear regression equation and summary statistics of the model. The equation is just filling in the regression equation using the intercept and slope values generated from the model. Recall the equation from the previous sections:

$$Y_i = \beta_0 + \beta_1 X_i + \epsilon_i$$

where

Y_i = the predicted value

β_0 = the intercept

β_1 = the slope

X_i = the actual value

ϵ_i = the error between the predicted value and the actual value

Now you just fill in the intercept, slope, and actuals to get a predicted value:

$7,149.15 + $2,647.53 × discrete month of order date = predicted value Y_i

Let's say you want to see what the profit would be on average for the month of October. If you add 10 into the equation you get the following predicted value:

$7,149.15 + $2,647.53 × 10 = $33,624.45

Let's review what the other values are as well. From the tooltip, you also have an R-squared and a p-value.

R-squared

The *R-squared* is a statistical measure that ranges from 0 to 1 and indicates the proportion of the variance in the dependent variable that is explained by the independent variables in a regression model. So, for this example, it is the percentage of profit that is explained by which sales month you are predicting. The closer to 1 the R-squared value, the better your model fits the data.

If you think through this logically, it makes sense. The R-squared value in the tooltip for this model is 0.6731 or roughly 67%. This means that about 67% of the profit made is explained by the month. This tells me that sales for this business are very seasonal, and the time of year plays a large role in the overall profit that is made. Here are some examples of other variables that could affect profit: shipping costs, labor costs, sales price, new products entering the market, manufacturing costs, etc. As you can see, there are many other variables we should consider, but being able to quantify the relationship between these variables using regression models is important in itself.

P-value

A p-value (probability value), as we discussed in Chapter 1, is a fundamental concept in statistical hypothesis testing. It quantifies the strength of the evidence against the null hypothesis. In hypothesis testing, we start with a null hypothesis (H_0), which typically represents the default or no-effect assumption. You then collect data and perform a statistical test to determine whether the data provides enough evidence to reject the null hypothesis in favor of an alternative hypothesis (H_a).

In simpler terms, it answers the question, "If the null hypothesis were true, how likely is it to observe that in the data you have?" The exact p-value you need to show that your hypothesis test fails to reject the null hypothesis is up to you. Normally, anything less than 0.05 is considered; however, if you wanted to be more confident in your results, you could set it to 0.01, or if you had room for interpretation, you could set it to 0.1.

Interpreting the Detailed Summary Statistics

I have shown you how to implement a linear regression model in Tableau using the Analytics pane. I also covered some brief information you can get from the tooltip and how to make some assumptions of the model. However, let's take a deeper look at the model and the full summary statistics. If you right-click on the regression line in the view, you will see several new options appear, as shown in Figure 8-11.

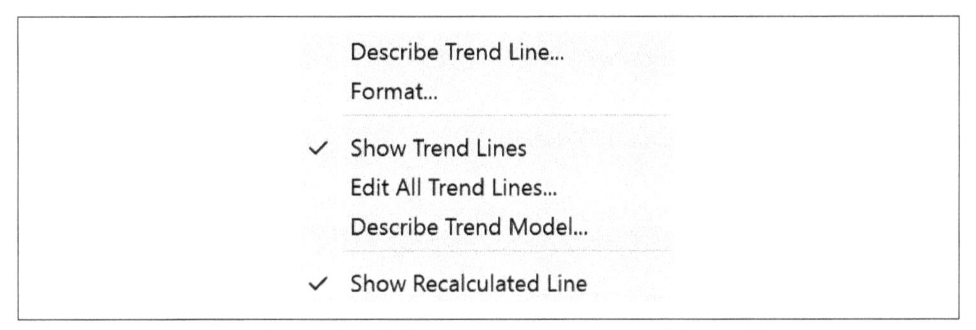

Figure 8-11. Options when right-clicking on a regression model in Tableau

From this menu select Describe Trend Model. This will open a new menu that has the full summary statistics within it, as shown in Figure 8-12.

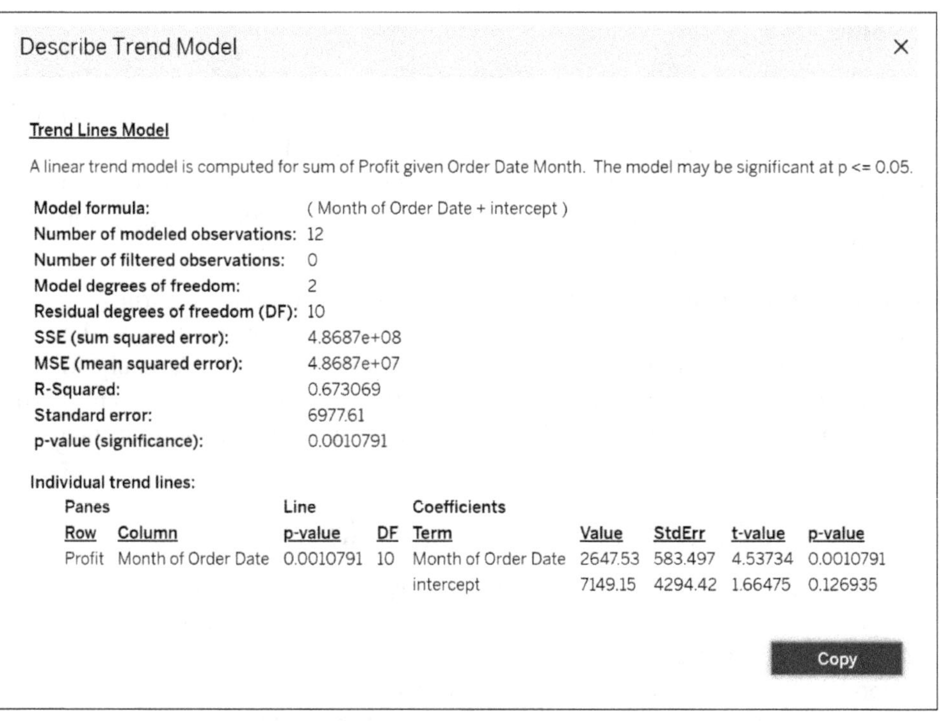

Figure 8-12. Describe Trend Model summary of statistics

There is so much information within this view that it would be impossible to cover it all in detail within a single chapter. However, here is a highlight of information you can find when describing the model:

Number of modeled observations
: How many data points there are in the model

Number of filtered observations
: How many observations you have filtered from the view and from the model

Mean squared error (MSE)
: This is a common metric used to evaluate the performance of a predictive model, particularly in regression analysis. It measures the average squared difference between the predicted values and the actual values in a dataset.

R-squared
: This is a figure used to evaluate the performance of the model and is a value between 0 and 1. The closer to 1 the R-squared is, the better your model fits the data.

P-value

This is used to determine if the model passes or fails your hypothesis tests. There is also a model p-value and independent variable p-values.

These statistics are the foundational components of understanding the outcome and quality of your model. Let's dive into the lower section and break out each variable, as shown in Figure 8-13.

Individual trend lines:

Panes		Line		Coefficients				
Row	Column	p-value	DF	Term	Value	StdErr	t-value	p-value
Profit	Month of Order Date	0.0010791	10	Month of Order Date	2647.53	583.497	4.53734	0.0010791
				intercept	7149.15	4294.42	1.66475	0.126935

Copy

Figure 8-13. Summary statistics of individual variables

This section can give you so much insight into your data and help you populate the linear regression equation. Again, recall the linear regression equation:

$$Y_i = \beta_0 + \beta_1 X_i + \epsilon_i$$

where

Y_i = the predicted value

β_0 = the intercept

β_1 = the slope

X_i = the actual value

ϵ_i = the error between the predicted value and the actual value

From the Value column in Figure 8-13, you can find the numbers to populate the equation, and it is also the same values that were shown in the tooltip when hovering over the model.

Summary

In this chapter, you were introduced to linear regression models and how to implement them in Tableau. You also learned how to interpret the results of the model. Throughout the chapter, I also gave you several ways you could use the model to either make assumptions about your data or predict future values.

CHAPTER 9

Polynomial Regression in Tableau

Linear regression can be a powerful tool in your toolkit, as you saw in the last chapter. However, sometimes it's not the best tool for the job. In this chapter, you will be introduced to another regression model called *polynomial regression*. This model allows your predictions to move along with the data rather than being a rigid line. There are many advantages of this model, but you have to be careful not to overfit the model.

In this chapter, you will be introduced to polynomial regression, learn the pros and cons of the model, and learn about overfitting a model.

What Is Polynomial Regression?

As with linear regression, polynomial regression is a predictive model that has many use cases:

Physics
: Polynomial regression is often used for curve fitting when the relationship between the independent and dependent variables appears to follow a curve or a nonlinear pattern. For example, in physics, it can be used to model the trajectory of a projectile.

Economics and finance
: In economics and finance, polynomial regression can be used to analyze the relationship between variables like gross domestic product (GDP) and time or stock prices and time, where linear models might not capture the underlying trends accurately.

Medicine and biology

In medical research and biology, polynomial regression can be used to model growth curves, drug concentration–response relationships, or the relationship between age and various physiological parameters.

Marketing and sales

Polynomial regression can help analyze consumer behavior, price elasticity, and market demand, especially when the relationships are nonlinear and exhibit saturation or diminishing returns.

Quality control

In manufacturing and quality control, polynomial regression can be used to model the relationship between process variables and product quality to optimize production processes.

It's important to note that while polynomial regression can be useful for capturing nonlinear relationships, it also comes with some limitations, such as overfitting when using higher-degree polynomials. Care should be taken to choose an appropriate degree of the polynomial and to evaluate the model's performance to avoid overfitting and ensure its reliability for making predictions or drawing conclusions. I will cover overfitting later in this chapter.

Let's look at an example of how polynomial regression differs from linear regression. To begin, let's fit a linear regression model to some data, as shown in Figure 9-1.

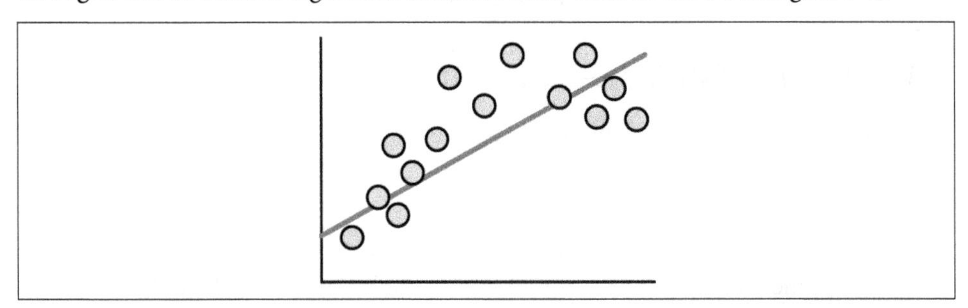

Figure 9-1. Plotting a linear regression

If you recall from Chapter 8, the purpose of a regression model is to fit a line that minimizes the distance between each data point and the line itself. The less distance between the data (marks in the view) and the line, the more accurate you could say your predictions could be.

Analyzing Figure 9-1, you can see that the data in the view arcs slightly up toward the middle, then it trends down toward the end. This is common in data, and when this pattern is observed over time, it is often referred to as seasonality because you will see peaks and valleys repeated over and over. While you can fit a linear line to data like this, you may not get the most accurate predictions. If you add lines between the points and the regression line you can see that there are certain data points that are close but most are pretty far off, as shown in Figure 9-2.

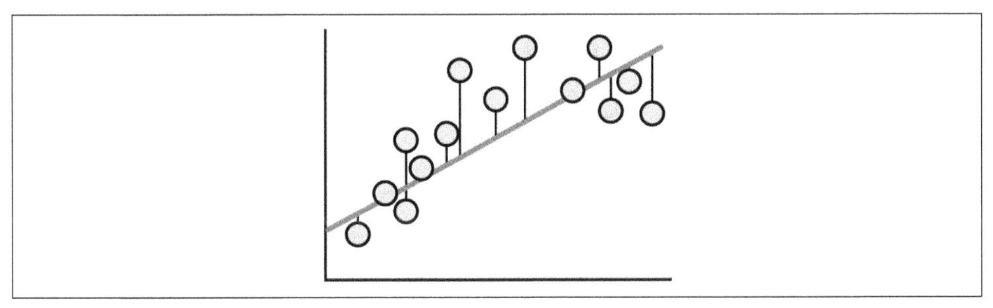

Figure 9-2. Drawing a line between the data points and the linear regression line

You can clearly see the areas where this model is not going to perform well to give you accurate predictions. In these situations, you need to be able to create a line that bends with the data to account for seasonality better. This is exactly what a polynomial regression model does. In Figure 9-3, you can see a polynomial regression line fit to the data represented by the dashed line.

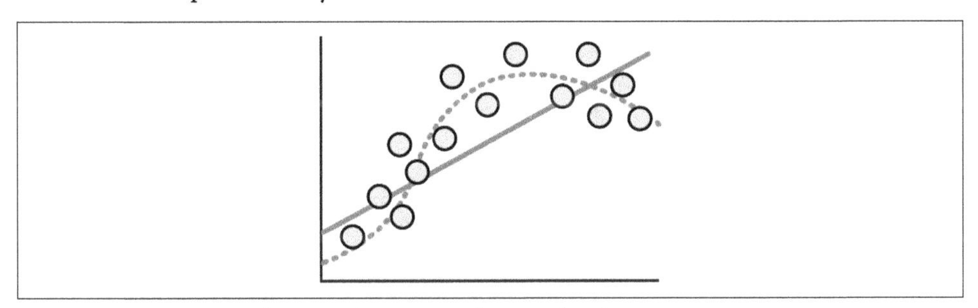

Figure 9-3. Fitting a polynomial regression line to the data

You can see that this model fits the data much better than the linear regression model. If you remove the linear regression line and draw lines between the polynomial regression line and the observations, as shown in Figure 9-4, you can see this more clearly.

As you can see, by choosing the polynomial regression model, you can reduce the distance between the line and observations in the visual. If you were to do an actual comparison of the models, you would most likely see a shift in the summary statistics, such as the R-squared and p-value of the model. I will show you actual examples of this comparison later in the chapter, and you will see how you can use the summary statistics to choose the best fit model.

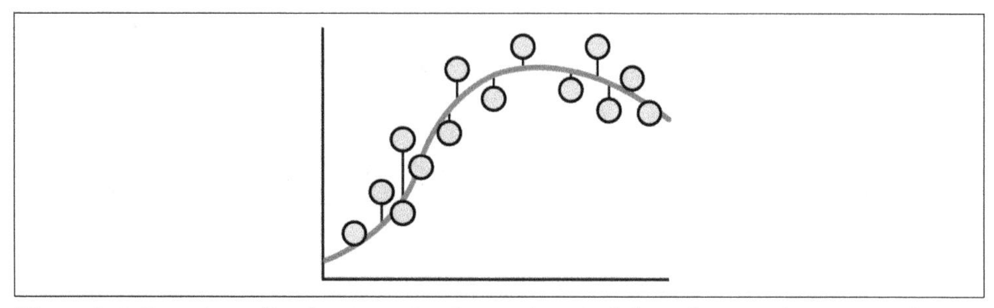

Figure 9-4. Drawing a line between the data points and the polynomial regression line

Polynomial Regression Equation

Mathematically, the polynomial regression model doesn't differ from the linear regression model all that much. However, to truly understand how the model works, I want to start with another basic example before I dive too far into the details.

If you were to plot the equation y = x onto a graph, you would get the results shown in Figure 9-5.

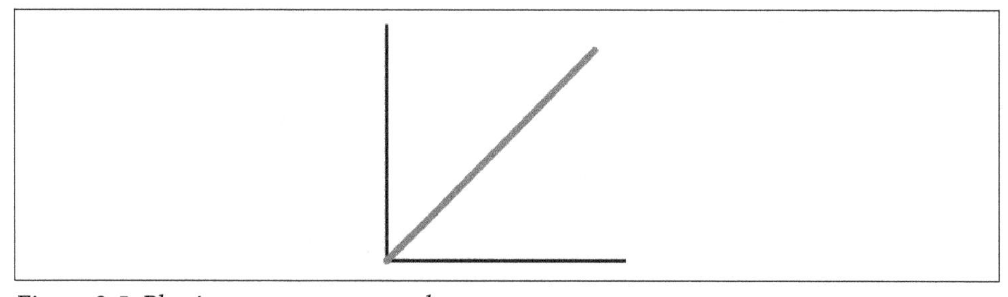

Figure 9-5. Plotting $y = x$ onto a graph

You can see that this results in a linear line, and if you recall back to Chapter 8, you will begin to see how this simple equation is familiar to a linear regression. Now, if you plot the equation $y = x^2$ into a graphing calculator, you will end up with results similar to Figure 9-6.

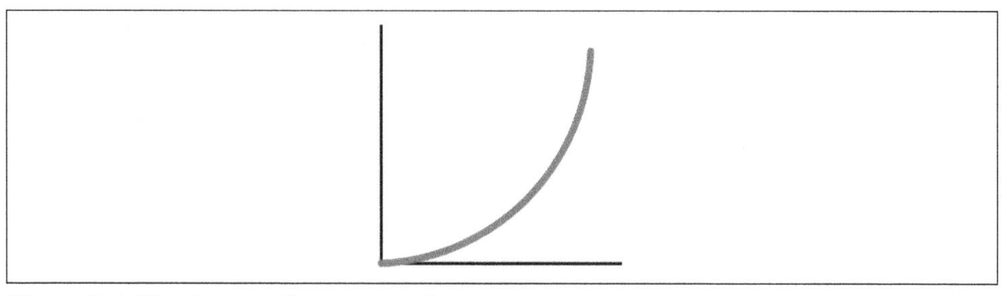

Figure 9-6. Plotting $y = x^2$ onto a graph

Now you can see that by adding a power function to the variable, you have introduced a curved line. This is how polynomial regressions curve as well, by introducing power functions into the linear regression equation.

Polynomial Regression Expressed Mathematically

The actual equation for polynomial regression is as follows:

$$y_i = \beta_0 + \beta_1 x_i + \beta_2 x_i^2 + \ldots + \beta_n x_i^n + \epsilon_i$$

where

y_i = the value of the dependent value

β_0 = the intercept

β_1 = the coefficient calculated by the model

x_i = a known value from the dataset at the ith point

ϵ_i = a random error that occurs

While it may look intimidating, you can compare it to the linear regression model, and you'll find that it isn't all that different. Remember, just like you saw from the plotting example in the previous section, you have only introduced a new part to the equation that has some power functions. The number of degrees to include in the equation is a parameter that you can control. For instance, if you wanted to add one slight curve, you could take the model to the second degree. If you wanted to add a second curve in the line you could set the model to the third degree, and so on to the *n*th degree.

This is where you have to worry about overfitting the model. In Figure 9-7, you can see there is a plotted linear regression, which is considered the first degree, and a polynomial regression to the third degree, represented by a solid and dashed line, respectively.

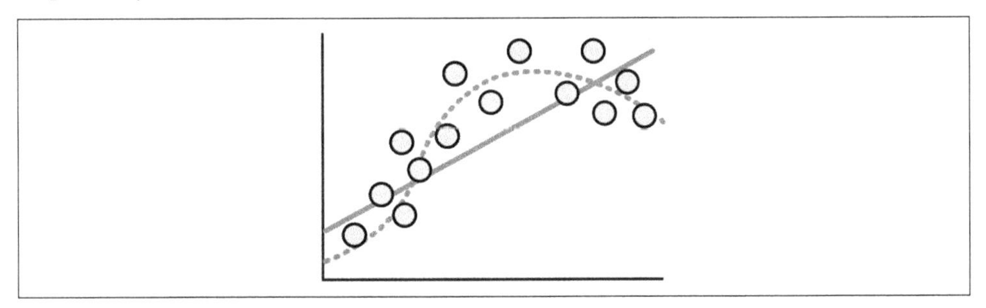

Figure 9-7. Explaining overfitting and underfitting a dataset

As explained earlier, the linear regression model is underfitting the data; it doesn't fit the trend of this data. However, the third-degree polynomial model fits the data well. If you changed your model to show a ten-degree polynomial model, you would get a line that passes through almost every data point in the view, as shown in Figure 9-8.

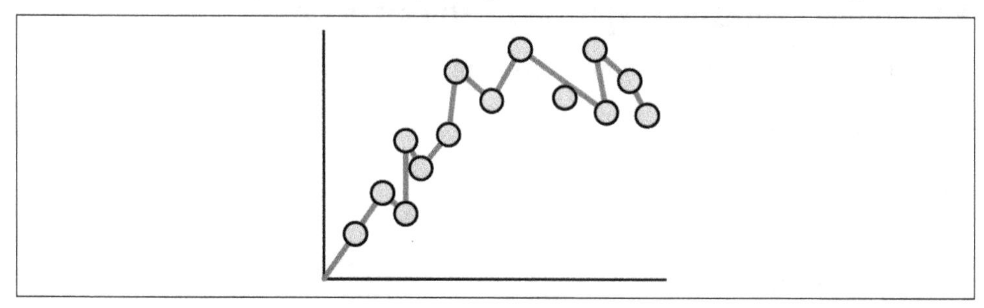

Figure 9-8. Overfitting a polynomial regression model

If you put this into practice, you would most likely observe a very high R-squared, which could lead you to believe that this model is very accurate. However, think about this logically for a moment and you can begin to see the fallacy with the accuracy of this model.

Take a moment and ask yourself, "If another data point was collected, where would it fall on this chart?" It is trained so precisely on what we have already observed that it would perform poorly if you received new data that wasn't the exact same as the test data. This problem as a whole is referred to as *overfitting the data.*

Choosing the Right Model

There is definitely a lot to think about when determining the best model to choose. You just learned about overfitting and underfitting a model, and you have learned about accuracy and summary statistics like R-squared and p-values. However, what about explainability? What if you had to take this model to your stakeholders and explain it to them or implement the model in a production environment? Model complexity is definitely another factor to consider.

Polynomial regression does add complexity to the model, and as you raise the degrees within the model, it becomes harder and harder to explain and interpret. Because of this, it is best to keep the degrees in the model as low as you can and weigh your options very carefully when making your decision.

Compare the three models side by side in Figure 9-9. If you take off the statistical hat for a moment and just think through it visually, which model would you prefer to use to predict the next data point?

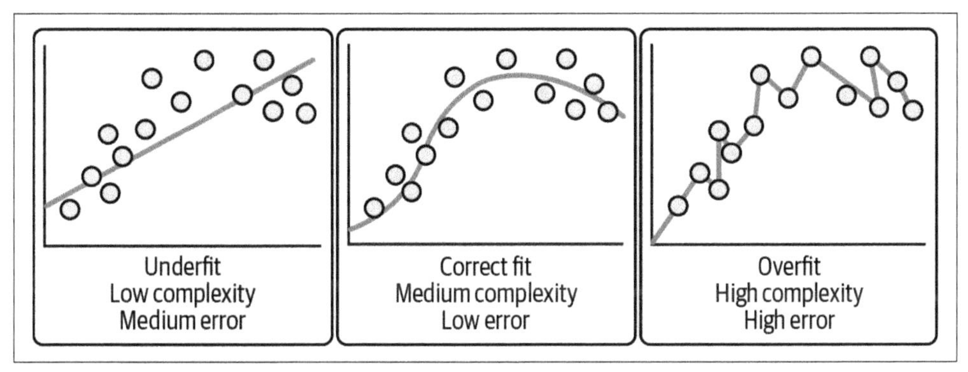

Figure 9-9. Comparing each model side by side

The middle chart best fits the pattern of the data and would convey the story of the data better than the others. Of course, in terms of model performance, you would need to analyze the summary statistics, train and test the model, and understand the underlying assumptions before putting any model into a production environment. However, adding this model to the data visualization can provide the stakeholders with more context and reduce the time to understand the data.

How to Implement Polynomial Regression in Tableau

To get started, create a dual-axis line chart using the sum of sales and month of order date, as shown in Figure 9-10.

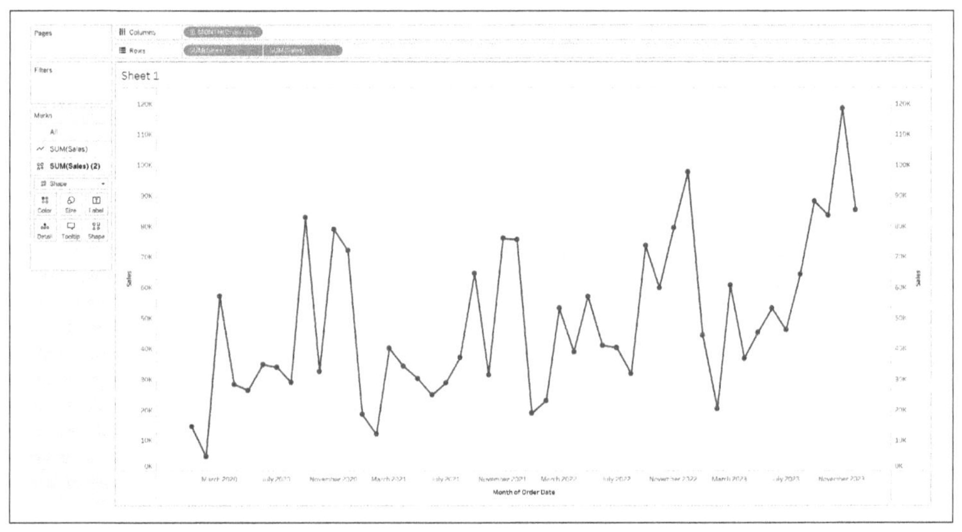

Figure 9-10. Dual-axis line chart of the sum of sales by month of order date

Next, add a linear regression from the Analytics pane. To do this, toggle to the Analytics pane and drag Trend Line onto the canvas, dropping it on Linear, as shown in Figure 9-11.

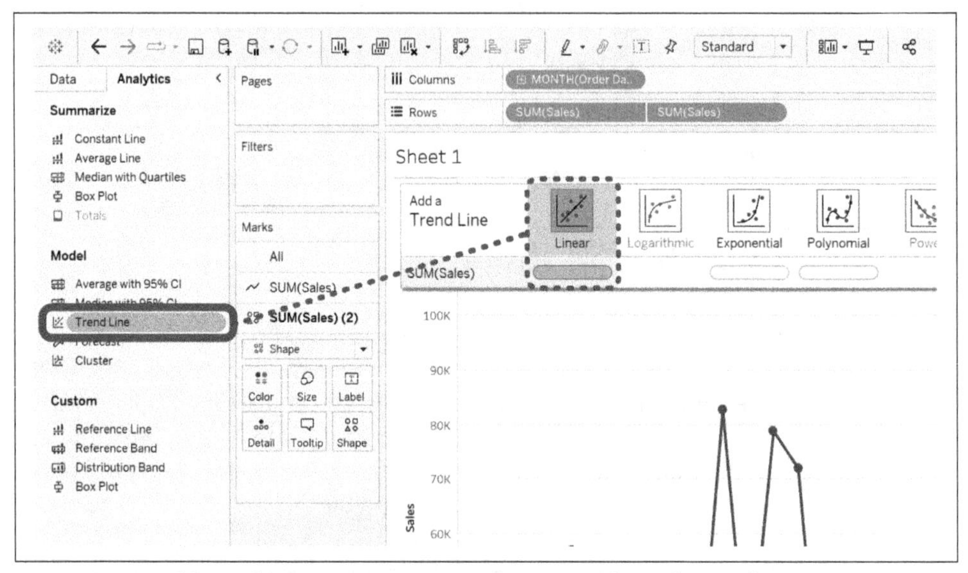

Figure 9-11. Adding a linear regression onto the canvas from the Analytics pane

This will add a straight line onto the view that represents a linear regression (covered in Chapter 8). As you can see in Figure 9-12, the linear regression does a good job of showing you the overall trend of this data. This shows that sales are increasing over time.

If you hover over the line to analyze the summary statistics, you can see that the p-value is less than 0.05 and the R-squared is about 0.25, as shown in Figure 9-13. This means that this model would be considered statistically significant.

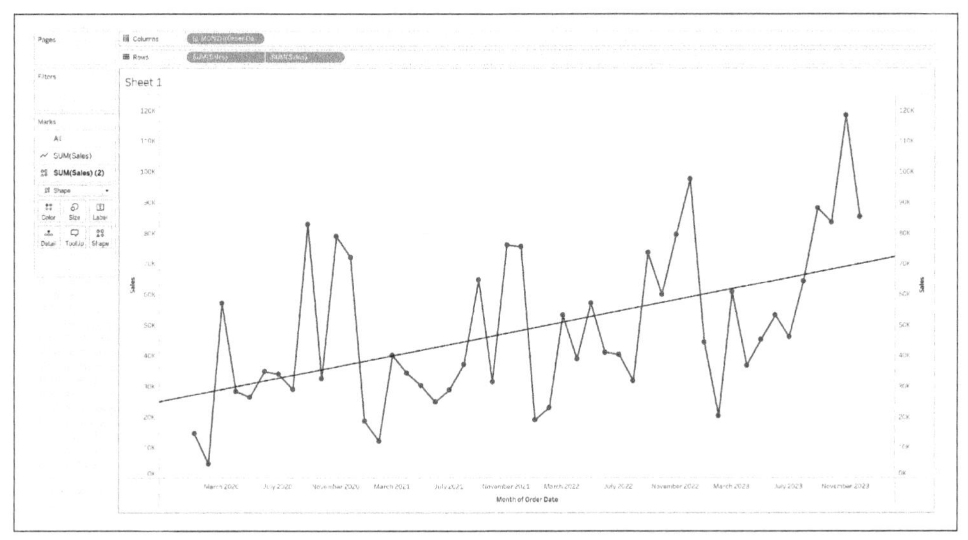

Figure 9-12. Linear regression of the sum of sales by month of order date

Sales = 29.9639*Month of Order Date + -1.2863e+06
R-Squared: 0.252128
P-value: 0.0002765

Figure 9-13. Summary statistics of the linear regression of month of order date

On the surface, this seems like it is underfitting the data. You can make that assumption by just observing the trend line on the view with the data. Let's see how a polynomial model performs with this data and compare the models. To add a polynomial regression to the view, toggle to the Analytics pane and drag Trend Line onto Polynomial, as shown in Figure 9-14.

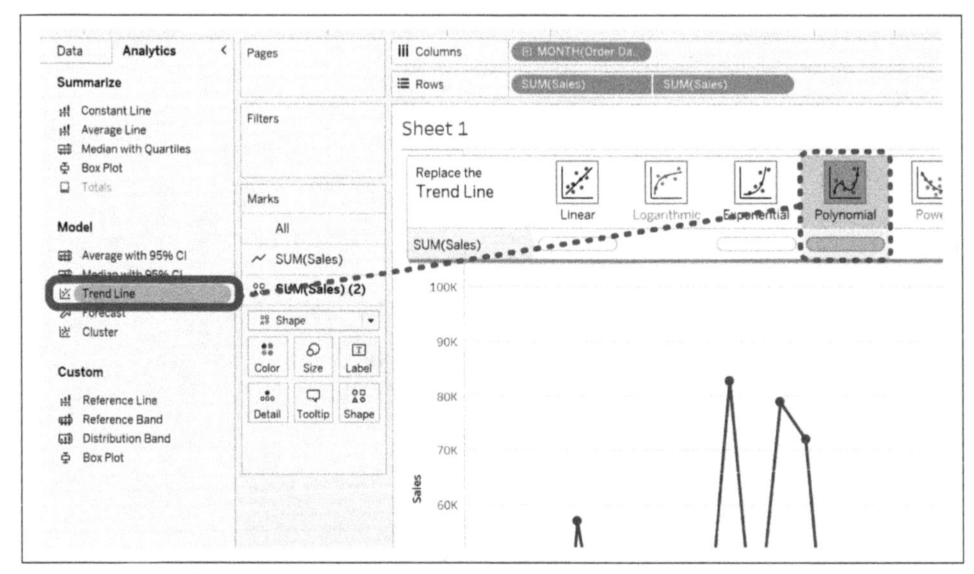

Figure 9-14. Adding a polynomial regression onto the view from the Analytics pane

You can see visually that this model fits the data much better. Rather than a straight line, you can see that the polynomial trend curves up quickly, then flattens out for a couple of years before trending up almost exponentially in the more recent months, as shown in Figure 9-15.

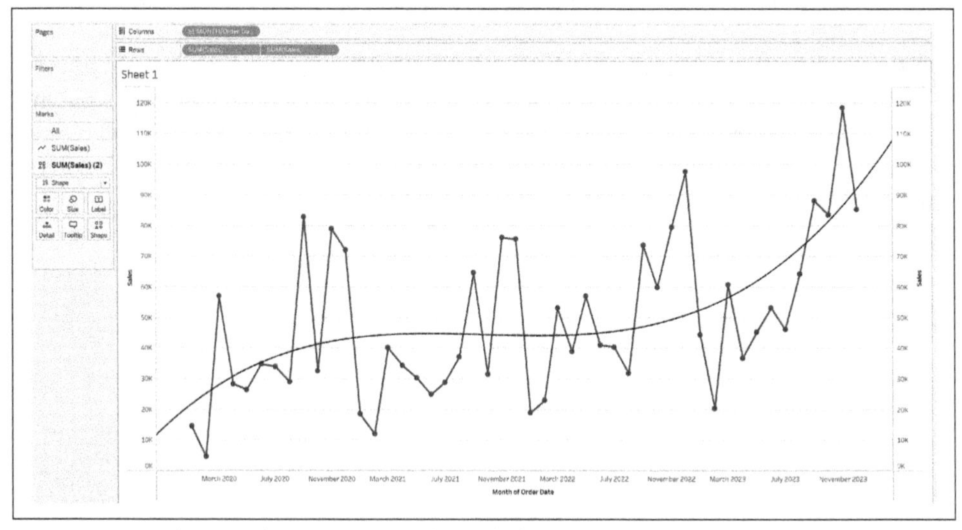

Figure 9-15. Polynomial regression model added to the view

Hovering over the trend line, as shown in Figure 9-16, you can immediately see that the model equation is much more complex. As I mentioned in the previous section, this model can lead to better results, but it becomes more difficult to interpret.

```
Sales = 9.99641e-05*Month of Order Date^3 + -13.3341*Month
of Order Date^2 + 592871*Month of Order Date + -8.78678e+09
R-Squared: 0.331254
P-value: 0.0004621
```

Figure 9-16. Summary statistics of the polynomial regression model

Looking at the p-value, you get a value less than 0.05 and an R-squared of 0.33. This means that the model could be considered statistically significant and that just knowing the month could explain about 33% of sales. Compare that to the values you got from the linear regression model, and you can see that mathematically, this model performs better.

Tableau is extremely flexible and allows you to easily change the parameters of the native models seamlessly. As mentioned in the previous sections, a polynomial model can be tuned by adding or removing degrees or powers from the equation. By default, Tableau starts a polynomial model with 3 degrees but you can adjust this setting from 2 degrees to a maximum of 8 degrees.

To edit this parameter, right-click on the trend line in the view and select Edit All Trend Lines from the menu, as shown in Figure 9-17.

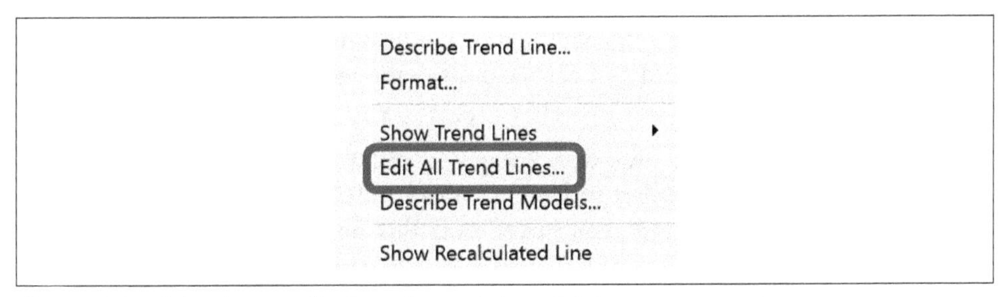

Figure 9-17. Selection to edit the polynomial model degrees parameter

You will see the Trend Lines Options menu appear, as shown in Figure 9-18. From this menu, you will see that the radio button for the model is selected for Polynomial and the Degree is 3, meaning the equation is using a third term. By clicking on the up and down arrows, you can change the model and see it update live in the view.

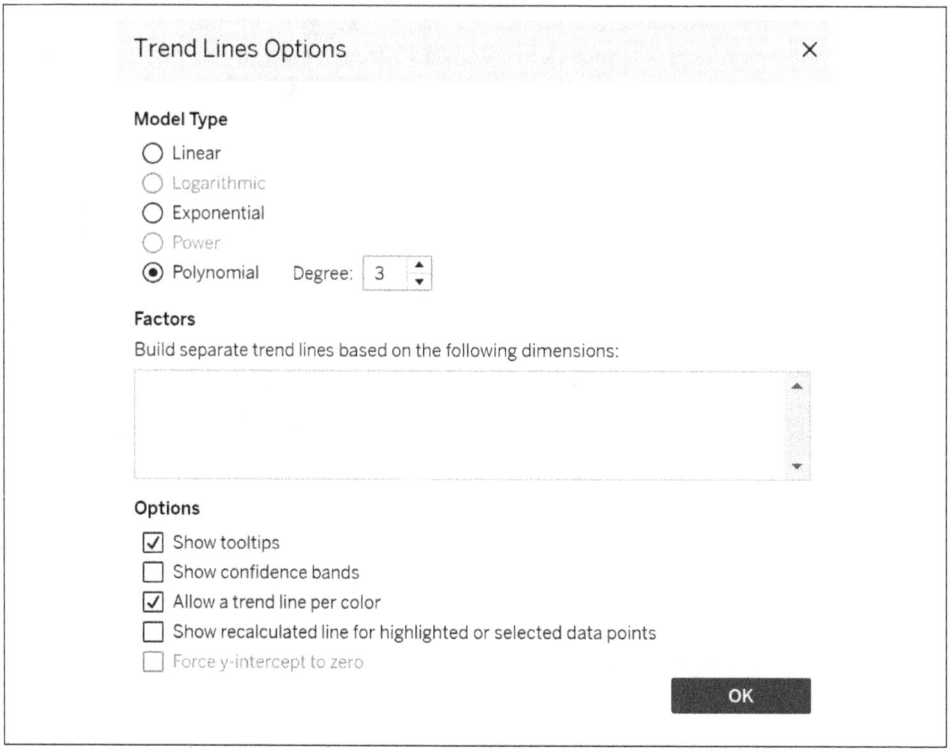

Trend Lines Options ×

Model Type

Linear
Logarithmic
Exponential
Power
Polynomial Degree: 3

Factors

Build separate trend lines based on the following dimensions:

Options

Show tooltips
Show confidence bands
Allow a trend line per color
Show recalculated line for highlighted or selected data points
Force y-intercept to zero

OK

Figure 9-18. Trend Lines Options menu

Let's see what happens when I update the model to just have 2 degree functions in the equation. I will click on the down arrow and see the view update live, as shown in Figure 9-19.

You can see what the model was with 3 degrees, and a new line appears that represents the results of the model at 2 degrees. When I click OK from the Trend Lines Options menu to close it and then hover over this new model, I can analyze the results from the tooltip, as shown in Figure 9-20.

You can see that this model is still statistically significant with a p-value less than 0.05 and an R-squared value at 0.27. With that said, this model does fit the data better than a linear regression model, but the results were slightly better with 3 degrees.

What if you adjust the model to the maximum degrees that Tableau allows? Edit the Trend Lines Options menu again and change the degrees from 2 to 8. As shown in Figure 9-21, you can see, interestingly enough, that you actually get a trend line that is almost the exact same as the model with 3 degrees.

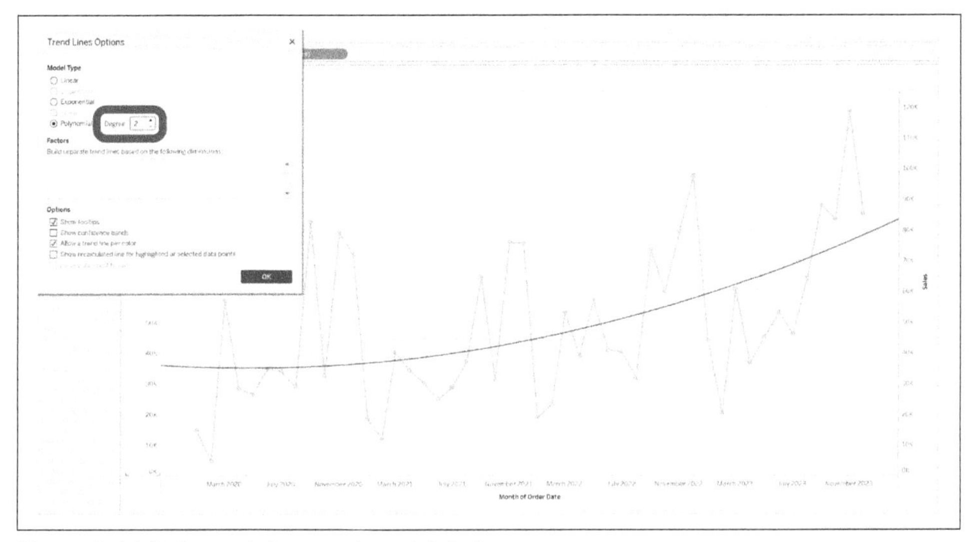

Figure 9-19. Polynomial regression with 2 degrees

Sales = 0.0248602*Month of Order Date^2 + -2184.88*Month of Order Date + 4.80403e+07
R-Squared: 0.276777
P-value: 0.0006818

Figure 9-20. Summary statistics of a polynomial model with degrees set at 2

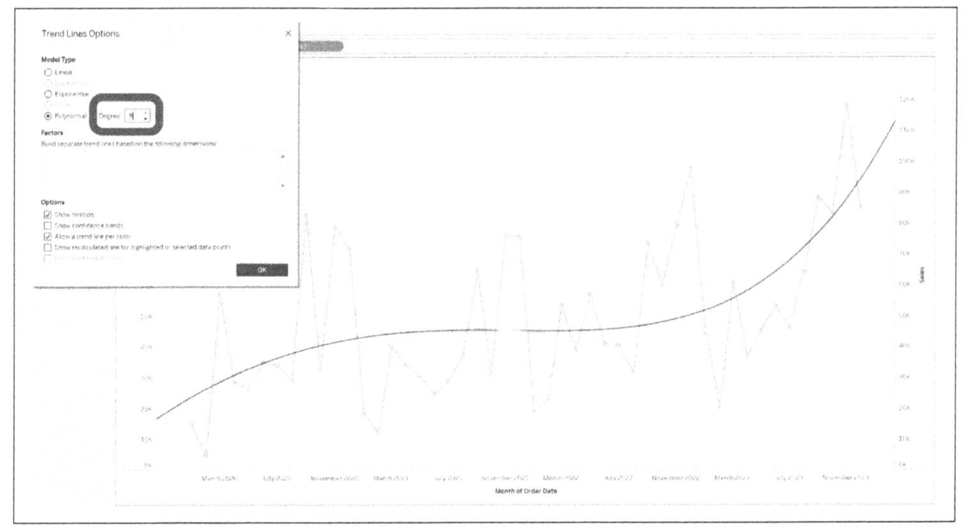

Figure 9-21. Polynomial regression with 8 degrees

If you hover over that new trend line, you can see you have almost the same results as well. The p-value is still less than 0.05 and the R-squared value is about 0.33; however, look at the model equation (see Figure 9-22).

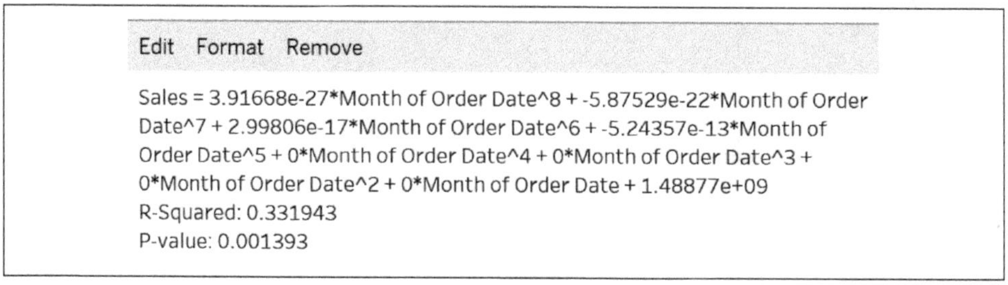

Edit Format Remove

Sales = 3.91668e-27*Month of Order Date^8 + -5.87529e-22*Month of Order Date^7 + 2.99806e-17*Month of Order Date^6 + -5.24357e-13*Month of Order Date^5 + 0*Month of Order Date^4 + 0*Month of Order Date^3 + 0*Month of Order Date^2 + 0*Month of Order Date + 1.48877e+09
R-Squared: 0.331943
P-value: 0.001393

Figure 9-22. Summary statistics of polynomial regression at 8 degrees

Putting something like this into practice or even trying to explain it to a stakeholder is going to be difficult. For the small change it made to the accuracy, it wouldn't make sense to go with this model that is more complex.

Summary

In this chapter, you learned about polynomial regression and how to implement it in Tableau. You were shown different elements of the summary statistics that could help you decide which model to use. Through that discussion, you learned about overfitting and underfitting a model. You were also introduced to how more complex models are harder to explain and interpret. Polynomial regression can be complex, but you should have enough knowledge from this chapter to know when to implement it and how to interpret the results using Tableau.

CHAPTER 10

Forecasting in Tableau

Forecasting is one of the best models to use in a business environment. There are many reasons why forecasting is popular, but the main reason is familiarity. You can see forecasts everywhere: in finance, banking, economics, healthcare, retail sales, supply chain, and much more.

The term *forecasting* itself can be rather vague from a statistical point of view because there are a lot of methods that can be used to forecast. In Tableau, when you use the built-in forecast model from the Analytics pane, it uses a method called *exponential smoothing*. Throughout this chapter, you see the term *forecasting* used interchangeably with *exponential smoothing*.

In this chapter, you will be introduced to exponential smoothing, its several methods of implementation, and the robustness of the model, and you will learn how to use it effectively in Tableau.

What Is Exponential Smoothing?

Exponential smoothing is a versatile technique that can be applied to various use cases across different industries. Here are five common use cases for exponential smoothing forecasting:

Retail demand forecasting
: Retailers often use exponential smoothing to predict future demand for products. By analyzing historical sales data, they can forecast product sales for different time periods, helping with inventory management, supply chain optimization, and stock replenishment strategies.

Financial forecasting
: Financial institutions and analysts use exponential smoothing to predict financial metrics like stock prices, exchange rates, and interest rates. This method can help investors make informed decisions and develop trading strategies.

Energy consumption forecasting
: Utility companies use exponential smoothing to forecast energy consumption patterns. By understanding how energy demand fluctuates over time, they can efficiently allocate resources, plan for infrastructure upgrades, and implement demand-response programs.

Staffing and workforce planning
: Workforce management departments use exponential smoothing to forecast workforce requirements. This helps organizations ensure they have the right number of employees available to meet varying workloads, seasonal demands, or project needs.

Website traffic and user engagement forecasting
: Online businesses and digital marketers use exponential smoothing to predict website traffic, user engagement, and conversion rates. This information is valuable for content planning, advertising budget allocation, and optimization of digital marketing campaigns.

These are just a few examples, and exponential smoothing can be applied to many other scenarios where time-bound data needs to be forecast. It's a valuable tool for organizations seeking to make data-driven decisions and improve their planning and resource allocation processes.

Tableau's forecasting tool is extremely robust and offers eight different algorithms for forecasting. Each of these algorithms is a different variation of exponential smoothing, but they all have a similar foundation, which is the simplest form of exponential smoothing. This foundational equation is called *simple exponential smoothing* and is what I will use to explain how the formula works.

Before jumping into the equations and how to implement these models in Tableau, let's break it down visually. Consider the data trended over time in Figure 10-1.

This data has no clear seasonality or trend and appears almost random. With data like this, it's not uncommon for analysts to take the average or median and use those summary statistics to predict values into the future. This is an example of where simple exponential smoothing excels, though. What sets the model apart from the more basic approaches (such as taking the average or median) is that exponential smoothing does not assign equal importance to all the data points. For instance, in getting a median of the data, I am assuming that the first data point recorded has the same weight as the last data point.

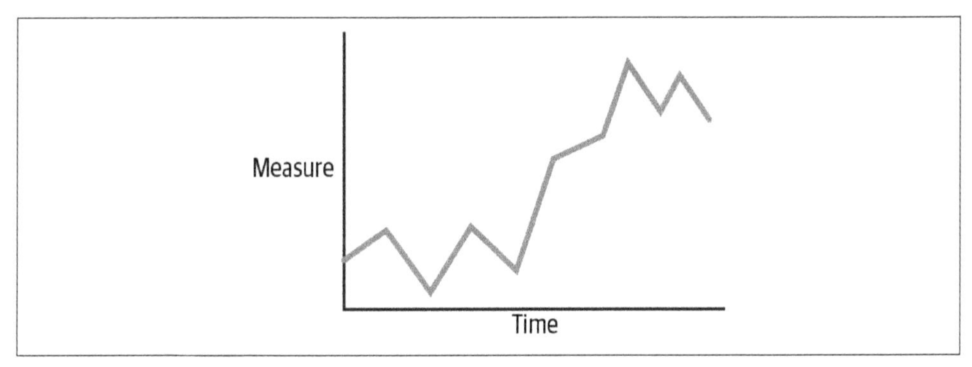

Figure 10-1. Exponential data example trended over time

Simple exponential smoothing actually assigns weights to each data point, and these weights increase exponentially as you move closer to the most recent data. The smallest weights are assigned to the oldest data points or observations. After all, if you were managing a retail store, more recent sales would be a better indicator of future sales due to continuous improvement in business. These weights are controlled by a value in the equation donated as α and is a value from 0 to 1. This value controls the rate at which the values decrease, as shown in Table 10-1.

Table 10-1. Simulating different alpha values and the weight applied at different data in a time series

	α = 0.2	α = 0.4	α = 0.6	α = 0.8
YT	0.2000	0.4000	0.6000	0.8000
YT-1	0.1600	0.2400	0.2400	0.1600
YT-2	0.1280	0.1440	0.0960	0.0320
YT-3	0.1024	0.0864	0.0384	0.0064
YT-4	0.0819	0.0518	0.0154	0.0013
YT-5	0.0655	0.0311	0.0061	0.0003

Consider the values in Table 10-1. You can see that as α moves closer to 1, more weight is given to more recent observations. You can also conceptualize that as α moves closer to 0, the smoother the fitted line would be for the observations within the time series. If you plotted a fitted line for different levels of α for our dataset, you would observe something like what is shown in Figure 10-2.

You can see that as α gets lower, the line becomes "smoother." However, as you move forward in time, because the points in the past have more weight, it causes the line to adjust to changes slower; the α at 0.8 reacted to the changes with the more recent data quicker.

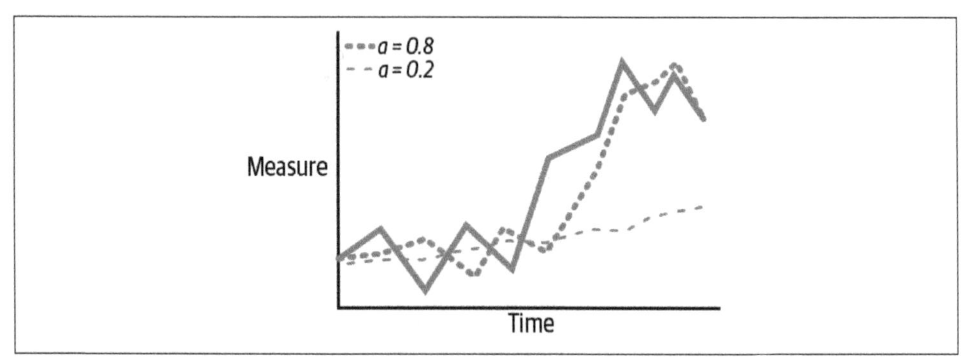

Figure 10-2. Simple exponential smoothing for different levels of α

Exponential Smoothing Equations

As mentioned earlier in this chapter, Tableau actually has eight algorithms built in natively with the forecasting model from the Analytics pane. Each of these models has its own pros and cons. However, when it comes to choosing the best model, Tableau makes it easy for you. When you use the Forecast model in Tableau, it runs through each of the models and selects what it thinks is the best fit for your data, giving you added confidence in your analysis. We will get more into this in "How to Implement Forecast Models in Tableau" on page 170. For now, let's focus on the equations.

To keep it simple, there will not be an explanation of each equation in detail. Just recall back to the visual examples we just explored. This is the foundation of each equation. Where they differ is how they react to trends and seasonality in the data. Exponential smoothing methods combine trend, seasonal, and error components together in a smoothing calculation. Each of those three components can be combined multiplicatively, combined additively, or left out of the model.

To break this down further, let's consider error, trend, and season (ETS). Exponential smoothing equations break the data into these three components, and that is how it forecasts the data, as shown in Figure 10-3.

You can see how each of these components relates back to the data. The *Trend* is a representation of how the data has trended overall. You can see from the top graph that the data trends from the bottom left and has slowly trended upward over time, matching the Trend visual.

The *Seasonality* data points are the peaks and valleys of the data over time. If you were to add up Trend and Seasonality, the difference between the sum of those values and the original data would be the *Error* term. If you summed Trend, Seasonality, and Error on any given increment of time, they would sum to the value of the original data on that date.

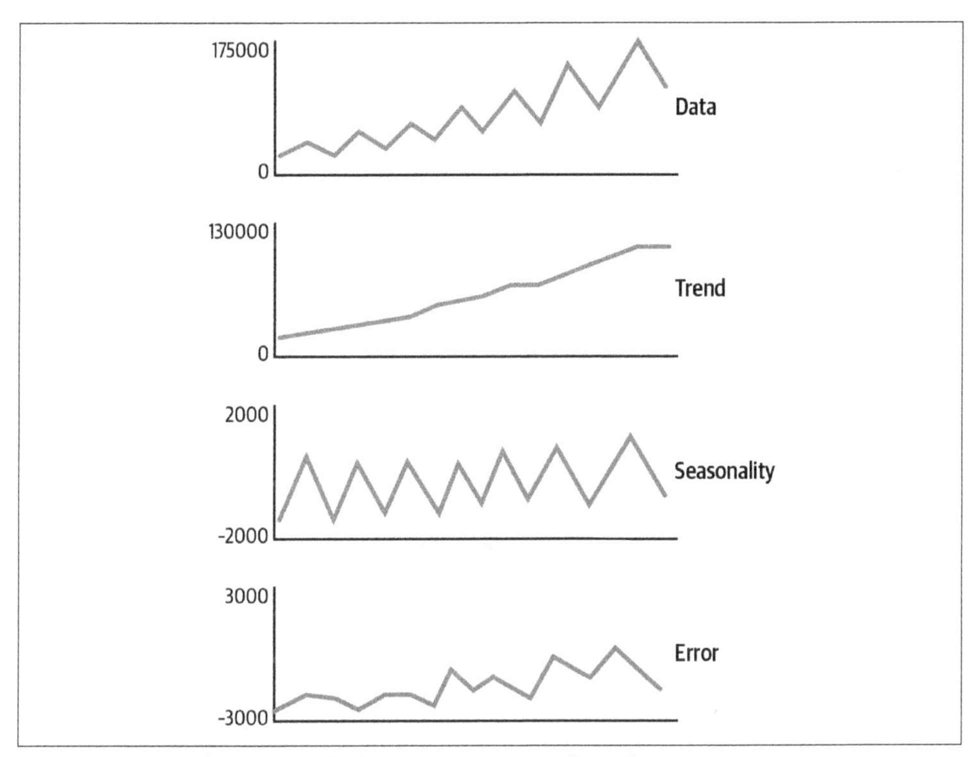

Figure 10-3. Breaking example data into error, trend, and season

The ultimate goal of exponential smoothing equations is to reduce the error component and better understand the trend and seasonality components. To achieve this, each equation manipulates these values by making them additive, multiplicative, or left out of the equation. Let's consider these different combinations for Trend and Seasonality, as shown in Table 10-2.

Table 10-2. Combinations of Trend and Seasonality being added, multiplied, or left out

	Seasonal component		
Trend component	**None (N)**	**Additive (A)**	**Multiplicative (M)**
None (N)	N,N	N,A	N,M
Additive (A)	A,N	A,A	A,M
Multiplicative (M)	M,N	M,A	M,M

Each of these combinations represents a different equation. As I mentioned earlier, I won't go into each in major detail, but here are some examples (for more detailed information, see the article on forecasting (*https://oreil.ly/BhiP7*) that Tableau links to in their own documentation on forecasting):

(N,N): Simple exponential smoothing

$$\hat{y}_{t+h|t} = \ell_t$$

(A,N): Holt's linear method

$$\hat{y}_{t+h|t} = \ell_t + hb_t$$

(A,A): Additive Holt-Winters' method

$$\hat{y}_{t+h|t} = \ell_t + hb_t + s_{t+h-m(k+1)}$$

(A,M): Multiplicative Holt-Winters' method

$$\hat{y}_{t+h|t} = \left(\ell_t + hb_t\right)s_{t+h-m(k+1)}$$

Just looking at each of these four methods expressed mathematically, you can see that they use the same foundation and then build on that using different techniques. To go full circle, these small changes to the equation help optimize how the weights are being calculated. These equations all have their own pros and cons, but that is where Tableau excels. When you use this model in Tableau, it runs through each and selects the best model for you. However, it is still flexible enough to allow you to tune these parameters yourself to manually select a model.

How to Implement Forecast Models in Tableau

Tableau makes it extremely easy for you to implement a forecast model. To get started, create a line chart of the sum of sales by month of order date, as shown in Figure 10-4.

From here, toggle to the Analytics pane on the top left of the authoring interface and drag Forecast onto the canvas, as shown in Figure 10-5.

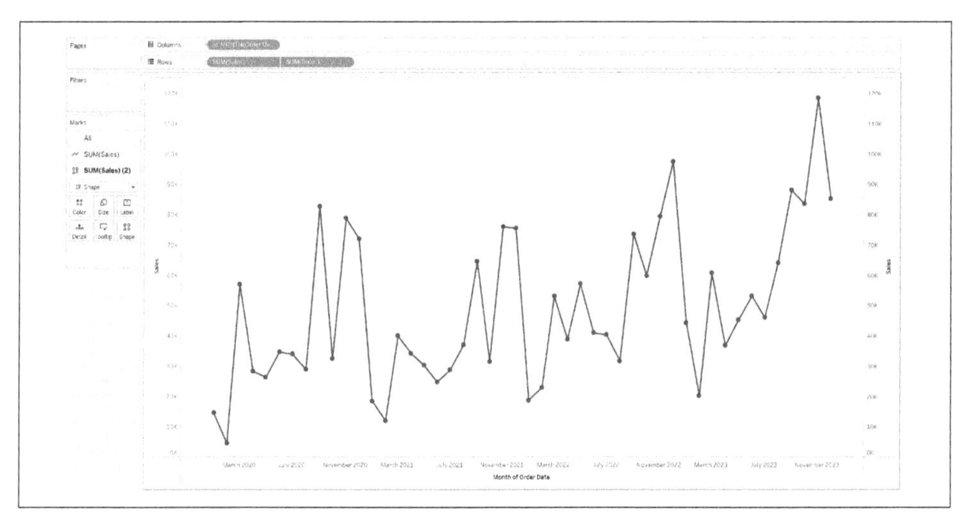

Figure 10-4. Sum of sales by month of order date

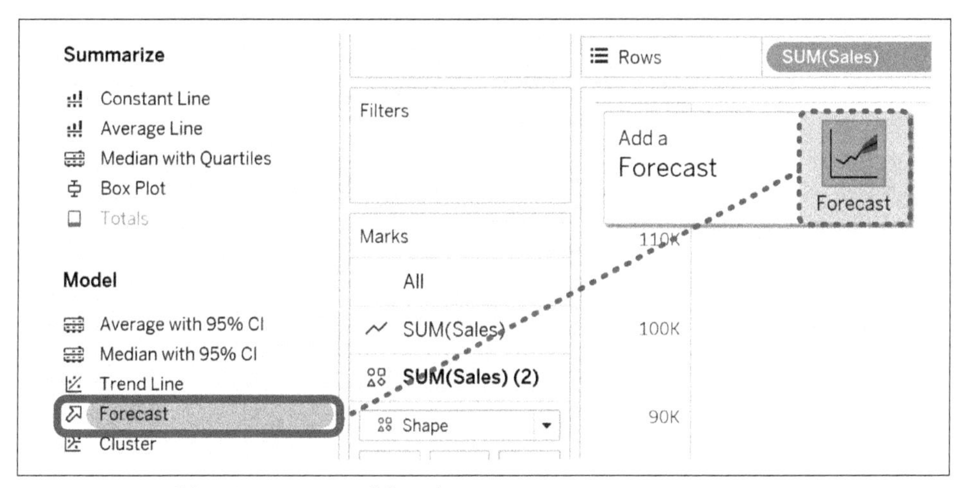

Figure 10-5. Adding Forecast model to the view

Immediately after dropping the model onto the canvas, you will see the Forecast model added to the Color property of the Marks card. This also gives you two indicators: Actual and Estimate. These will denote what your actuals are in your dataset and what the estimates are based on the forecast model, as shown in Figure 10-6.

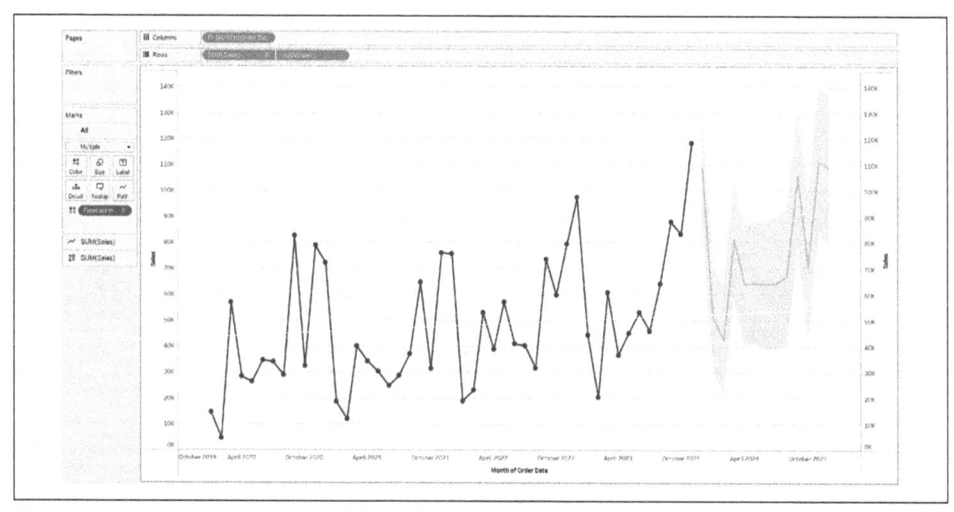

Figure 10-6. After Forecast model has been added to the view

It is important to note that this model has virtually generated future values for your dataset. This is important to call out because you can view the new data, but you won't be able to use it in calculated fields or reference them in any way. At most, you can "view the data" in Tableau and extract the estimates to Excel or another source. Conceptually, this is important because you have to realize that Tableau is a visualization tool and stays within the context of the data you have connected to. Unlike tools like Excel, where you can easily enter new data in the tool, there are limitations to doing this directly in Tableau Desktop.

This drag-and-drop approach is really helpful for you as a developer because it makes it so easy to implement these models, and you can feel confident in the results. However, what if someone asks you what type of algorithm you are using to calculate these values? To do some discovery of the model results and what model is being used, you simply right-click on the estimated values that were added to the view. In the menu that opens you will see a new option labeled Forecast. If you hover over Forecast you will see several other options: Show Forecast, Forecast Options, and Describe Forecast, as shown in Figure 10-7.

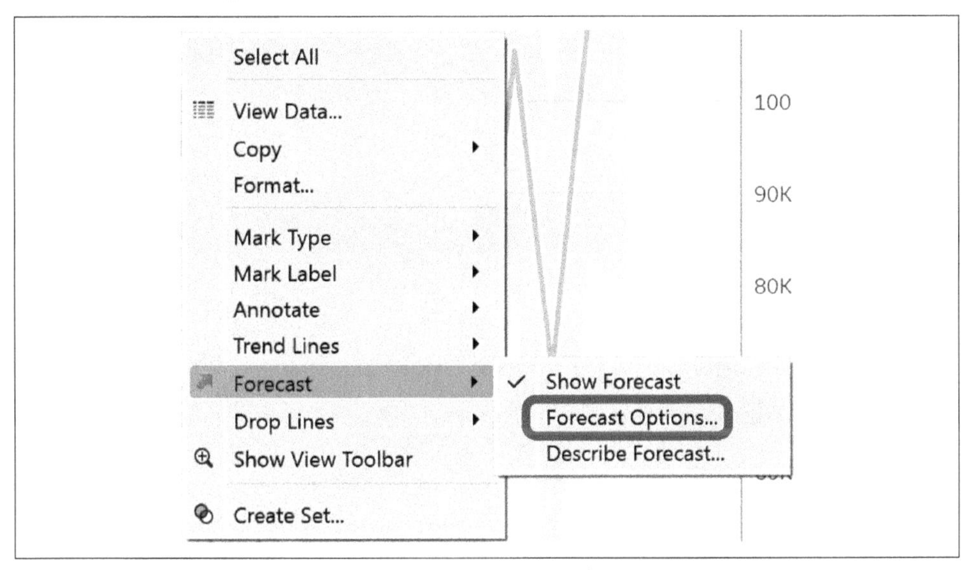

Figure 10-7. Forecast Options within the Tableau Desktop menu

The first option, Show Forecast, will remove the forecast from the view. Another option to remove the forecast is to drag the Forecast indicator pill that is in the Color property off the Marks shelf. Either of these approaches will remove the estimates from the view, and you will be left with the actuals.

Forecast Options

The second option is Forecast Options. When you click this, the Forecast Options window will appear, as shown in Figure 10-8. You can see that this menu is separated into three sections: Forecast Length, Source Data, and Forecast Model.

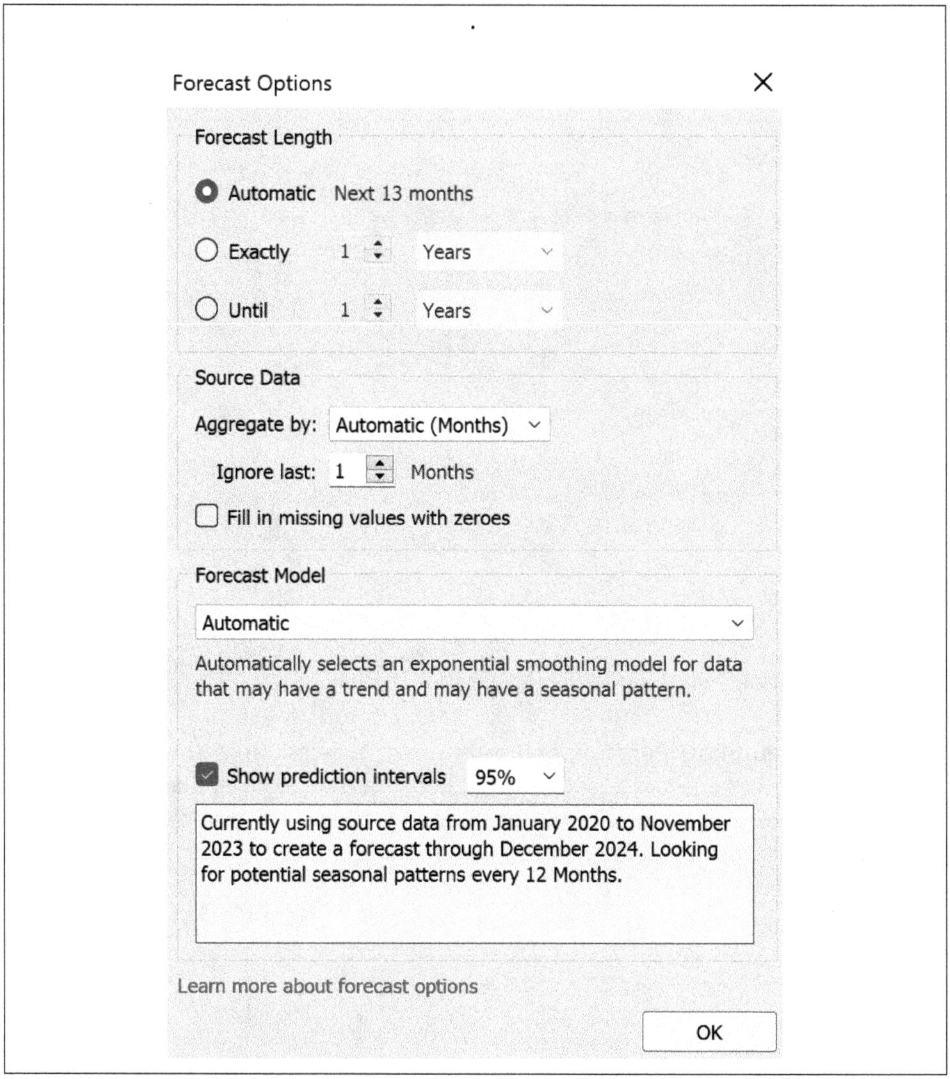

Forecast Options

Forecast Length

Automatic Next 13 months

Exactly 1 Years

Until 1 Years

Source Data

Aggregate by: Automatic (Months)

Ignore last: 1 Months

Fill in missing values with zeroes

Forecast Model

Automatic

Automatically selects an exponential smoothing model for data that may have a trend and may have a seasonal pattern.

Show prediction intervals 95%

Currently using source data from January 2020 to November 2023 to create a forecast through December 2024. Looking for potential seasonal patterns every 12 Months.

Learn more about forecast options

OK

Figure 10-8. Forecast Options menu

Forecast Length

There are three configurable options in the Forecast Length section. By default, Tableau will automatically assign a length of time for the forecast. In this example, you can see it selected "Next 13 months" automatically. However, you can define this length of time manually by selecting one of the other two available options. To demonstrate this section, select Exactly, which will enable the options to the right of the radio button selection, as shown in Figure 10-9.

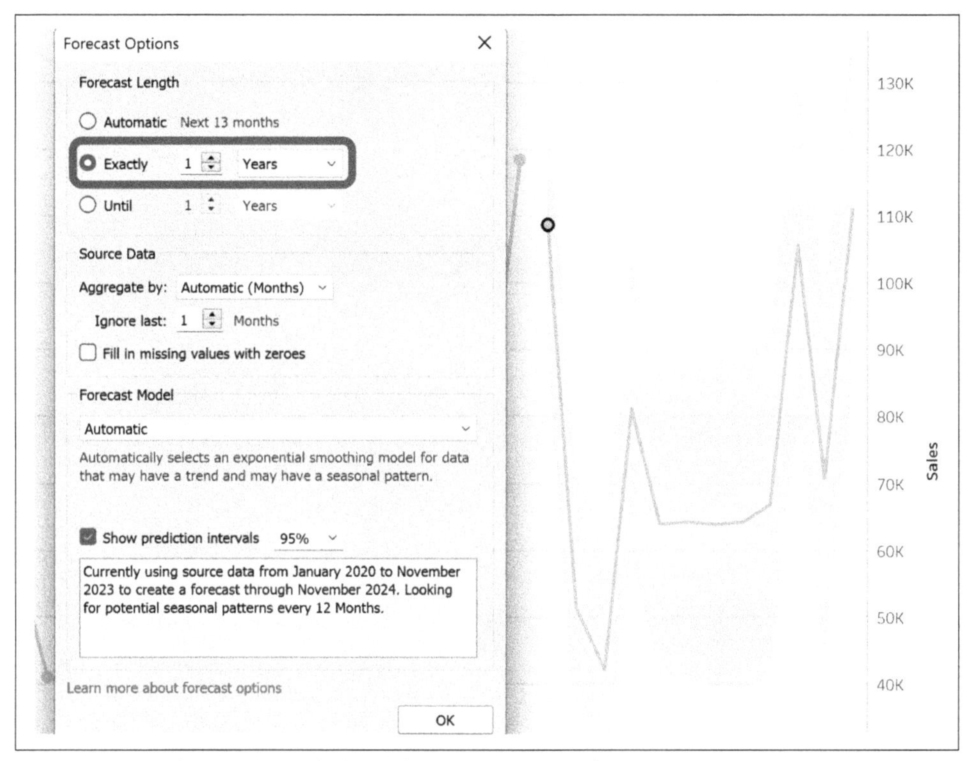

Figure 10-9. Selecting Exactly from the Forecast Length section

Beyond activating the two options on the right of the radio button, you can also observe a subtle shift on the estimates in the canvas as well, as it shifts from 13 months to 12 months of estimates. The two new options allow you to select how many time periods and the date part. So if you select 3 periods by quarters you will have 9 total months estimated in the view, as shown in Figure 10-10.

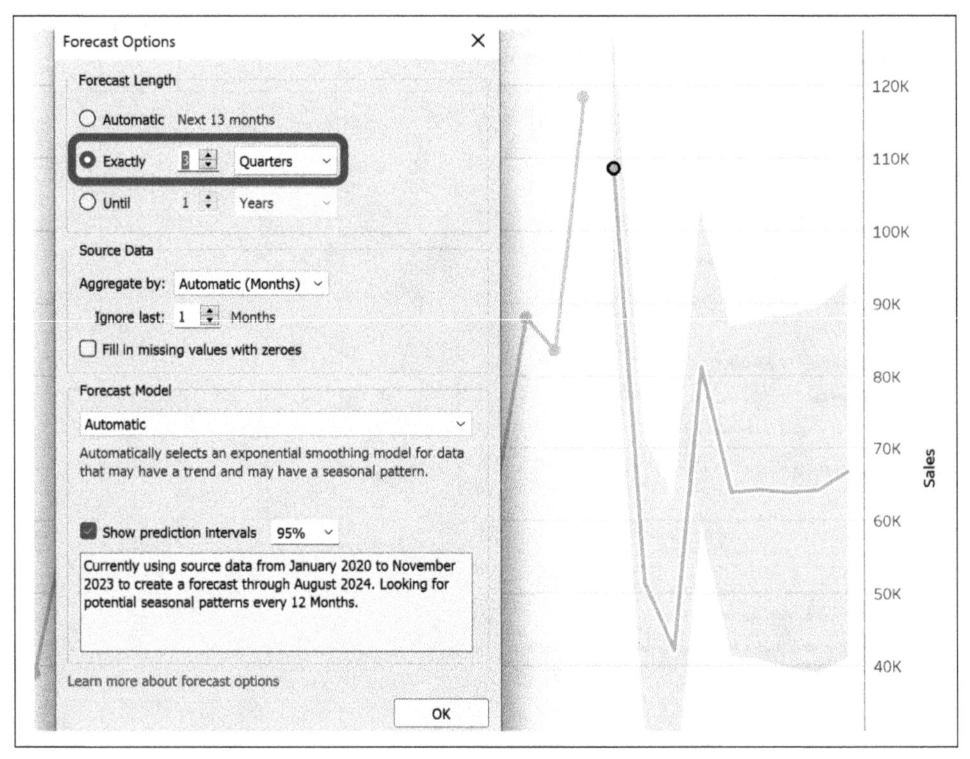

Figure 10-10. Showcasing the Exactly option in Forecast Length

If you select the Until option, you can see the difference immediately when you toggle the same options: 3 periods and date part of quarters. You can see in Figure 10-11 that there are only 7 periods that were estimated, and then it stopped. This gives you a bit more flexibility if you need a dynamic option that stops at certain milestones versus a rolling number of periods forward.

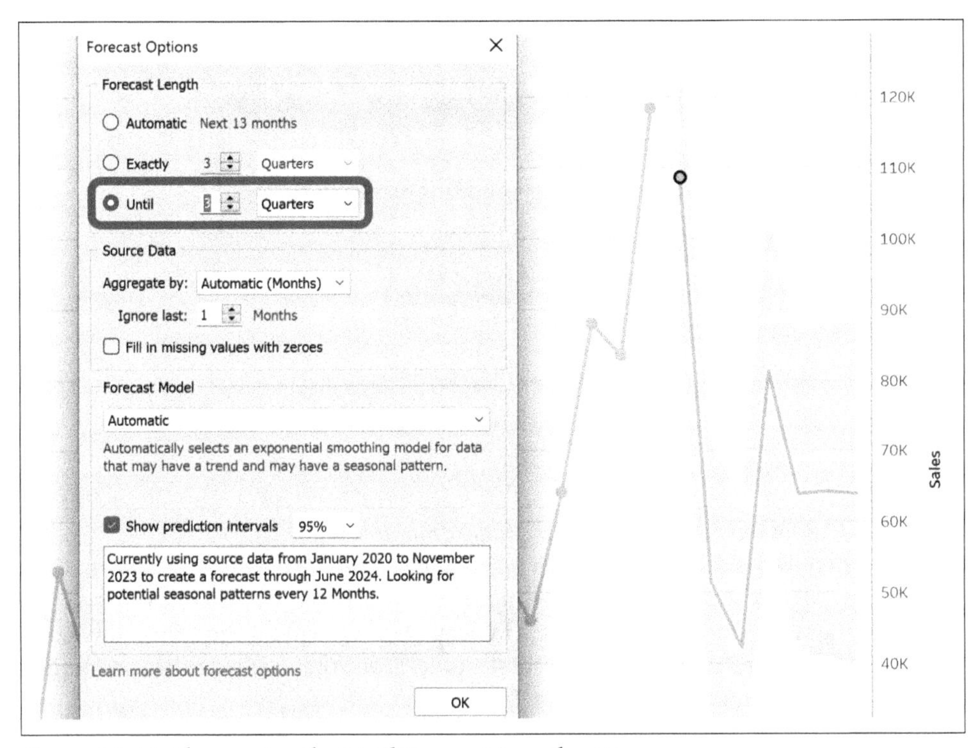

Figure 10-11. Showcasing the Until Forecast Length option

Source Data

The next section in Forecast Options is Source Data. This section gives you the ability to change the aggregation of the forecast model, an option to ignore the last *x* periods, and a toggle to fill in missing values with zeros.

If you adjust the "Ignore last" option, it will remove some of your actuals and put an estimate in their place. This is a really useful feature if you want to explore the accuracy of the model compared to some of the last few months of actuals. The last option to choose is "Fill in missing values with zeroes," if you have any in your data.

Forecast Model

The final section of the Forecast Options menu is the Forecast Model section. The first drop-down in this section will be set to Automatic by default; this is where you can manually change the exponential smoothing technique that is being used. If I expand that drop-down, as shown in Figure 10-12, you can see that there are three selectable options.

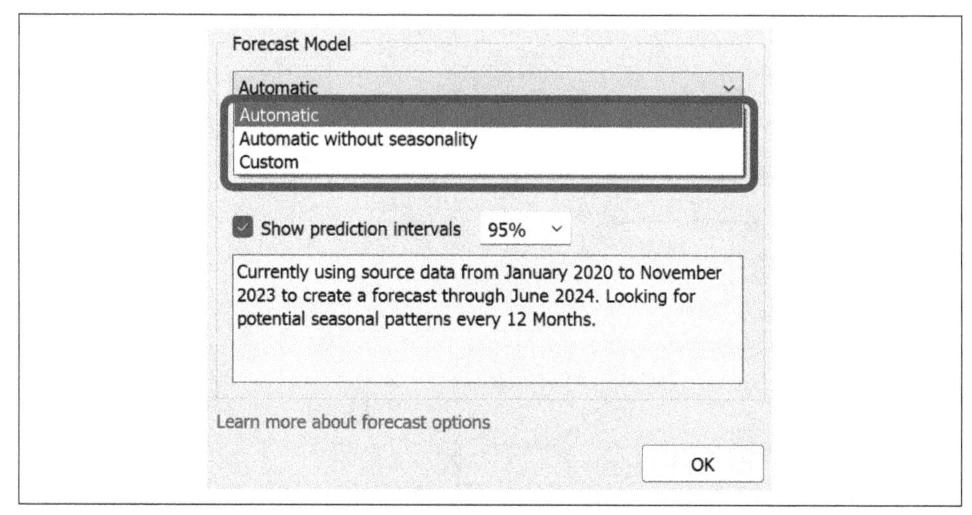

Figure 10-12. Options from the Forecast Model drop-down

Automatic will select the best possible model for you from the eight options. If you choose "Automatic without seasonality," Tableau Desktop will set the seasonality component to None and then choose the best model from the available three combinations, which you can see highlighted in Table 10-3.

Table 10-3. Highlighted available models with seasonality configured as None

	Seasonal component		
Trend component	None (N)	Additive (A)	Multiplicative (M)
None (N)	N,N	N,A	N,M
Additive (A)	A,N	A,A	A,M
Multiplicative (M)	M,N	M,A	M,M

The final option from that drop-down is Custom. This is where you can manually configure the Trend and Season components to select a specific exponential smoothing technique, as you can see in Figure 10-13.

With Custom selected, you can provision Trend and Season. Currently, in Figure 10-13, the simple exponential smoothing model is selected because Trend and Season are not considered in the model and are configured to None. Also notice in Figure 10-13 that the estimates in the view have changed to reflect this new model:

(N,N): Simple exponential smoothing

$$\hat{y}_{t+h|t} = \ell_t$$

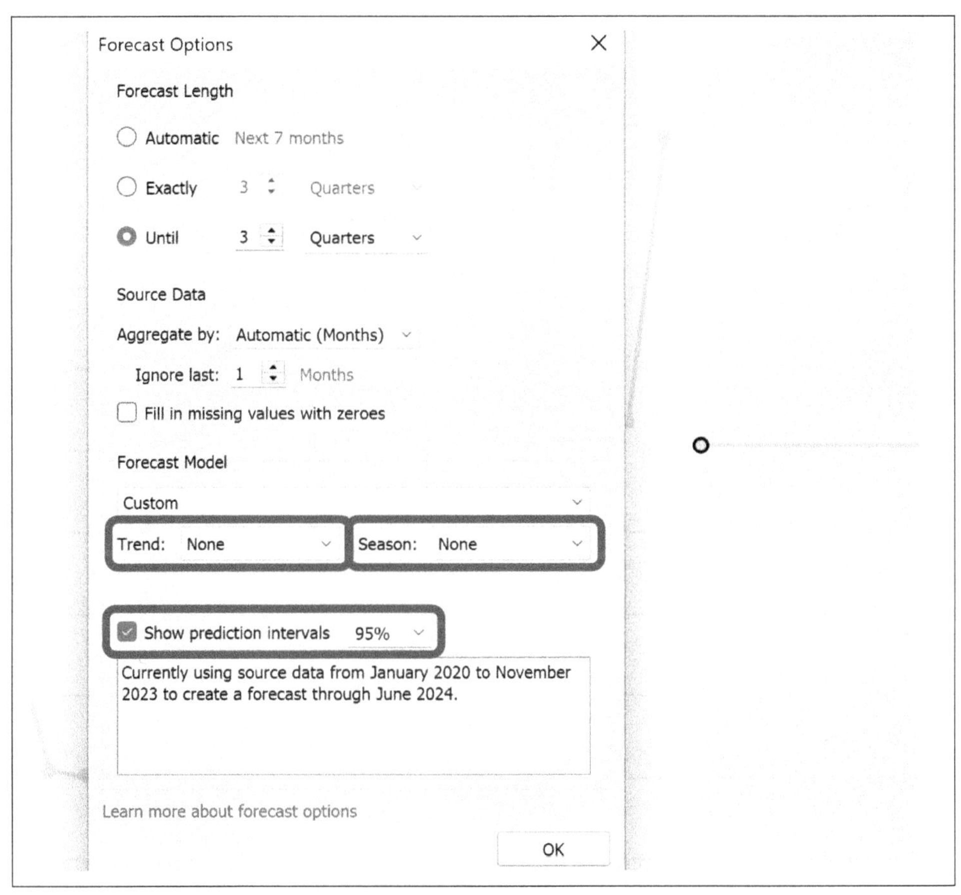

Figure 10-13. Simple exponential smoothing model selected

If you click the Trend drop-down and choose Additive, you will be toggling the model to Holt's linear method, and you will see the estimates update in the view (see Figure 10-14):

(A,N): Holt's linear method

$$\hat{y}_{t+h|t} = \ell_t + hb_t$$

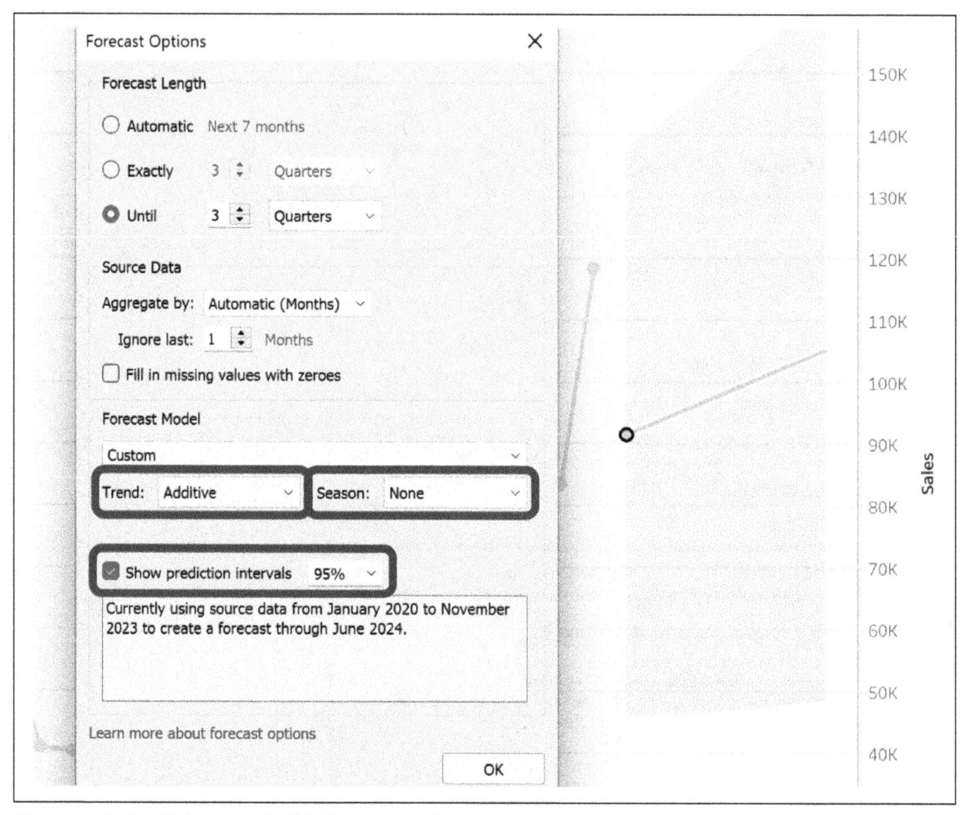

Figure 10-14. Selecting Holt's linear method

If you click the Season drop-down and select Additive, you will have selected the Additive Holt-Winters' method, as shown in Figure 10-15:

(A,A): Additive Holt-Winters' method

$$\hat{y}_{t+h|t} = \ell_t + hb_t + s_{t+h-m(k+1)}$$

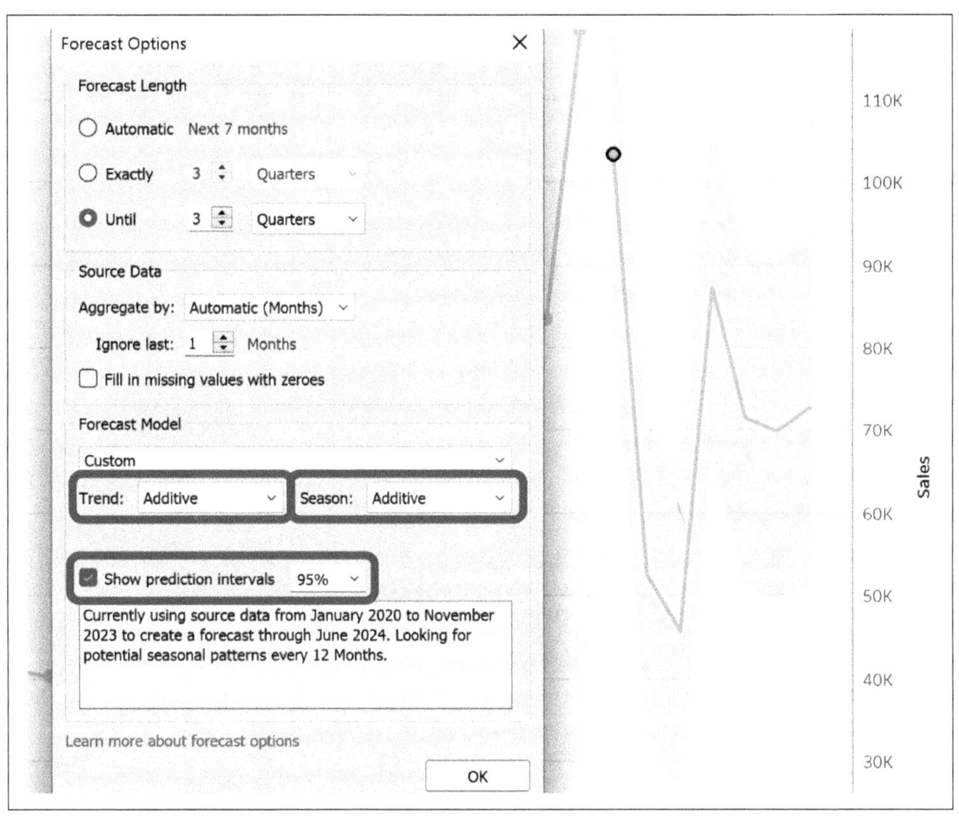

Figure 10-15. Additive Holt-Winters' method

The goal would be to toggle to the different models using your industry knowledge or context that is not within the data itself. Maybe you know that a new marketing campaign is about to occur, and therefore sales estimates will be higher. Maybe there is a new product launch or a downturn in the market; regardless of the reason, you can provision the method appropriately to account for them here.

You can also toggle the Custom options until you find the model that matched what Tableau chose by default with the Automatic setting. This is helpful when you go to present your findings and will build credibility and adoption of your model if you can explain and interpret the results of the model.

The last piece of the Forecast Options menu allows you to toggle the prediction intervals on and off. Prediction intervals are very similar to confidence intervals, but they define where your predictions will fall on average. Just like confidence intervals, your confidence level can be adjusted to be either more or less confident. With the dropdown to the right of the "Show prediction intervals" checkbox, you can adjust the confidence level to 90%, 95%, or 99%. If you toggle the confidence level to 90%, you can see that prediction intervals in the view get thinner, as shown in Figure 10-16.

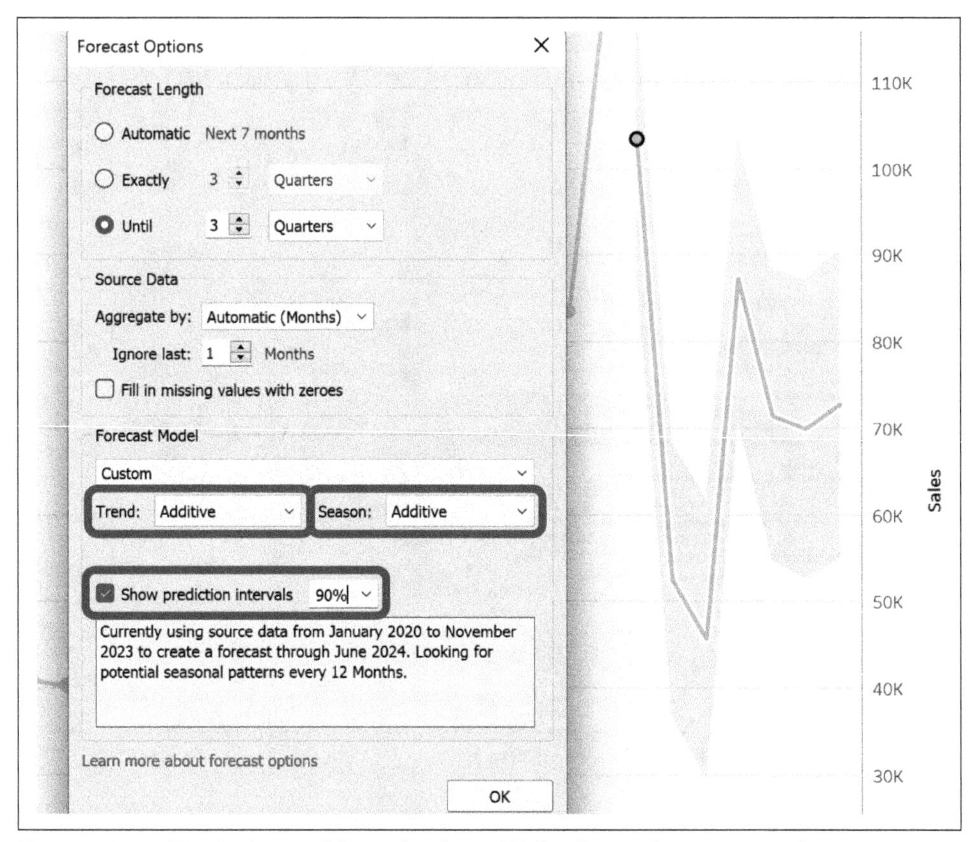

Figure 10-16. Toggle the confidence level to 90% for the prediction intervals

If I toggle the confidence level to 99%, the intervals become much wider, as you can see in Figure 10-17.

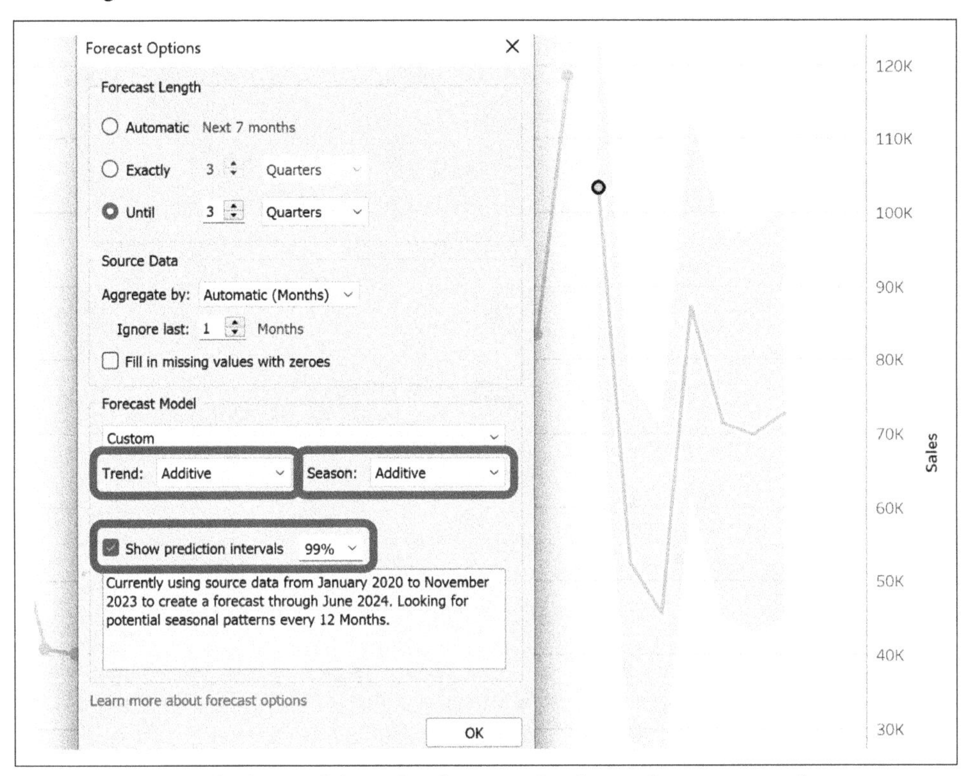

Figure 10-17. Toggle the confidence level to 99% for the prediction intervals

Below the options for the prediction intervals is a brief explanation of the model. I have always found this bit of text a great starting point for documentation or a write-up of the model. That text will dynamically update based on your selections you made.

Describe Forecast

Close the Forecast Options menu, right-click on the estimations again, and select the Describe Forecast option, as shown in Figure 10-18.

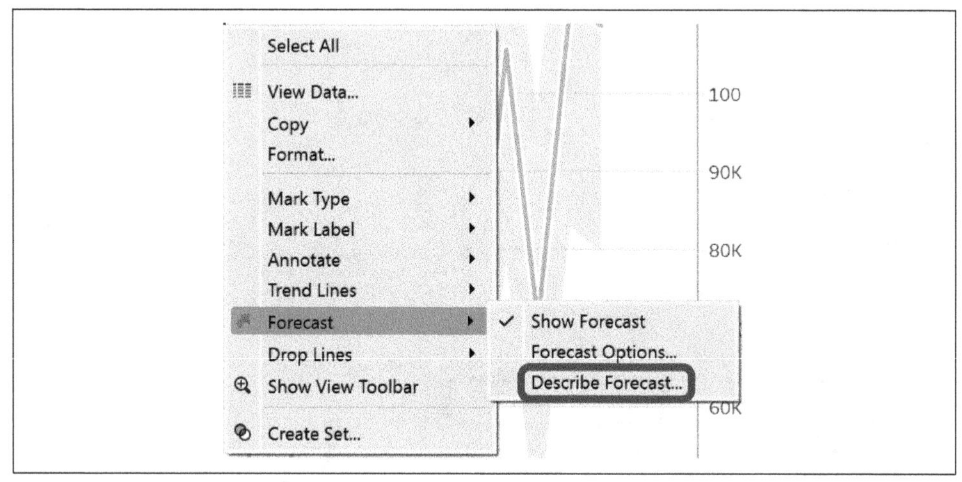

Figure 10-18. Selecting Describe Forecast from the menu

This will open the Describe Forecast window, which gives us the summary statistics of the currently selected model. There are two tabs in this window that you can access from the top left of the window pane: Summary and Models. The Summary tab is what will open by default, as shown in Figure 10-19.

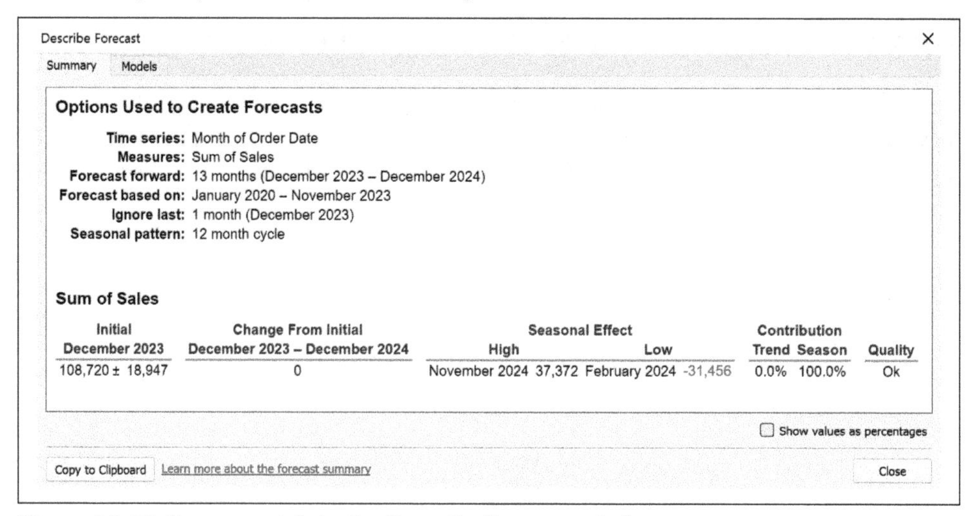

Figure 10-19. Summary tab in the Describe Forecast window

This tab has some very useful summary statistics that you can use to better understand the model. However, if you toggle to the Models tab, you will find more detailed information, as shown in Figure 10-20.

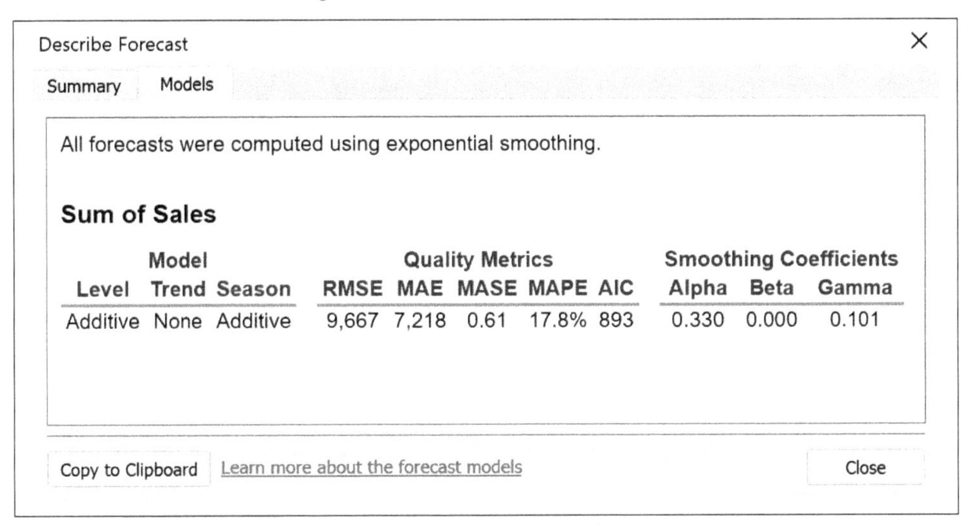

Figure 10-20. Models tab in the Describe Forecast window

Here you can find the components Model, Quality Metrics, and Smoothing Coefficients. You can use this information to better understand the accuracy of the model, which model is selected, and the coefficients of the model.

Another important point: for forecasting to be effective, you need to have enough historical data. If you have four points of data or observations, that is most likely not going to be enough. You need to consider this when implementing the model.

Summary

In this chapter, you were introduced to forecasting using exponential smoothing. You learned about different techniques that are available in Tableau and how to access them within the Forecast Options menu. You also learned how you can find the summary statistics of the model to better understand how the model is performing.

CHAPTER 11

Clustering in Tableau

Oftentimes, you will find yourself wanting to better understand how things relate to one another. What groups of products sell well when paired together? How should I market to certain groups of customers? Are there anomalies in my data? If you're asking these types of questions, then clustering is a great model to start finding answers. The primary objective of clustering is to partition a dataset into subgroups or clusters. The models achieve this by partitioning the data so that the data points in one cluster are more similar to each other than another cluster's data points.

There are many different clustering models, each with its own pros and cons. In Tableau, the algorithm that is built in for clustering is called *k-means*. K-means is a widely used model that provides an automated approach to grouping data. Unlike the other regression models, k-means is also an unsupervised model, which means having a normal distribution is not an assumption for this model.

In this chapter, you will learn how the k-means model works, the difference between supervised and unsupervised models, and how to implement k-means in Tableau.

What Is K-Means Clustering?

K-means clustering is a versatile, unsupervised technique that can be applied to various domains and problems. Examples of how you could use k-means clustering include:

Customer segmentation
: K-means clustering is used to group customers based on their purchasing behavior, demographics, or other relevant features. This can help businesses tailor marketing strategies, product recommendations, and customer support to different customer segments more effectively.

Recommendation systems

Use k-means clustering to create user or item profiles in recommendation systems. By clustering users or items with similar preferences, you can make personalized recommendations to users based on the preferences of other users in the same cluster. This is often referred to as *collaborative filtering*.

Anomaly detection

K-means clustering can identify anomalies in datasets by clustering data points into "normal" and "anomalous" groups. Data points that do not belong to any cluster or are distant from all clusters can be considered anomalies, making it useful for fraud detection, network security, and quality control.

Inventory management

K-means clustering can be applied to optimize inventory management by grouping products based on demand patterns. This helps in maintaining appropriate stock levels, reducing carrying costs, and minimizing stockouts, which can lead to improved supply chain efficiency.

Fraud detection

Financial institutions and ecommerce businesses can utilize k-means clustering to detect fraudulent activities. By clustering transaction data, anomalies can be detected more effectively, helping in the early identification of potentially fraudulent transactions or accounts.

These are just a few examples of how k-means clustering can be applied in various domains. Its simplicity and effectiveness make it a valuable tool for exploratory data analysis, pattern recognition, and data-driven decision making in many fields.

K-Means Conceptual Example

To understand how k-means clustering works, you will read a simple example and walk through it step by step with visual aids. The equation expressed mathematically is as follows:

$$J = \Sigma_{j=1}^{k} \Sigma_{i=1}^{n} ||x_i^{(j)} - c_j||^2$$

where

J = objective question

k = number of clusters

n = number of operations

$x_i^{(j)}$ = the ith observation

c_j = centroid for cluster j

On the surface, this model seems like one of the more complex ones that have been covered in this book. However, it can be broken down to a simple example and executed by hand. Here are the steps you need to take to calculate the results and assign each Row ID to a cluster:

1. Define how many *k* clusters there are and assign a centroid randomly.
2. Assign each data point to the nearest cluster centroid based on the Euclidean distance.
3. Recalculate the cluster centroids as the mean of all the data points assigned to the cluster.
4. Repeat steps 2–3 until there is no change to each cluster where Row ID is assigned.

I am using some new language here (Euclidean distance, centroid), and while these are important terms to understand, I don't want to lose you here. Really, *Euclidean distance* is just a calculation used to see which cluster centroid is closer to the data point or the shortest distance between two points. The *centroid* is the center point of a cluster that will be used as the point to measure the distance to the data point. As an example, here is a scatterplot (see Figure 11-1).

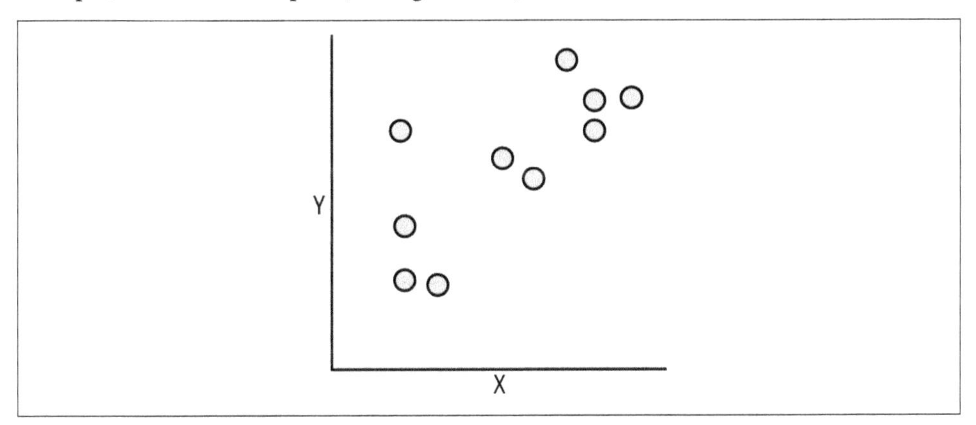

Figure 11-1. Example scatterplot for k-means cluster example

Step 1 is to determine how many clusters to use and then assign some random centroid points. For this example, you will use two clusters and assign some centroids (the star shapes) in the scatterplot (see Figure 11-2).

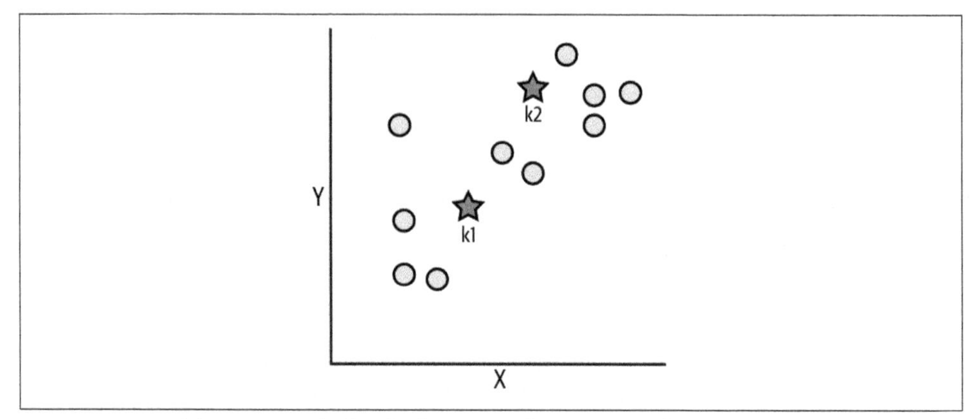

Figure 11-2. Adding centroids for the k-means cluster example

Step 2 is to assign a point to a cluster based on its Euclidean distance. You can do that by selecting a point and drawing a line to both of the centroid points, as shown in Figure 11-3.

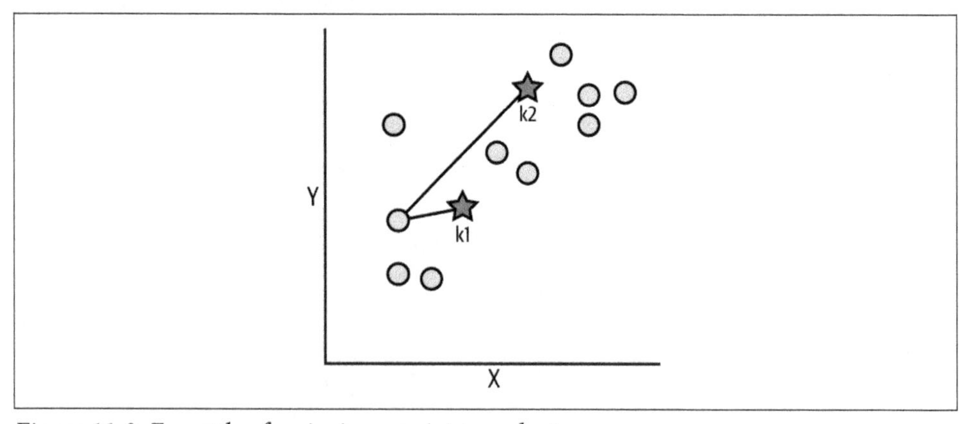

Figure 11-3. Example of assigning a point to a cluster

You can see visually that the point used as an example is closer to k1; therefore you would assign it to cluster 1. In this step, you would do that for all the data points in the view until every point was assigned to one of the clusters. Step 3 is to recalculate the centroid points as the mean of all the points assigned to that cluster. You can see this demonstrated in Figure 11-4.

From here, you repeat steps 2–3 by mapping the distance between each point and the centroid again, then assigning the point to one of the clusters, as shown in Figure 11-5.

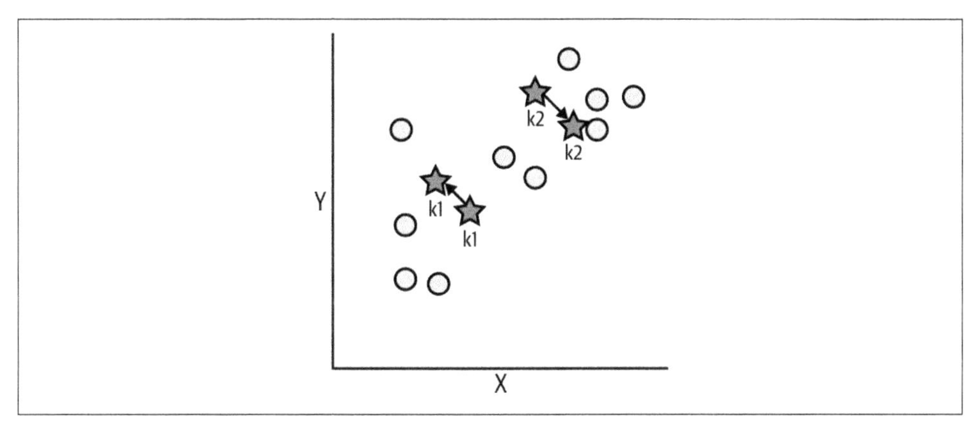

Figure 11-4. Assigning new centroids based on the mean of the clusters

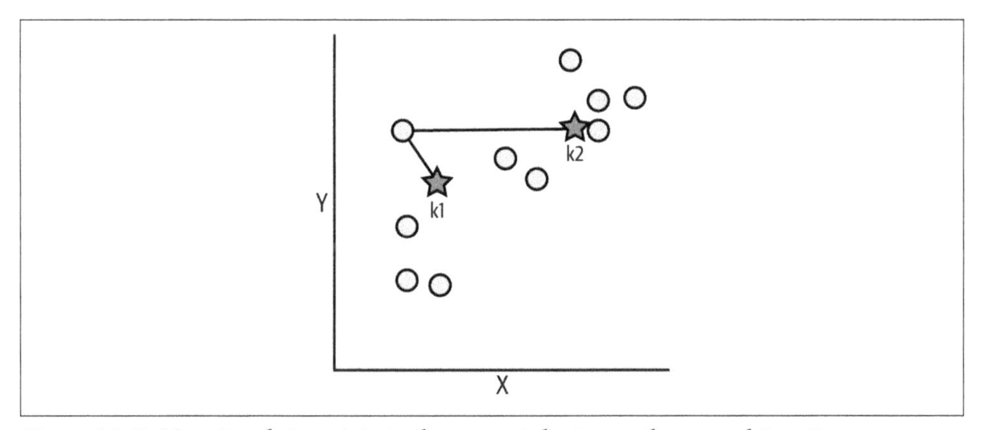

Figure 11-5. Mapping data points to the correct cluster on the second iteration

Once all of the data points are mapped to the clusters based on the new centroid location, you would see if any point changed from its assigned cluster from the last iteration. If there was a point that was reassigned to a different cluster, you would recalculate the new centroids and then repeat steps 2–3 again. If none of the points were reassigned, the iteration process would stop and that would be the final assigned cluster for the data points. In this example, you did not have to reassign any points to a different cluster. In Figure 11-6, you can see all the points assigned to cluster 1 denoted by a square shape and everything assigned to cluster 2 as a circle shape.

You can see an ellipse plotted from the center of the centroid expanding out. This can help provide context around the distance between the centroid and the points.

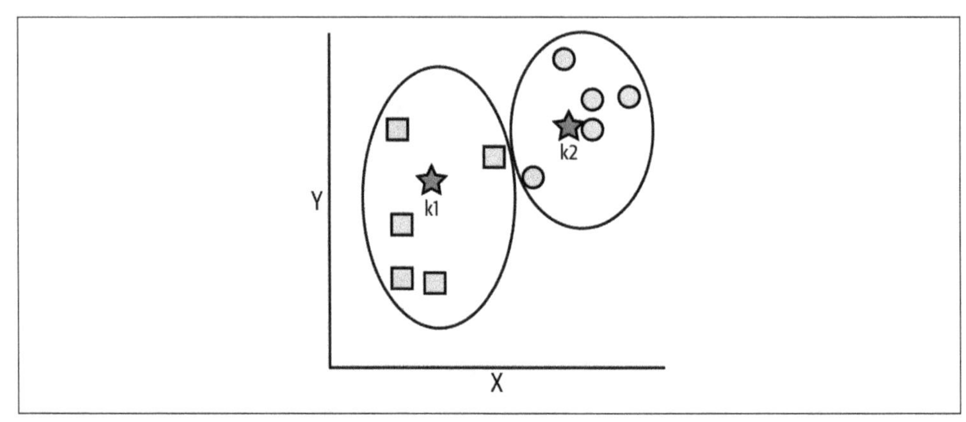

Figure 11-6. Final cluster in k-means example

K-Means Example with Real Values

You've just seen a conceptual example, but how does the actual equation apply to real data? Here is some mock data that represents the number of orders certain sales reps produced over the course of a month. You want to cluster the sales reps into two groups that represent high performance and average performance. You will use two clusters, and their initial centroids assigned randomly are 17 and 22.

The data is recorded in Table 11-1. You will see each sales rep denoted by their ID. The second column is the number of sales they obtained for the month. Columns three and four represent the centroid for each cluster. Columns five and six represent the absolute difference between the number of sales and the centroid. Column seven represents the nearest cluster based on the smallest number from columns five and six. Finally, column eight represents the new centroid calculated from the mean of the values assigned to that cluster.

Table 11-1. First iteration of k-means example

Rep ID	Sales	c1	c2	Sales – c1	Sales – c2	Nearest cluster	New centroid
001	18	17	22	1	4	1	17.33
002	30	17	22	13	8	2	36.83
003	51	17	22	34	29	2	36.83
004	22	17	22	5	0	2	36.83
005	47	17	22	30	25	2	36.83
006	59	17	22	42	37	2	36.83
007	23	17	22	6	1	2	36.83
008	19	17	22	2	3	1	17.33
009	27	17	22	10	5	2	36.83
010	15	17	22	2	7	1	17.33

Rep ID	Sales	c1	c2	Sales – c1	Sales – c2	Nearest cluster	New centroid
011	33	17	22	16	11	2	36.83
012	39	17	22	22	17	2	36.83
013	42	17	22	25	20	2	36.83
014	32	17	22	15	10	2	36.83
015	37	17	22	20	15	2	36.83

Now you reiterate based on the results of Table 11-1 to see if there are any shifts between clusters. The next interaction is calculated in Table 11-2.

Table 11-2. Second iteration of k-means example

Rep ID	Sales	c1	c2	Sales – c1	Sales – c2	Nearest cluster	New centroid
001	18	17.33	36.83	0.67	18.83	1	20.66
002	30	17.33	36.83	12.67	6.83	2	41.11
003	51	17.33	36.83	33.67	14.17	2	41.11
004	22	17.33	36.83	4.67	14.83	1	20.66
005	47	17.33	36.83	29.67	10.17	2	41.11
006	59	17.33	36.83	41.67	22.17	2	41.11
007	23	17.33	36.83	5.67	13.83	1	20.66
008	19	17.33	36.83	1.67	17.83	1	20.66
009	27	17.33	36.83	9.67	9.83	1	20.66
010	15	17.33	36.83	2.33	21.83	1	20.66
011	33	17.33	36.83	15.67	3.83	2	41.11
012	39	17.33	36.83	21.67	2.17	2	41.11
013	42	17.33	36.83	24.67	5.17	2	41.11
014	32	17.33	36.83	14.67	4.83	2	41.11
015	37	17.33	36.83	19.67	0.17	2	41.11

You can see that there were some sales reps reassigned to a different cluster, and the centroids recalculated once again. Since there were some shifts, you will reiterate again (see Table 11-3).

Table 11-3. Third iteration of k-means example

Rep ID	Sales	c1	c2	Sales – c1	Sales – c2	Nearest Cluster	New Centroid
001	18	20.66	41.11	2.66	23.11	1	22
002	30	20.66	41.11	9.34	11.11	1	22
003	51	20.66	41.11	30.34	9.89	2	42.5
004	22	20.66	41.11	1.34	19.11	1	22
005	47	20.66	41.11	26.34	5.89	2	42.5
006	59	20.66	41.11	38.34	17.89	2	42.5

Rep ID	Sales	c1	c2	Sales – c1	Sales – c2	Nearest Cluster	New Centroid
007	23	20.66	41.11	2.34	18.11	1	22
008	19	20.66	41.11	1.66	22.11	1	22
009	27	20.66	41.11	6.34	14.11	1	22
010	15	20.66	41.11	5.66	26.11	1	22
011	33	20.66	41.11	12.34	8.11	2	42.5
012	39	20.66	41.11	18.34	2.11	2	42.5
013	42	20.66	41.11	21.34	0.89	2	42.5
014	32	20.66	41.11	11.34	9.11	2	42.5
015	37	20.66	41.11	16.34	4.11	2	42.5

Once again, there were some shifts between clusters and a recalculation of the centroids. You will iterate a fourth time, as shown in Table 11-4.

Table 11-4. Fourth iteration of k-means example

Rep ID	Sales	c1	c2	Sales – c1	Sales – c2	Nearest Cluster	New Centroid
001	18	22	42.5	4	24.5	1	22
002	30	22	42.5	8	12.5	1	22
003	51	22	42.5	29	8.5	2	42.5
004	22	22	42.5	0	20.5	1	22
005	47	22	42.5	25	4.5	2	42.5
006	59	22	42.5	37	16.5	2	42.5
007	23	22	42.5	1	19.5	1	22
008	19	22	42.5	3	23.5	1	22
009	27	22	42.5	5	15.5	1	22
010	15	22	42.5	7	27.5	1	22
011	33	22	42.5	11	9.5	2	42.5
012	39	22	42.5	17	3.5	2	42.5
013	42	22	42.5	20	0.5	2	42.5
014	32	22	42.5	10	10.5	2	42.5
015	37	22	42.5	15	5.5	2	42.5

As you can see, there weren't any shifts between clusters from the third and fourth interactions. You would stop here, and you can see how k-means has grouped the sales reps into two clusters. This is why a tool like Tableau is so great for this kind of analysis because it iterates through this process very quickly and automatically.

Supervised Versus Unsupervised Learning Methods

In this chapter, you have been introduced to a new concept of supervised versus unsupervised learning. The easiest way to break down the differences between each technique is to look at how they train data, their overall tactics, and the feedback loop (interaction with you).

Supervised Learning

Supervised learning is characterized by the following factors:

Training data
: The algorithm is trained on a labeled dataset. If you recall back to Chapters 8 and 9 on regression models, you learned about dependent variables versus independent variables, where you select a target variable and run hypothesis testing on that variable using independent variable inputs. The algorithm learns to make predictions based on the labeled dependent variable and the corresponding independent variable.

Objective
: The primary objective of supervised learning is to learn how to predict unseen data accurately based on specific input and output variables. These variables are defined by you and require you to run the hypothesis tests with certain assumptions defined up front.

Feedback loop
: Supervised learning models are refined and improved based on the accuracy of their predictions and the results of the hypothesis testing. For this reason, you will need to be involved to define each test and then run the model again. This requires you to be heavily involved in defining the parameters, testing, interpreting the results, and then retesting.

Unsupervised Learning

Unsupervised learning is characterized by the following factors:

Training data
: The algorithm works with unlabeled data, meaning that there are no predefined dependent variables or independent variables. The algorithm aims to uncover the inherent structure or patterns in the data on its own.

Objective

Unsupervised learning is primarily used for exploratory data analysis, data reduction, and clustering. The main goal is to discover hidden relationships, group similar data points, or reduce the dimensionality of data.

Feedback loop

Unsupervised learning models often lack a direct feedback mechanism. This means that they are often harder to interpret, and evaluation can be more challenging. However, you will be less involved in the process, especially upfront.

How to Implement Clustering in Tableau

To start, connect to the Sample - Superstore dataset and create a jittered dot plot of the average discount by customer. To do this, drag Discount to the Columns shelf, right-click the pill and change the aggregation to AVG. Then drag Customer Name to the Detail property of the Marks card. You should have a dot plot similar to Figure 11-7.

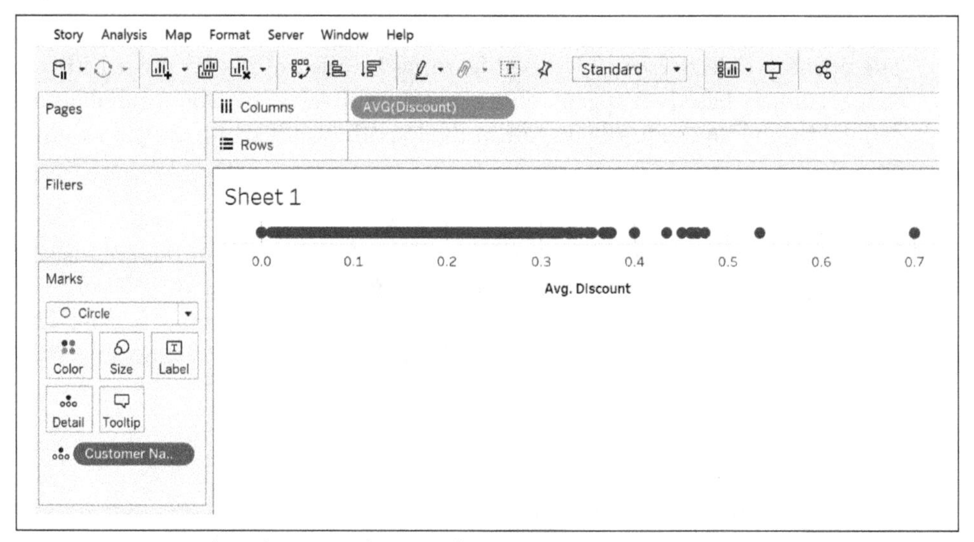

Figure 11-7. Dot plot of average discount by customer

Next, you need to jitter the dot plot. You can do this by right-clicking on the Rows shelf, typing RANDOM() in the blank pill, and then pressing Enter on your keyboard. This will jitter each mark in the view to a random point on the y-axis, as shown in Figure 11-8. The advantage of this is that you can hover over each mark now and analyze them within the view. It is also important to mention that you could implement this tactic using a calculated field.

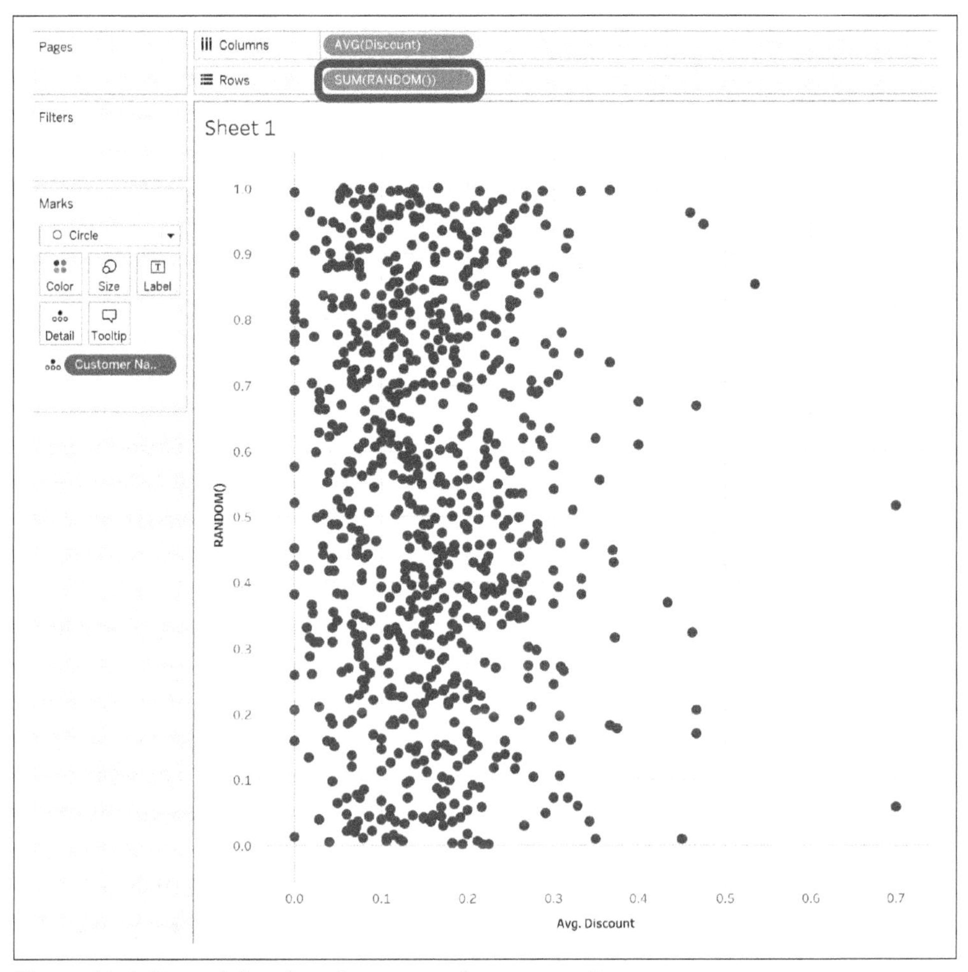

Figure 11-8. Jittered dot plot of customers by average discount

From here, implementing a clustering model is extremely easy. Toggle to the Analytics pane and drag Cluster into the view, as shown in Figure 11-9.

You will see a new field appear in the color property of the Marks card titled Clusters (1). In addition to the new field, you will see a window pop-up that shows you the variables in the current model; in the text box you can enter the desired number of clusters, as shown in Figure 11-10.

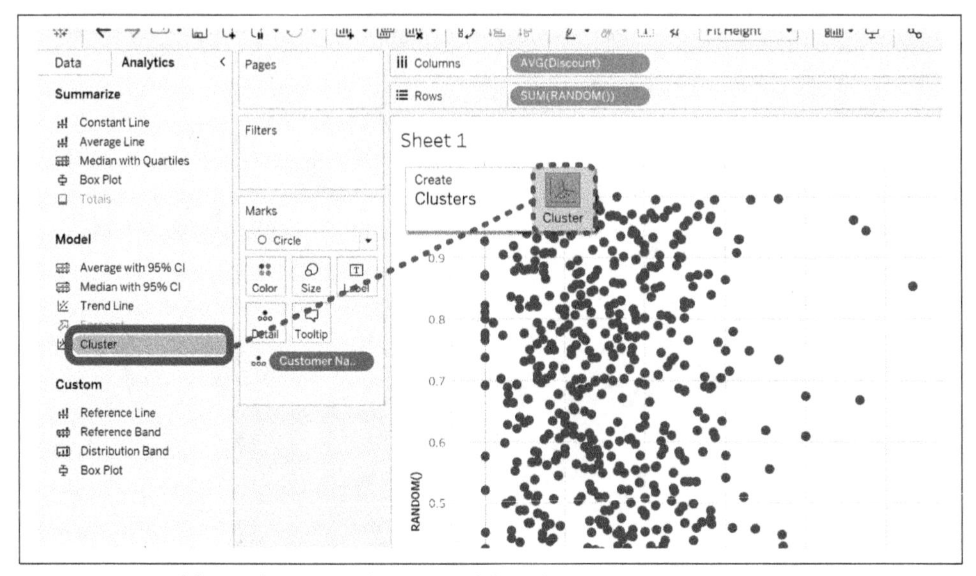

Figure 11-9. Adding a k-means cluster model to the view

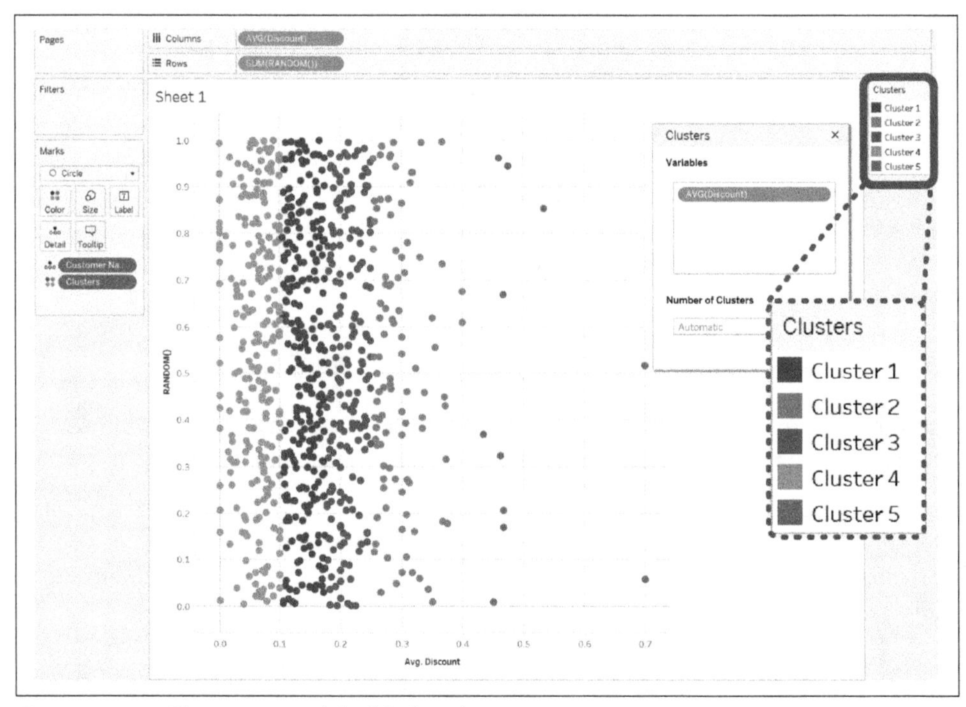

Figure 11-10. Clustering model added to the view

By default, Tableau will assign a cluster based on color. However, you can also change that to shape. To implement that, choose Shape from the Marks type drop-down and drag Clusters in the Marks shelf to the Shape property, as shown in Figure 11-11.

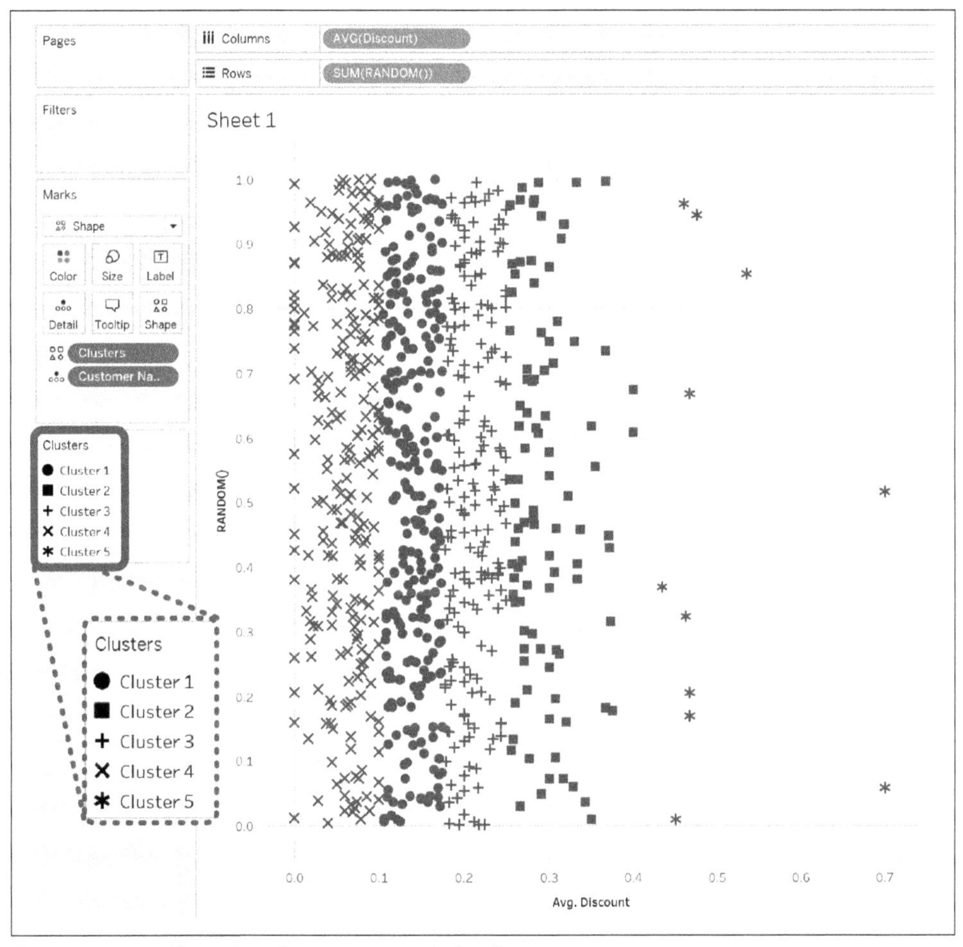

Figure 11-11. Changing cluster to encode by shape

You can see in this example that, by default, Tableau Desktop chose five clusters for this analysis. You can adjust this by right-clicking on Clusters in the Marks shelf and choosing Edit Cluster. From the menu section "Number of Clusters," enter 3 and close the Edit Cluster window. If you are following along, you will have results similar to what is in Figure 11-12.

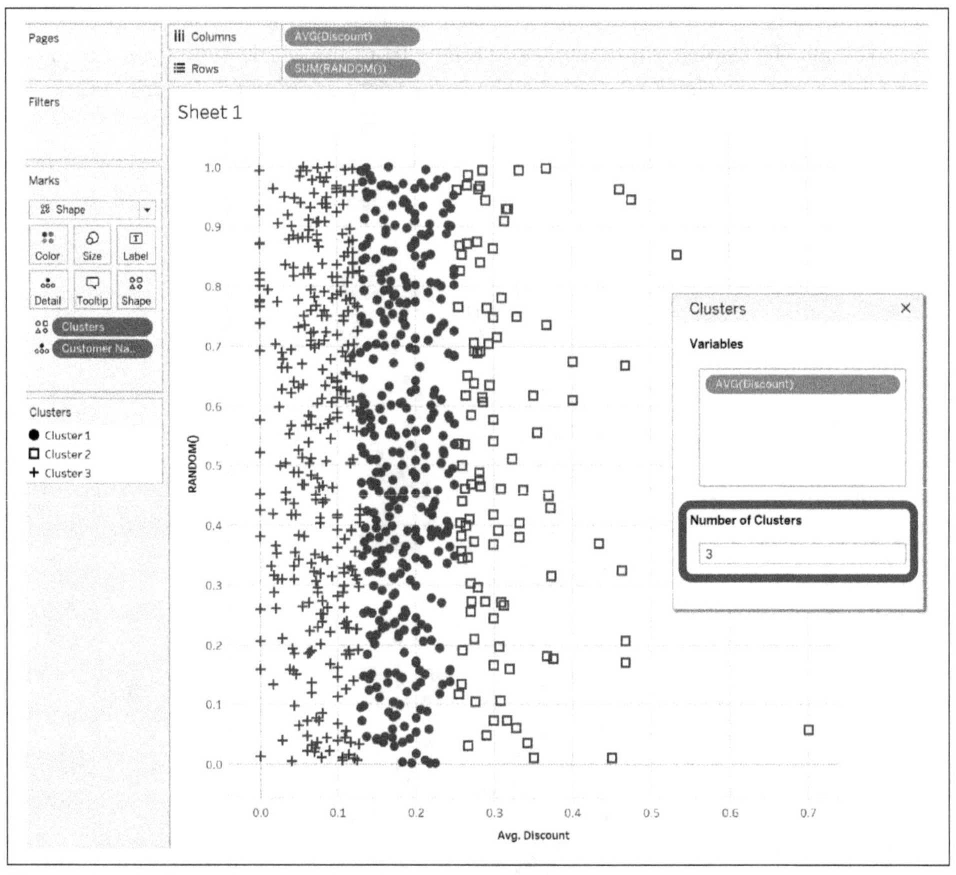

Figure 11-12. Three clusters selected

Now let's analyze the results of the model. If you right-click on the Clusters field in the Marks shelf, you will see two new options appear in the menu, namely "Describe clusters" and "Edit clusters," as shown in Figure 11-13.

If I click "Describe clusters," a new window will appear, as shown in Figure 11-14.

Similar to the window you saw in Chapter 10 (see Figure 10-19), Tableau Desktop gives you two tabs: Summary and Models. The Summary tab shown in Figure 11-14 gives you a lot of good details about the cluster model, including the centroids for each cluster. You can extract the centroids if you want to use those to build out and calculate fields using this analysis. The Summary tab also shows you how many marks are in the view, as well as how many of the total points are assigned to each cluster.

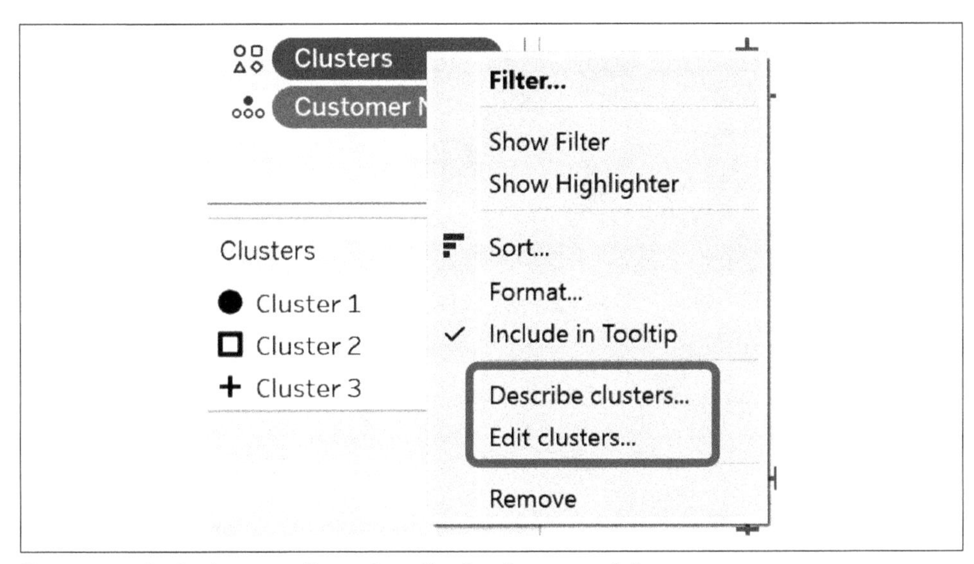

Figure 11-13. Options to edit or describe the cluster model

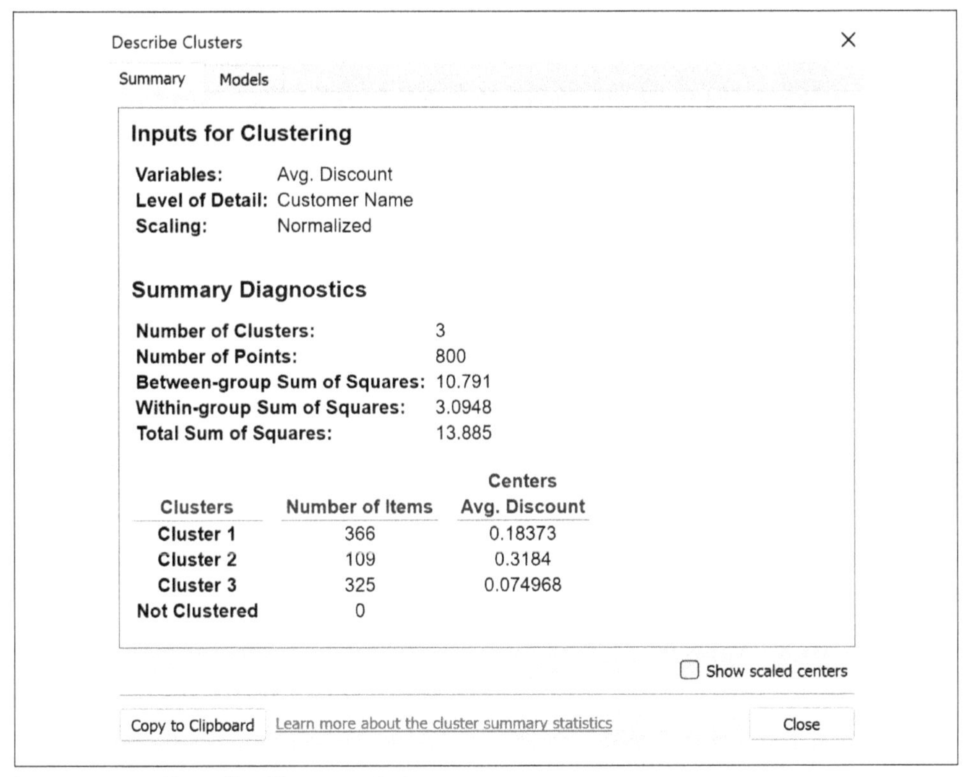

Figure 11-14. Describe Clusters window

Next, click the Models tab (see Figure 11-15).

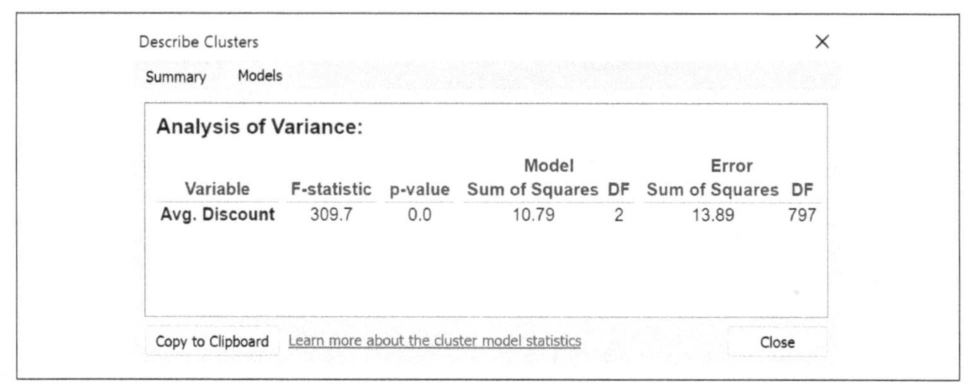

Figure 11-15. Models tab in "Describe clusters" window

The Models tab will show you all the detailed summary statistics for each variable you are using in the model. Now that you have viewed the statistics of the model, you want to add another variable to the analysis. Tableau Desktop makes this extremely easy for you. Close this window and open the "Edit clusters" menu by right-clicking the Clusters field and selecting "Edit clusters."

To add a new variable to the analysis, simply toggle to the Data pane and drag new variables into the window. To demonstrate this, drag *Orders (Count)* onto the menu, as shown in Figure 11-16.

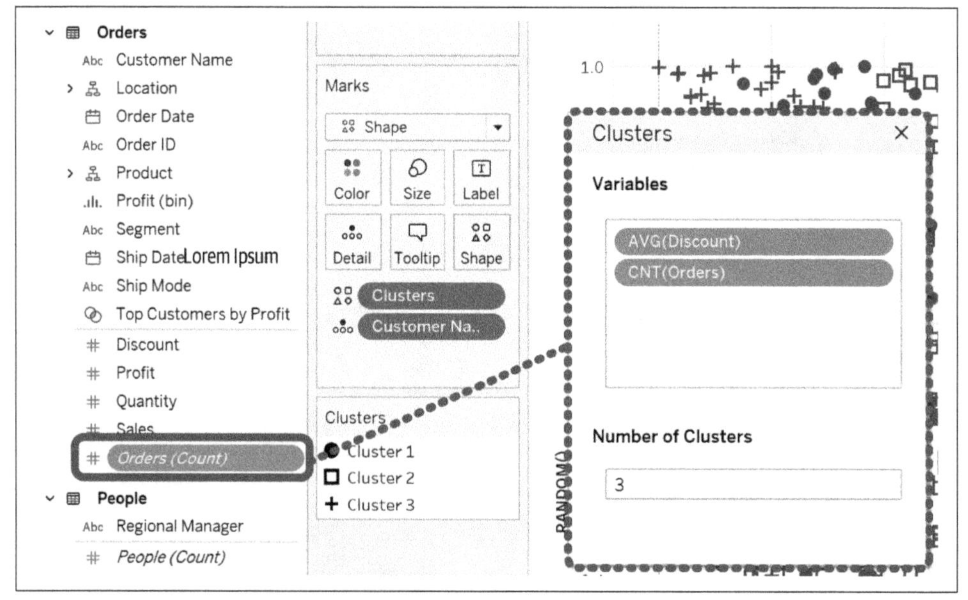

Figure 11-16. Adding the count of orders to the model

You can see in the view behind the "Edit clusters" menu that the marks are all jumbled up and scrambled together within the scatterplot. This is not going to be an effective way to present this information. I will replace RANDOM() in the Rows shelf with *Orders (Count)*, as shown in Figure 11-17.

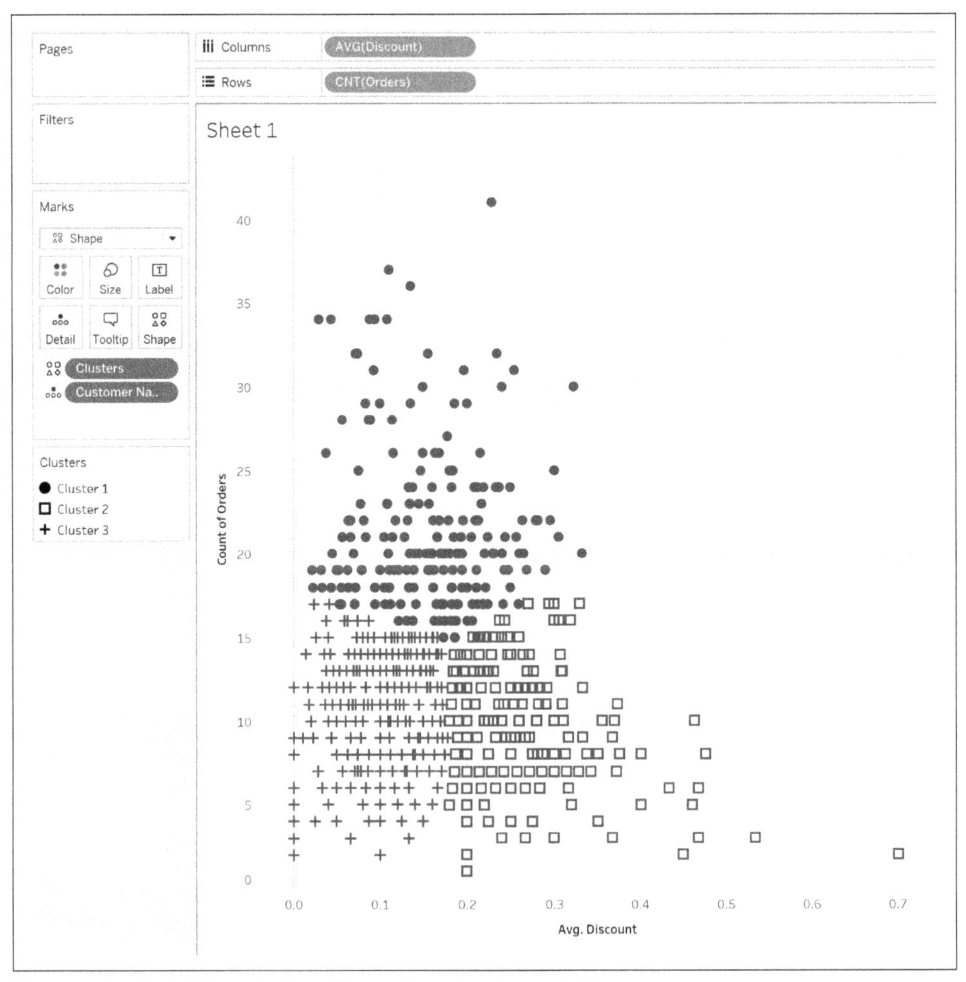

Figure 11-17. Adding Orders (Count) to the Rows shelf

You can see that this is a much better visualization to present the data, and it shows the three clusters clearly separated.

Next, close the "Edit clusters" menu and open the "Describe clusters" window again to view how the model has changed (see Figure 11-18).

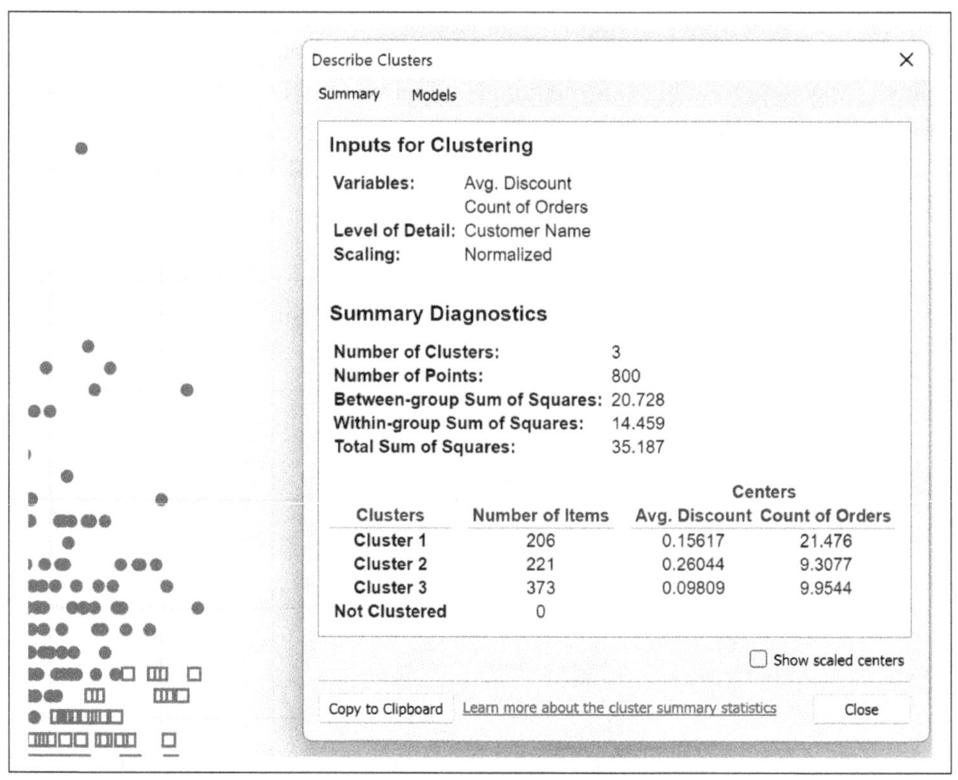

Figure 11-18. "Describe clusters" window after changing the clusters

You can see some interesting insights emerge by interpreting the results of the clusters. Looking at Cluster 1, it seems there is a sweet spot for these customers and the number of orders they have in relation to the average discounts. Using this information, you would be able to dive in deeper to see if this is a strategy that the business could adopt to optimize sales. However, clusters can be harder to interpret due to this being an unsupervised model. For instance, what is the exact relationship between the count of orders and the average discount? You don't have the exact coefficients; you would have them if this was a regression model, so that relationship becomes harder to determine.

Every technique is going to come with its own pros and cons.

Summary

In this chapter, you learned how to calculate k-means clustering by hand. While the formula can be intimidating, the method is relatively straightforward and easy to explain. Explaining models is half the battle, making k-means an ideal technique to have in your toolkit. Also, you learned the difference between supervised and unsupervised methods. Ultimately this stems from how much you need to do throughout the training process when implementing a model. Last, you saw how easy it is to implement and interpret this model within Tableau.

This is the last chapter on the native models within Tableau Desktop. In the next chapters, you will learn how to connect to and implement a model using the external connections to R and Python.

CHAPTER 12

Creating an External Connection to R Using Tableau

When working with statistical models within Tableau, there are limitations to what you can do, given the models that come native in the tool. However, Tableau has incorporated ways for you to go beyond what is available in the tool and establish external connections. One of the external connections that is available within Tableau is R.

In this chapter, you will learn how to download and install R as well as RStudio. You will also learn how to establish a connection to R through Tableau and use R within Tableau.

What Is R Code?

R is a programming language and open source software environment primarily used for statistical computing and graphics. It was developed by statisticians and data analysts to provide a versatile and powerful platform for data analysis, visualization, and statistical modeling. R is widely used in academia, industry, and research fields for tasks such as data manipulation, exploratory data analysis, hypothesis testing, machine learning, and more.

Key Features

Some key features of R include:

Data analysis
: R offers a wide range of functions and packages for data manipulation, transformation, and summarization. It allows users to handle and process data efficiently.

Graphics and visualization

R provides extensive capabilities for creating various types of high-quality graphs and visualizations, enabling users to effectively present their findings and insights.

Statistical modeling

R is equipped with a rich collection of statistical functions and libraries, allowing users to perform a wide variety of statistical analyses, including regression, hypothesis testing, time series analysis, and more.

Machine learning

R has a growing ecosystem of packages for ML and predictive modeling. It supports popular algorithms and techniques for classification, regression, clustering, and other ML tasks.

Community and packages

R has a vibrant and active community of users and developers who contribute to the creation of packages (libraries) that extend its functionality. These packages cover a wide range of topics, from specialized statistical methods to domain-specific analyses.

Open source

R is open source software, which means it is freely available for anyone to use, modify, and distribute. This has contributed to its widespread adoption and continuous development.

What Is RStudio?

RStudio is an integrated development environment (IDE) that was launched in 2011 and improves the way you can write R code. RStudio and RStudio Server have both a free and fee-based version of the tool. In the next section, you will read how to install both R and RStudio on your machine. For now, you need to understand what an IDE is and how it differs from base R.

There are a lot of advantages to coding in an IDE like RStudio. Here are a few key features you will get in RStudio:

Code editing

RStudio offers code editors with features like syntax highlighting, autocompletion, code suggestions, and error detection. These features help developers write code more accurately and quickly.

Code navigation

RStudio makes it easier to navigate through codebases, allowing developers to jump to function or variable definitions, find and replace code, and browse project files and directories effortlessly.

Debugging
: RStudio comes with built-in debugging tools that help developers identify and fix issues in their code. It provides features like breakpoints, variable inspection, call stack visualization, and step-by-step execution.

Compilation and execution
: RStudio can automate the process of compiling and running code, saving developers from the need to switch among multiple tools or command-line interfaces. This accelerates the feedback loop during development.

Integrated terminal
: RStudio includes an integrated terminal that allows developers to run command-line tasks and scripts directly within the IDE, streamlining tasks like package installation or server setup.

Documentation and help
: RStudio has access to documentation, language references, and help resources directly within the interface. This makes it convenient for developers to quickly look up information or learn about specific programming concepts and packages.

Integration with third-party tools
: RStudio supports integration with third-party tools, libraries, and plug-ins, allowing developers to extend the functionality of the IDE to suit their specific needs and download packages quickly.

Ultimately there are many advantages to coding within an IDE like RStudio versus the base R graphical user interface (GUI).

Installing R and RStudio

In this section, you will download and install both R and RStudio on a Windows machine. It will generally be the same process for Mac computers with slight differences in the UX. If you already have R and RStudio installed, feel free to skip to "Establishing an External Connection in Tableau" on page 220.

Installing R

To begin, you need to download base R from the Comprehensive R Archive Network (CRAN). You can access CRAN by navigating to its installation site (*https://oreil.ly/DwBmK*). You will see a page similar to Figure 12-1.

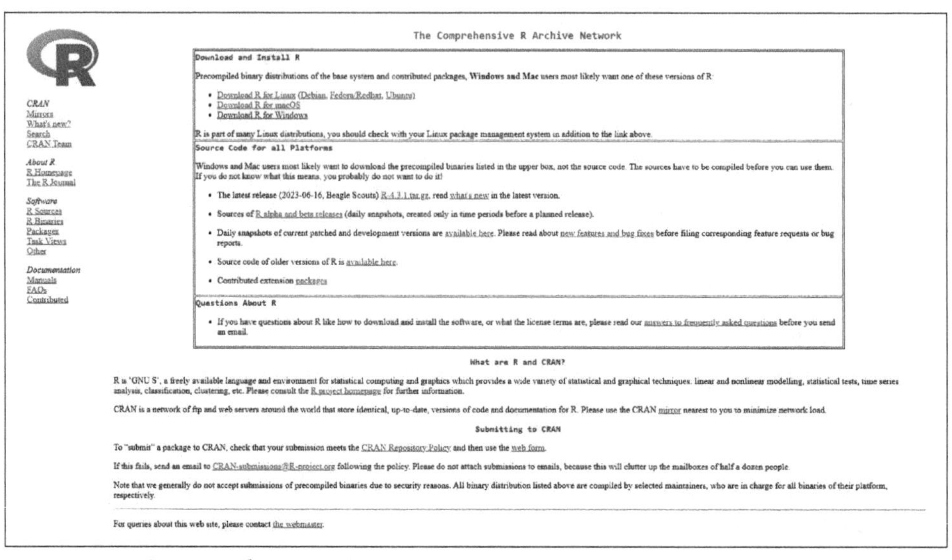

Figure 12-1. CRAN website

There is a lot of useful information on this site, and it is highly recommended that you click the links to learn more about R. At the top of the page, you will find links to download R for the different operating systems, as shown in Figure 12-2.

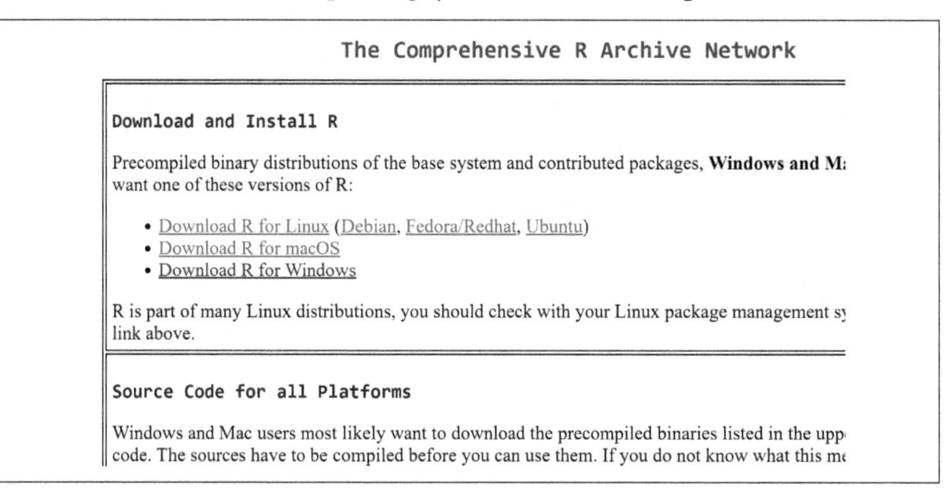

Figure 12-2. Download and Install R section on the CRAN website

For this section, you will see the installation process step by step through a Windows operating system. Click "Download R for Windows." This will take you to another screen of subdirectory options, as shown in Figure 12-3.

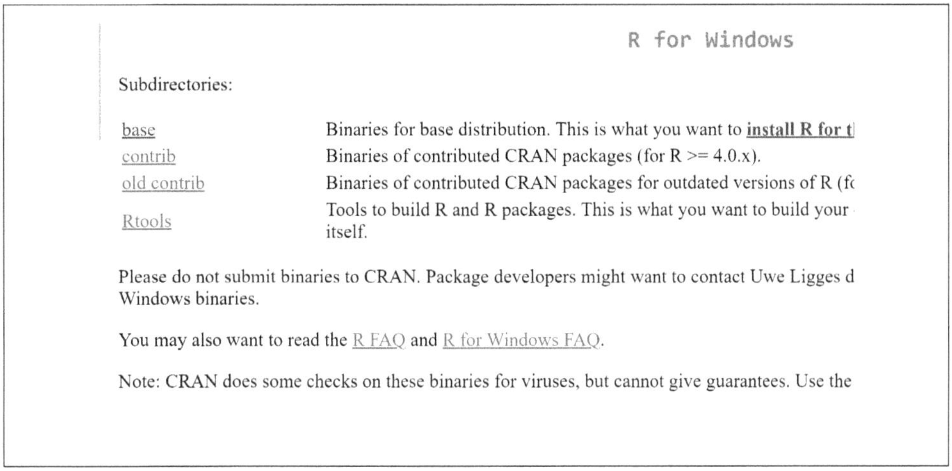

R for Windows
Subdirectories:
base
Binaries for base distribution. This is what you want to install R for t
contrib
Binaries of contributed CRAN packages (for R >= 4.0.x).
old contrib
Binaries of contributed CRAN packages for outdated versions of R (fo
Rtools
Tools to build R and R packages. This is what you want to build your itself.
Please do not submit binaries to CRAN. Package developers might want to contact Uwe Ligges d Windows binaries.
You may also want to read the R FAQ and R for Windows FAQ.
Note: CRAN does some checks on these binaries for viruses, but cannot give guarantees. Use the

Figure 12-3. Subdirectory R for Windows

From here, you can see the differences among each download. If this is the first time you are installing R on your computer and you want the latest version, click "base." The other options will allow you to download older versions of R or root packages and source code of R. Clicking "base" will lead you to another screen, as shown in Figure 12-4.

R-4.3.3 for Windows
Download R-4.3.3 for Windows (79 megabytes, 64 bit)
README on the Windows binary distribution
New features in this version
This build requires UCRT, which is part of Windows since Windows 10 and Windows Server 201 from here.
If you want to double-check that the package you have downloaded matches the package distribut to the fingerprint on the master server.
Frequently asked questions
Does R run under my version of Windows?
How do I update packages in my previous version of R?
Please see the R FAQ for general information about R and the R Windows FAQ for Windows-spe

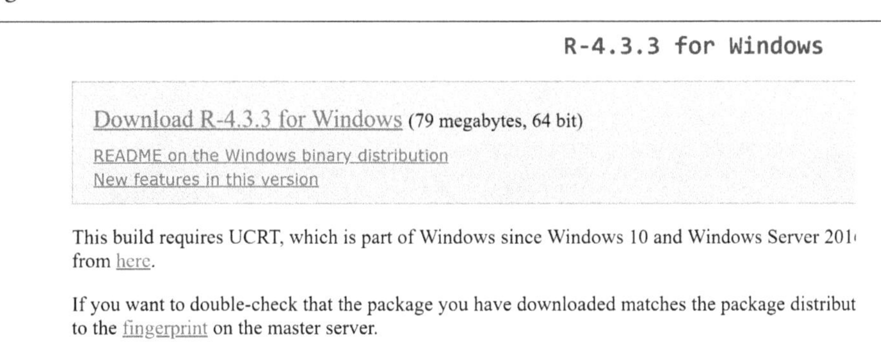

Figure 12-4. Latest version of R ready to download

At the time of this writing, R-4.3.3 is the latest version of base R. Click "Download R-4.3.3 for Windows" from the top of the screen to begin downloading the executable file needed for installation.

After the executable file is downloaded, double-click it to launch the setup process from your downloads folder in Windows. You will see a prompt asking you to allow or deny the program from making changes to your computer. Click Allow and you will launch the setup. The first option wants you to choose a language for the setup process, as shown in Figure 12-5.

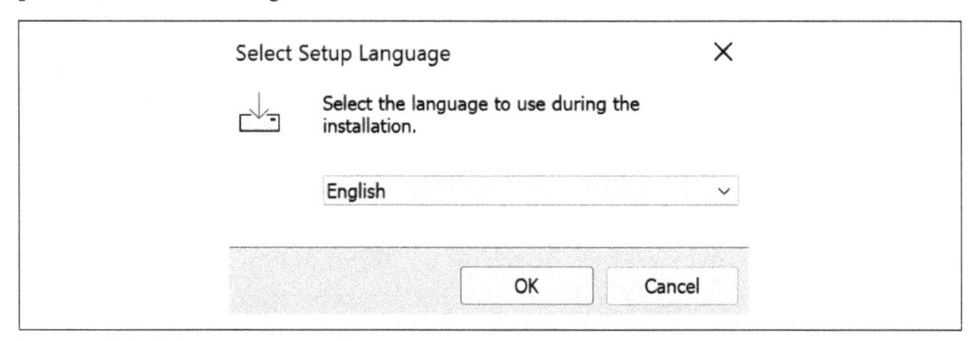

Figure 12-5. Select your setup language

Continue in English by clicking OK. On the next screen, you will see the GNU General Public License agreement, as shown in Figure 12-6.

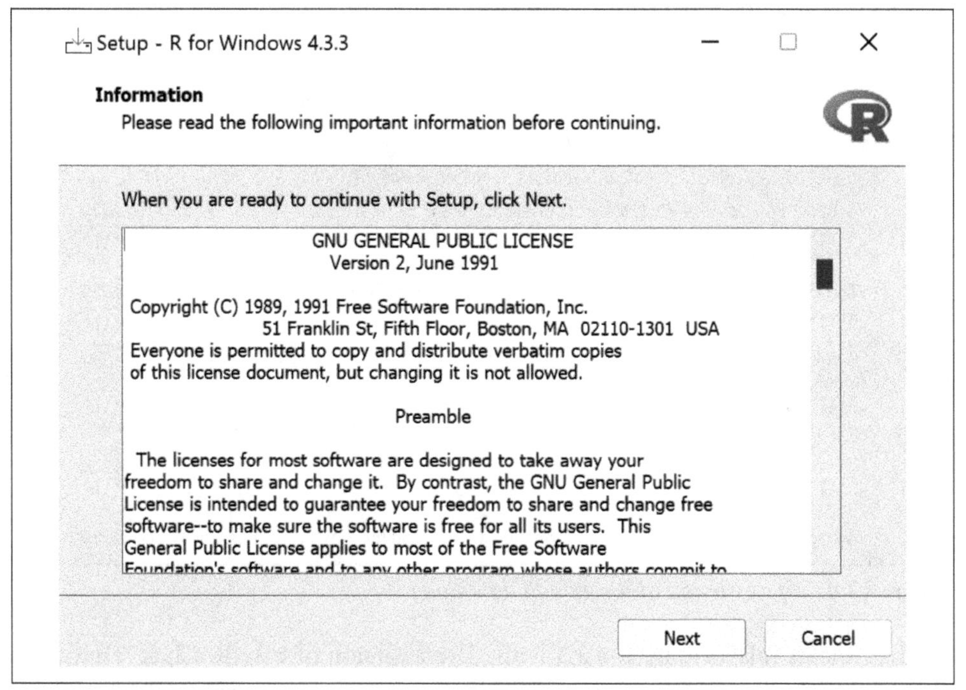

Figure 12-6. GNU General Public License agreement

After you read the information, click Next to continue. The next screen prompts you to choose a destination location to install R to Windows. By default, Windows will suggest creating a new folder directory that ends with the version you are installing. It is recommended to keep this default location. Later on, if you decide to download another version of R, this will help you understand what version you have. Leave the suggested default location, as shown in Figure 12-7, and then click Next.

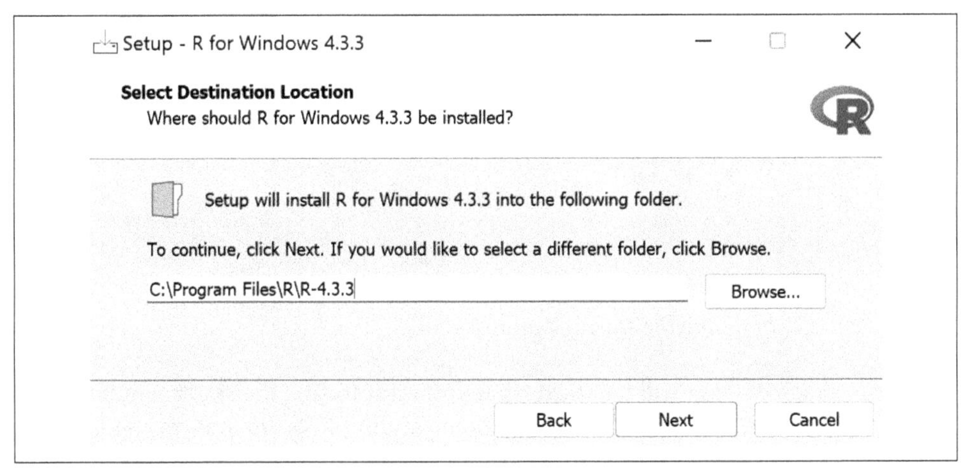

Figure 12-7. Destination location for R to be installed on Windows

The next screen will prompt you to select different components to install (see Figure 12-8). Leave the default selections checked and then click Next to continue.

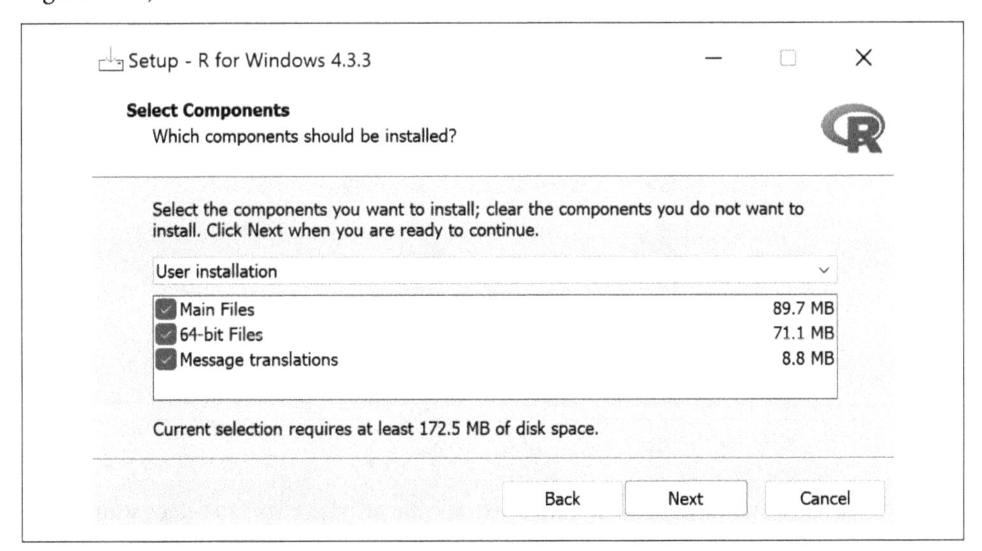

Figure 12-8. Select components to install for R

After continuing, you will be asked if you want to customize the startup options, as shown in Figure 12-9. This is truly up to you. For this writing, select the default options and then click Next.

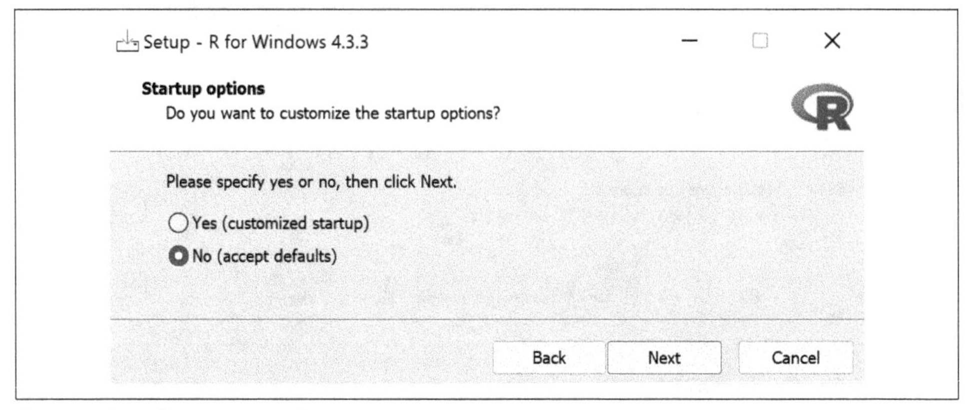

Figure 12-9. Customizing the startup options

On the next screen, you will be asked what you want to title the start menu folder, as shown in Figure 12-10. Since you will be downloading RStudio and coding from the IDE, you could select the checkbox at the bottom and opt to not create a start menu folder. This could help you later from becoming confused on which is which. Keep the default and create the start menu folder.

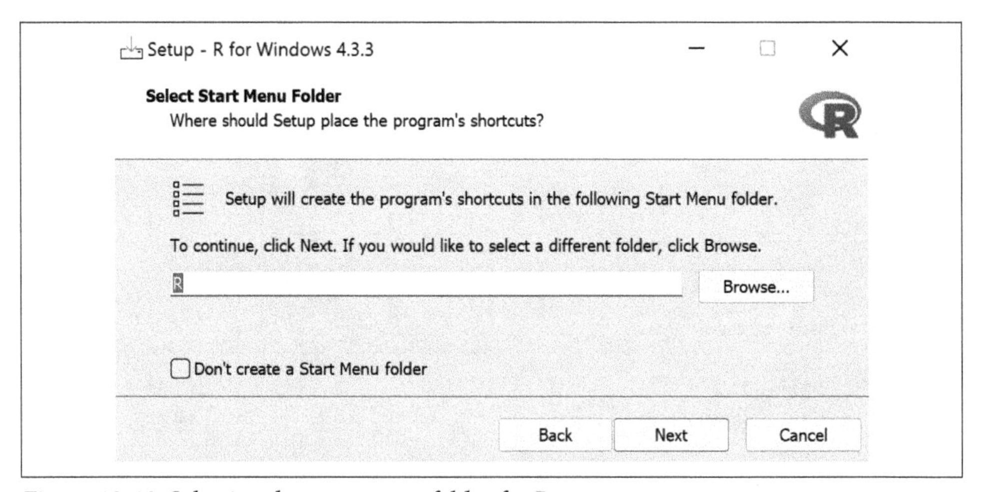

Figure 12-10. Selecting the start menu folder for R

After choosing a start menu folder, the next screen prompts you to select some additional tasks. The shortcuts are up to you; however, it is recommended to leave the registry entries selected. These will be important when you install RStudio. Leave the default selections, as shown in Figure 12-11, and then click Next.

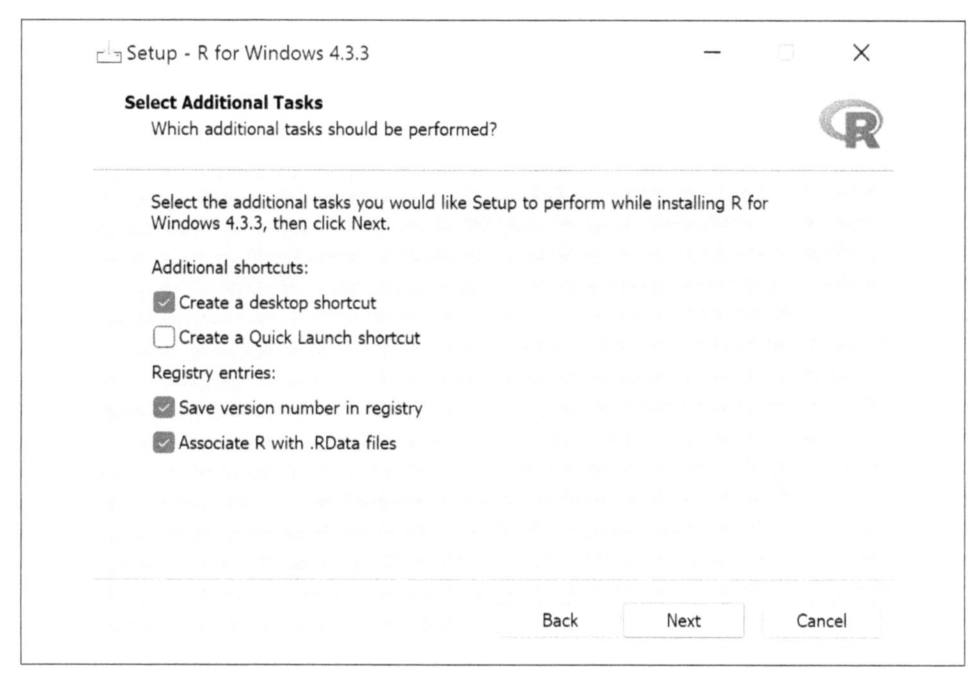

Figure 12-11. Selecting additional tasks while installing R

The next screen will begin the installation. You will see a progress bar, as shown in Figure 12-12.

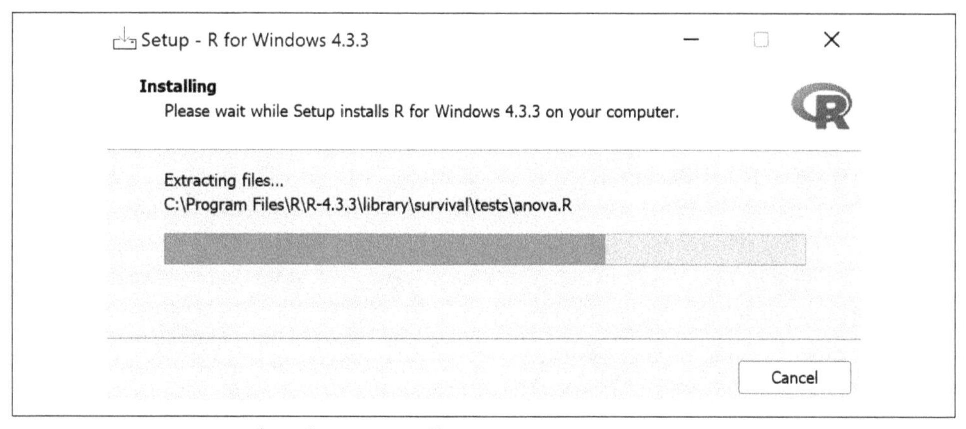

Figure 12-12. Progress bar during installation

Depending on your machine, the installation should take 30 seconds or less. If the installation is successful, you will see a screen similar to Figure 12-13. To close the installation progress, select Finish.

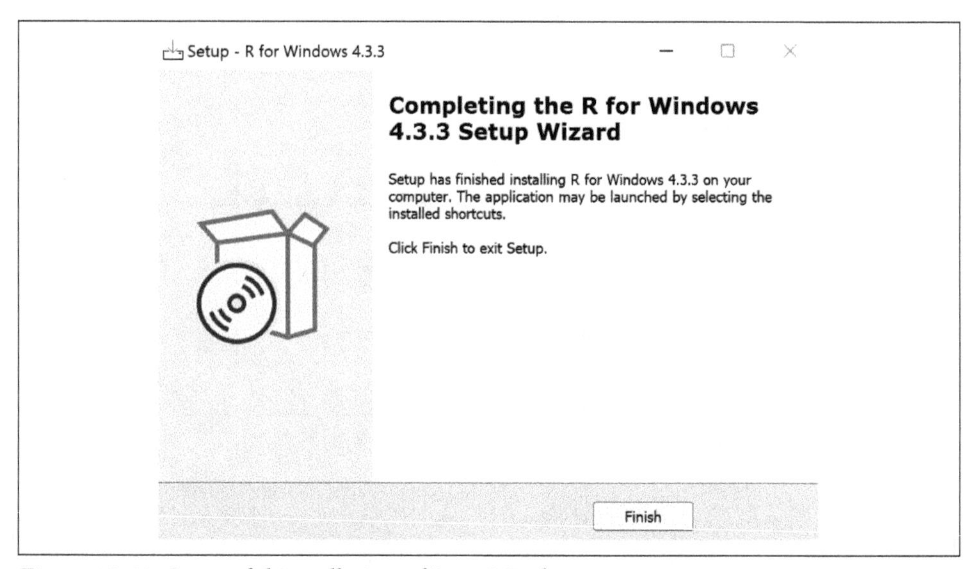

Figure 12-13. Successful installation of R on Windows

That concludes the installation of base R. In the next section, you will download and install RStudio.

Installing RStudio

After installing base R, you can install RStudio Desktop. To begin installation, go to the Posit website (*https://oreil.ly/oYpAw*). You will see a window similar to Figure 12-14.

Click DOWNLOAD RSTUDIO DESKTOP FOR WINDOWS to download the executable file for installation. This website will detect which operating system you have, so if you have a Mac, the button will be DOWNLOAD RSTUDIO DESKTOP FOR MAC. Once the file is downloaded, launch the executable from the downloads folder in Windows. You will be asked if you want to allow this app to make changes to your computer. Select Allow to continue. You will then see the start screen of the installation wizard, as shown in Figure 12-15.

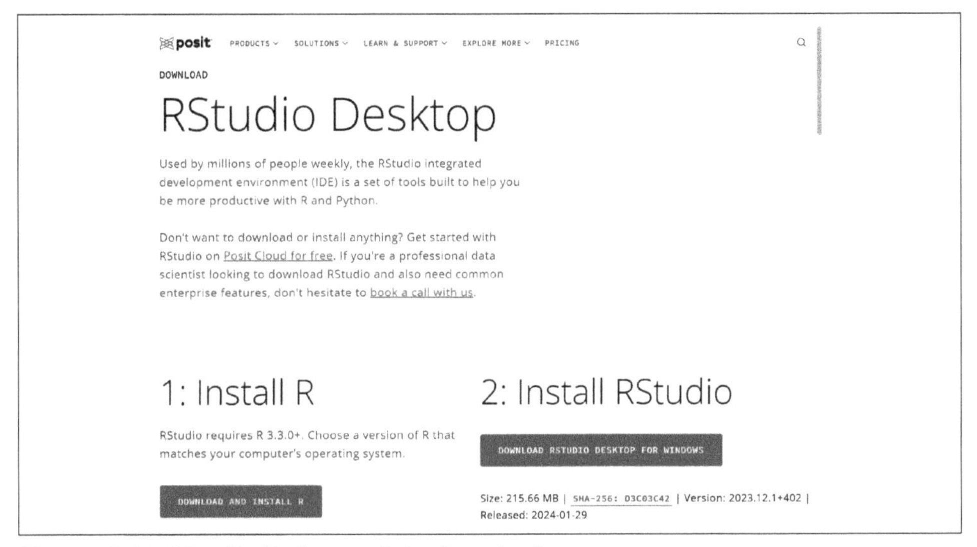

Figure 12-14. RStudio Desktop website download

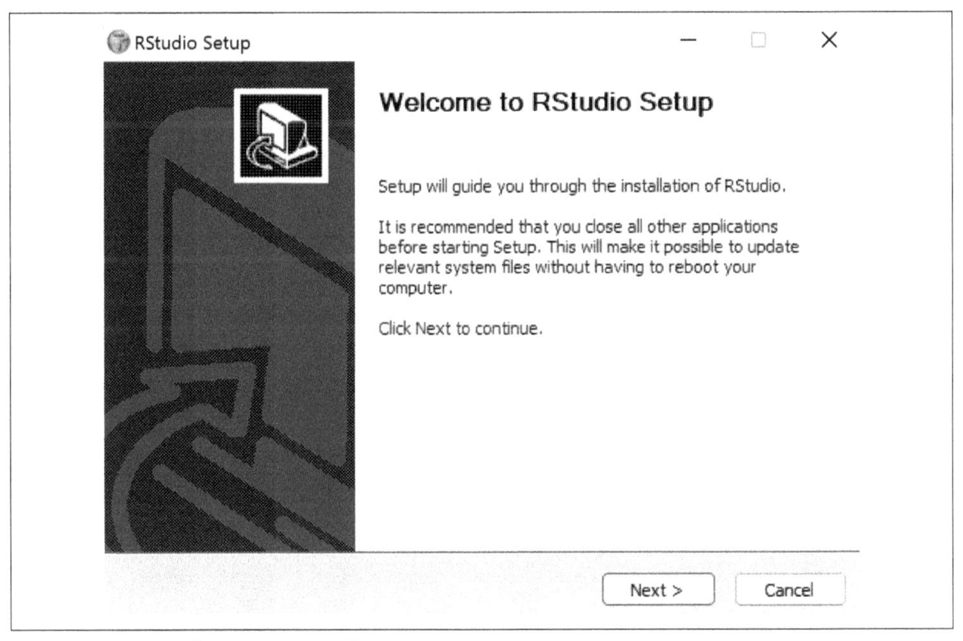

Figure 12-15. Welcome to RStudio Setup

Click Next to continue. On the following screen, you will be asked to choose a destination location for RStudio. Leave the defaults, as shown in Figure 12-16, and then click Next.

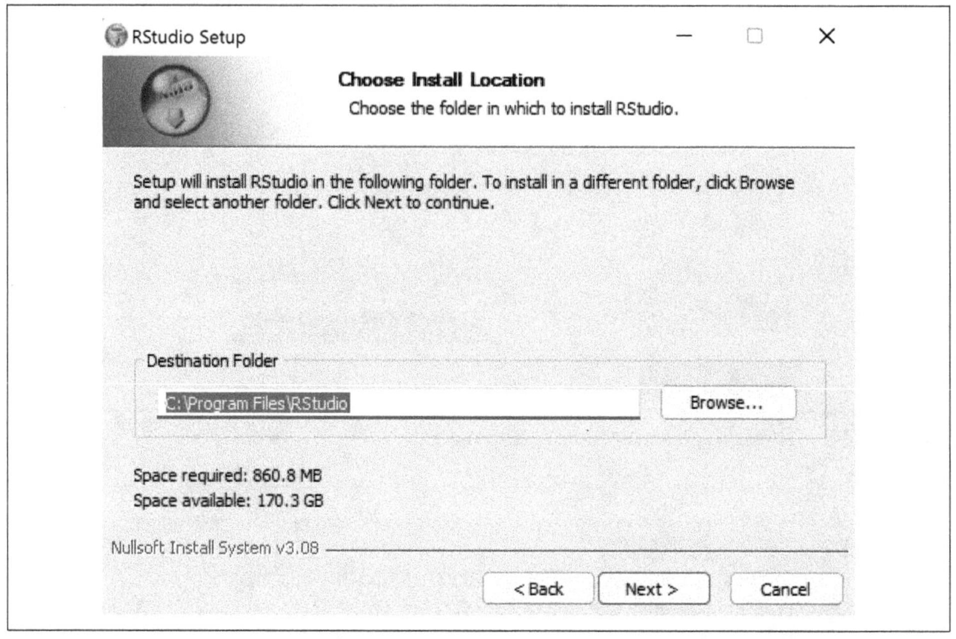

Figure 12-16. Choose a destination location for RStudio

On the next screen, you will be prompted to select a start menu folder, as shown in Figure 12-17. Leave the default for this and then click Install.

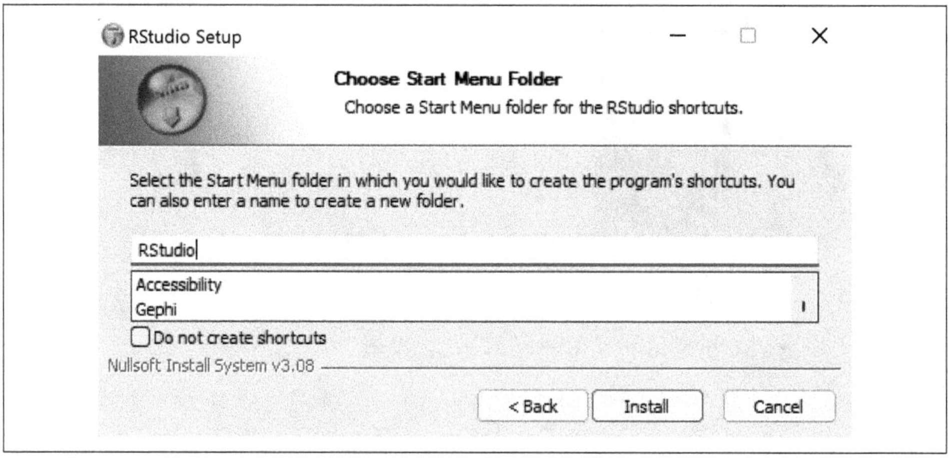

Figure 12-17. Choosing a start menu folder

After clicking Install, you will see a progress bar appear. The installation progress normally takes slightly longer than base R. After the installation is complete, you will see a finish screen, as shown in Figure 12-18.

Figure 12-18. Successful installation of RStudio

The next step is to launch the program. Find RStudio and launch it. When the program opens, it will ask you to connect to R. From the radio button options, select "Choose a specific version of R:" and select the version of R you installed in the previous section, as shown in Figure 12-19.

You now have RStudio and R installed on your machine. You are now ready to establish a connection to those tools through Tableau.

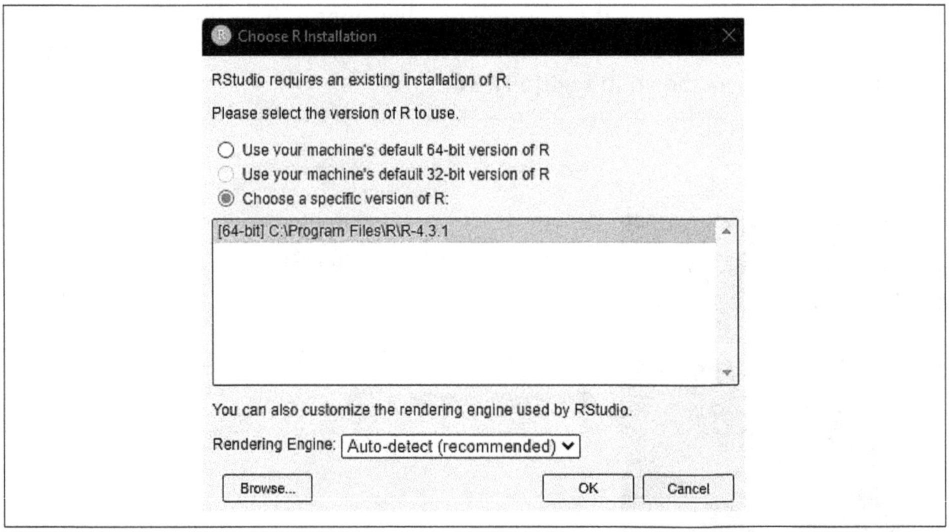

Choose R Installation

RStudio requires an existing installation of R.

Please select the version of R to use.

- Use your machine's default 64-bit version of R
- Use your machine's default 32-bit version of R
- Choose a specific version of R:

[64-bit] C:\Program Files\R\R-4.3.1

You can also customize the rendering engine used by RStudio.

Rendering Engine: Auto-detect (recommended)

Browse... OK Cancel

Figure 12-19. Choose R Installation window for RStudio

Establishing an External Connection in Tableau

To establish an external connection from Tableau to R, you will first launch RStudio. To give you a brief overview of the interface, you can see four distinct sections, as shown in Figure 12-20.

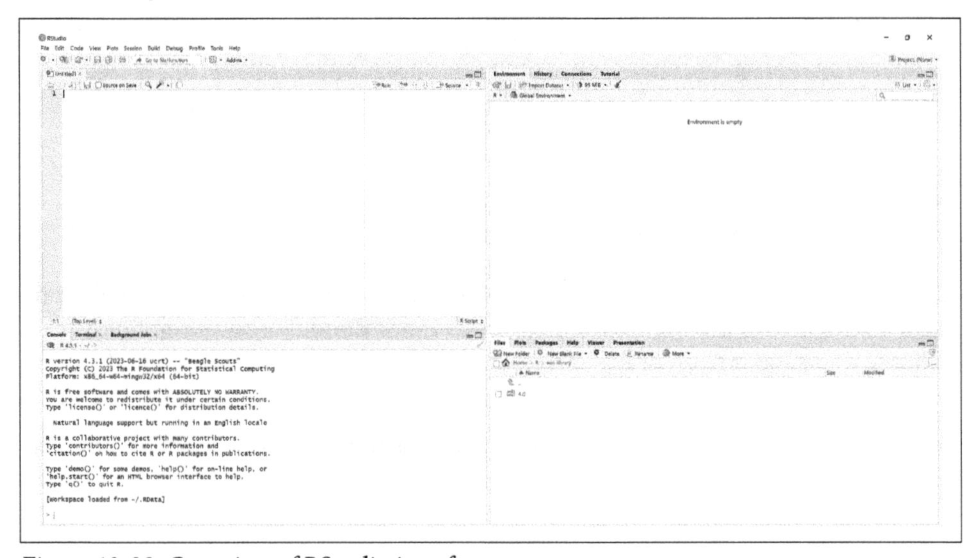

Figure 12-20. Overview of RStudio interface

The top-left box is the coding interface, where you will type all of your code into the editor. From the bottom-left box you can see a console where you can open a terminal command interface. The top-right section tracks different variables, objects, arrays, and data frames you create as you code. The last section is the bottom right, which allows you to access files, view plots, view documentation for packages, and several other options.

This is a brief overview of the RStudio interface. Covering every aspect of RStudio is beyond the scope of this writing. However, from here you will learn each step needed to launch and connect to R through Tableau. The first step is to click on Tools from the top navigation bar and then select Install Packages, as shown in Figure 12-21.

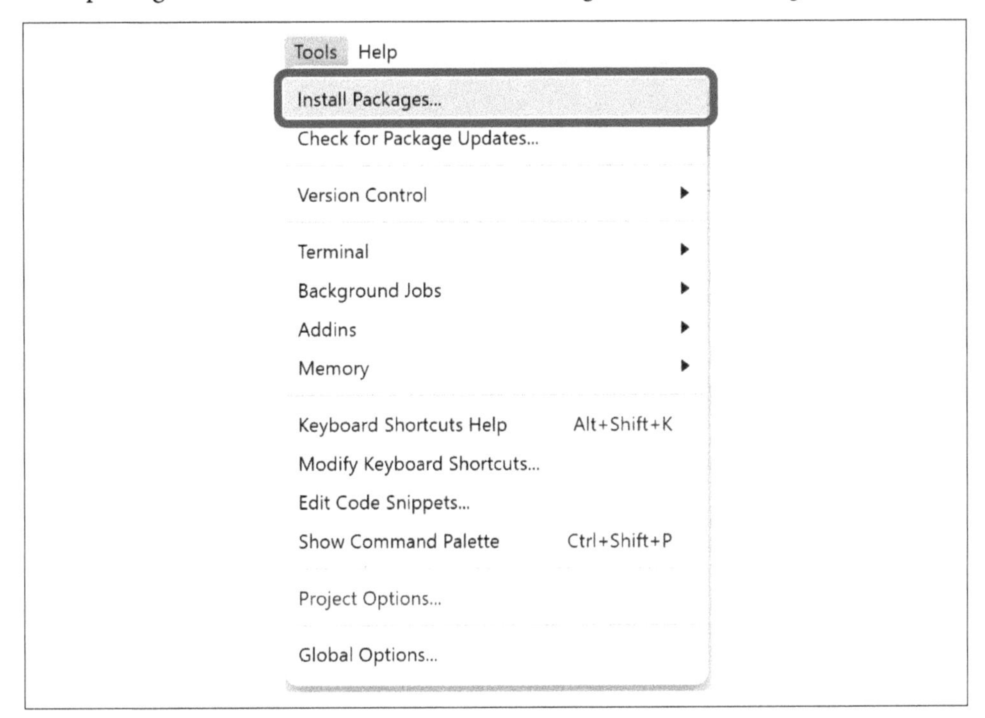

Figure 12-21. First step to installing Rserve in RStudio

From the menu, type Rserve into the search bar and then click Install, as shown in Figure 12-22.

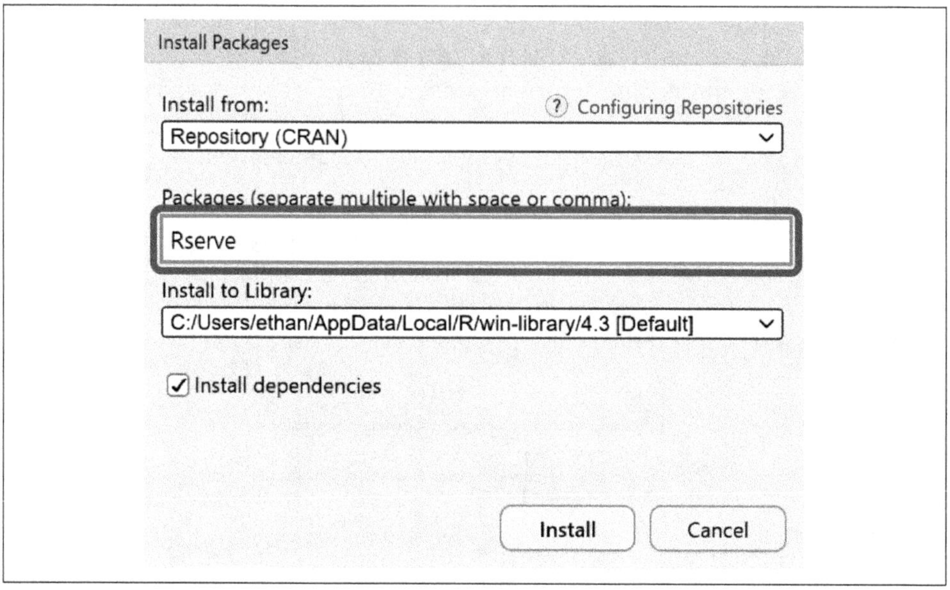

Install Packages

Install from: (?) Configuring Repositories

Repository (CRAN)

Packages (separate multiple with space or comma):

Rserve

Install to Library:

C:/Users/ethan/AppData/Local/R/win-library/4.3 [Default]

Install dependencies

Install Cancel

Figure 12-22. Installing Rserve in RStudio

In the console, you should see a note that says the installation was successful. Now from the console, type `Rserve::Rserve()`, as shown in Figure 12-23, and then press Enter on your keyboard. You will see a message in the console that says `Starting Rserve`.

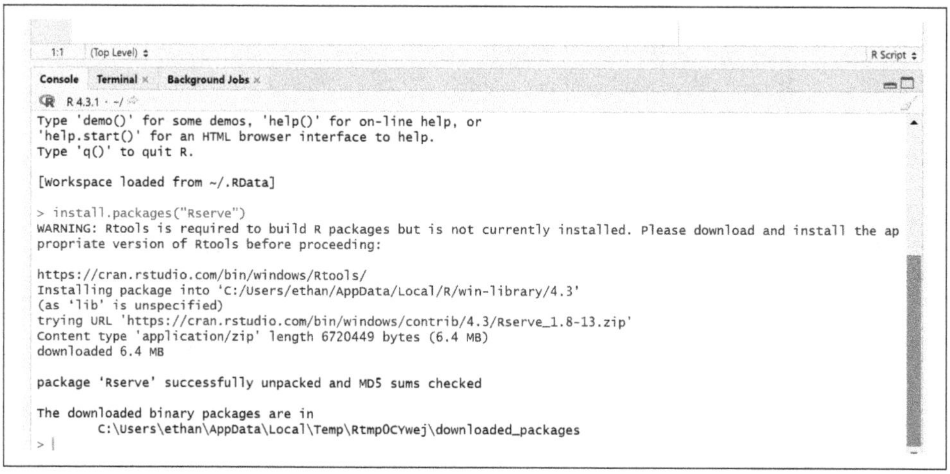

```
1:1   (Top Level) ÷                                                    R Script ÷
Console   Terminal ×   Background Jobs ×
R 4.3.1 · ~/
Type 'demo()' for some demos, 'help()' for on-line help, or
'help.start()' for an HTML browser interface to help.
Type 'q()' to quit R.

[Workspace loaded from ~/.RData]

> install.packages("Rserve")
WARNING: Rtools is required to build R packages but is not currently installed. Please download and install the ap
propriate version of Rtools before proceeding:

https://cran.rstudio.com/bin/windows/Rtools/
Installing package into 'C:/Users/ethan/AppData/Local/R/win-library/4.3'
(as 'lib' is unspecified)
trying URL 'https://cran.rstudio.com/bin/windows/contrib/4.3/Rserve_1.8-13.zip'
Content type 'application/zip' length 6720449 bytes (6.4 MB)
downloaded 6.4 MB

package 'Rserve' successfully unpacked and MD5 sums checked

The downloaded binary packages are in
        C:\Users\ethan\AppData\Local\Temp\RtmpOCYwej\downloaded_packages
> |
```

Figure 12-23. Launch Rserve in the console of RStudio

Now launch Tableau and click on Help from the top navigation bar. Select "Settings and Performance" > Manage Analytics Extension Connection, as shown in Figure 12-24.

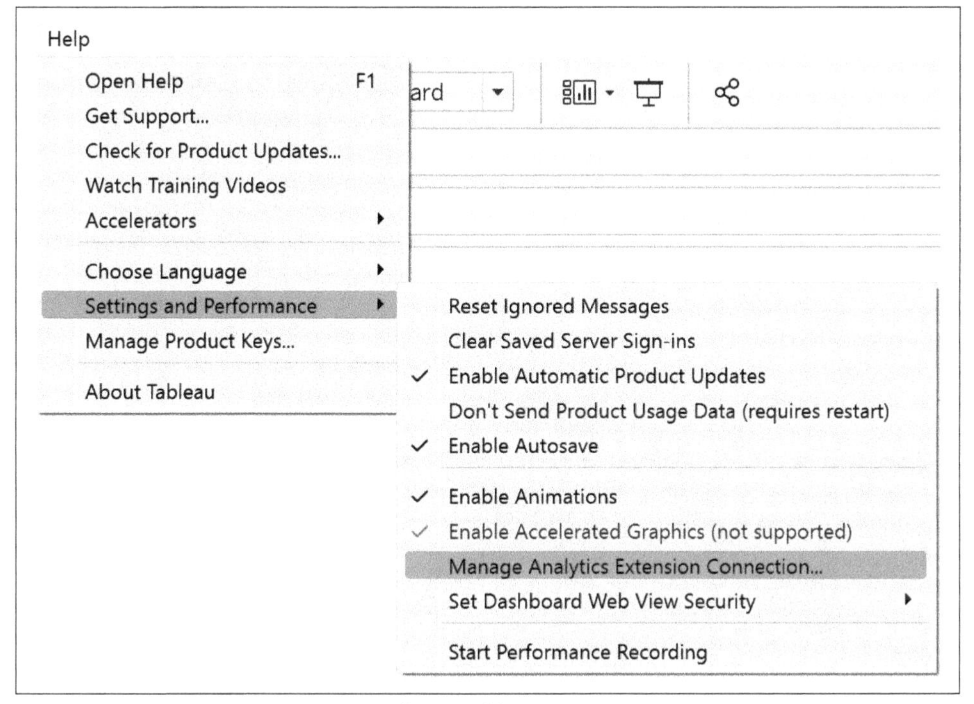

Figure 12-24. Connecting to Rserve from Tableau

There will be a menu that appears, asking you to configure the settings of the Rserve you are attempting to connect. Rserve is a server software that allows the R statistical computing environment to be accessed and controlled remotely. Since you established and launched Rserve from RStudio locally on your machine, you can simply leave the Hostname as "localhost" and the Port as 6311. Click Test Connection. You will see the Action Completed pop-up that appears and says the connection was established successfully, as shown in Figure 12-25.

Close the Action Completed pop-up and then click Save on the Manage Analytics Extensions Connection page. Now from Tableau, test the connection by creating a new calculated field (see Figure 12-26):

```
SCRIPT_INT("as.integer(.arg1 * 5)",SUM([Profit]))
```

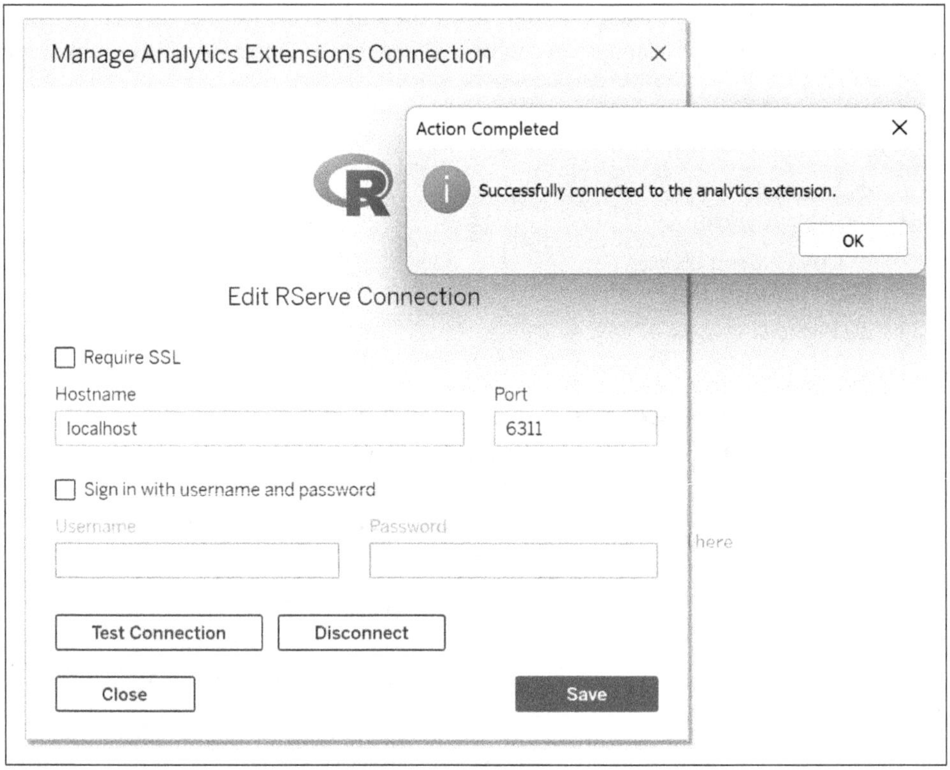

Figure 12-25. Successful connection to Rserve from Tableau

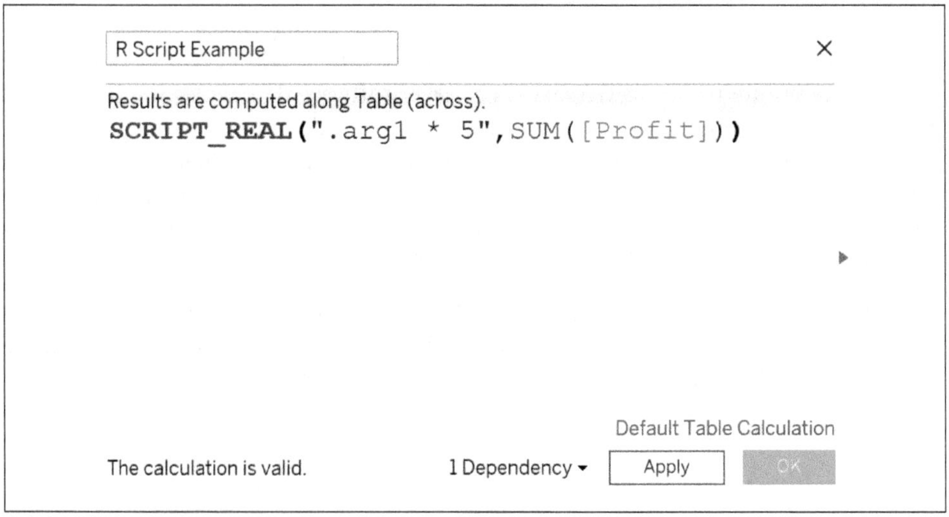

Figure 12-26. SCRIPT_INT example in Tableau

This is a simple formula that will take the sum of the profit in Tableau and then send that value to R to apply some basic arithmetic operations. In this example, it will multiply the value by 5 and then return those values back to be visualized in Tableau. To demonstrate that function, add Category to the Rows shelf and SUM(Profit) to the Text property of the Marks card. Then add the R Script Extension calculated field to the Text property as well, as shown in Figure 12-27.

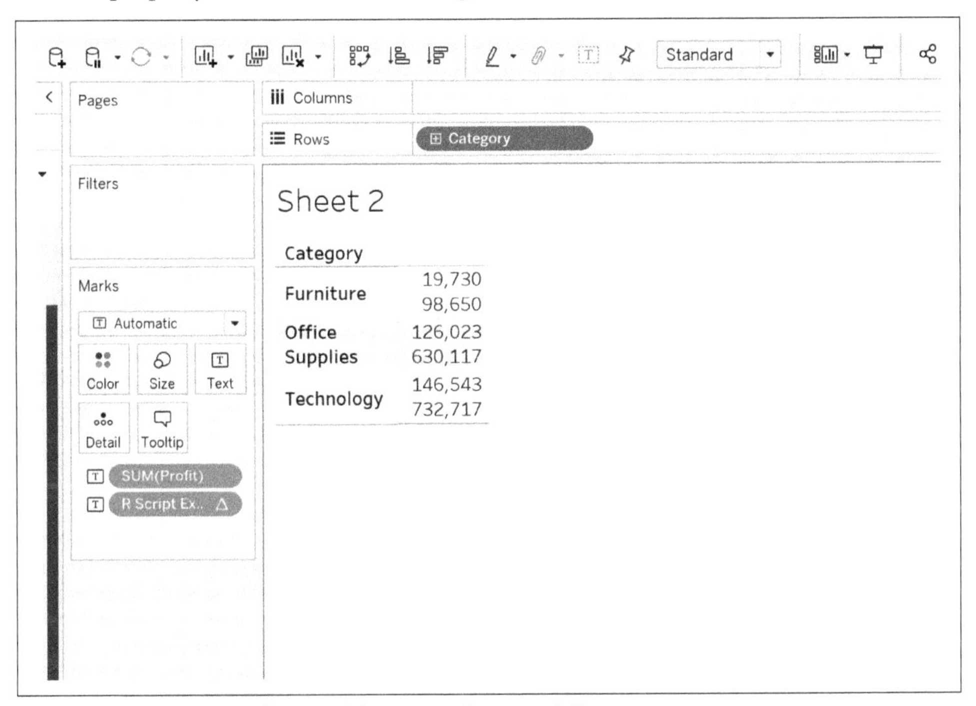

Figure 12-27. R Script from Tableau tested successfully

You can see that the calculated field, which takes the profit value and multiples it by 5, is working. Now that you have tested the connection, it is always best practice to shut down and close the connections. Select Help in the top navigation, hover over "Settings and Performance," and click Manage External Analytics Extensions. When the menu to edit the Rserve connection appears, select Disconnect. You will see a pop-up that alerts you that disabling the connection will break any visuals that have been created using the connection. Select Disable, and the connection will close.

Summary

In this chapter, I showed you how to download and install R and RStudio. Then I showed you how to launch Rserve locally on your machine and connect to Tableau. Although I simply touched on an example calculation that showed the connection was working, I will cover script functions in more depth in Chapter 15.

CHAPTER 13

Creating an External Connection to Python Using Tableau

When working with statistical models within Tableau, there are limitations to what you can do, given the models that come native in the tool. However, Tableau has incorporated ways for you to go beyond what is available in the tool and establish external connections. One of the external connections that is available within Tableau is Python.

In this chapter, you will learn how to download and install Python and Anaconda. You will also learn how to establish a connection to Python through Tableau and use Python within Tableau.

What Is Python?

Python is a popular language in various fields, including web development (using frameworks like Django and Flask), data science and machine learning (with libraries like NumPy, pandas, and scikit-learn), and automation (with tools like Selenium and Beautiful Soup). Its versatility and simplicity have contributed to widespread adoption in the software development community.

Key Features

Some key features of Python include:

Data visualization
: Python offers a range of data visualization libraries, such as Matplotlib and Seaborn, which allow you to create a wide variety of static and interactive visualizations to explore and present data effectively.

Jupyter Notebook
: Jupyter Notebook provides an interactive and literate programming environment for data analysis and visualization. Notebook allows you to combine code, data, and documentation in a single document.

Machine learning libraries
: Python has popular ML libraries such as scikit-learn, TensorFlow, Keras, and PyTorch, which are widely used for building and training ML models.

Statistical analysis
: Python provides support for statistical analysis through libraries like SciPy and statsmodels, allowing you to perform statistical tests and hypothesis testing.

Data integration
: Python can easily integrate with various data sources, including databases (e.g., through SQLAlchemy or Django ORM), web scraping (e.g., using Beautiful Soup or Scrapy), and data extraction from various file formats (e.g., CSV, Excel, JSON).

What Is Anaconda?

Anaconda is a popular open source distribution of Python for data science, machine learning, and scientific computing. It is designed to simplify package management and deployment for data scientists, researchers, and developers working on data-related projects. In the next section, I will walk you through installing Anaconda, which comes with the Python language.

There are a lot of advantages to using the Anaconda environment. Here are a few key features:

Preinstalled packages
: Anaconda comes with many preinstalled packages that will help jump-start your development and is great for folks just getting started.

Environments
: Anaconda allows you to create environments and install different packages to independent environments. This helps with development across multiple operating systems or development with different versions of Python.

Anaconda Navigator
: This is a GUI program that allows you to launch different popular applications for coding such as Jupyter, VS Code, and even RStudio.

Bypass computer admin rights
: Using Conda (the Anaconda environment manager), you can install packages completely independent of the system libraries and admin rights. This allows you

to install what you need to continue development, even with limited access to the permissions on your machine.

Installing Python and Anaconda

In this section, you will learn how to download and install Anaconda. If you already have Anaconda installed, feel free to skip to "Establishing an External Connection to Python in Tableau" on page 236.

Anaconda is going to come with everything you need to get started. It will have most of the core packages, different coding applications, and the latest versions of Python you can choose from for coding.

To begin the installation process, you need to first download the program. The easiest way is to go to the Anaconda site (*https://oreil.ly/4_gnO*), as shown in Figure 13-1. Click the download button for your operating system. This demonstration will be conducted with Windows. If you are using a Mac, it will be the same process but will look slightly different than the figures.

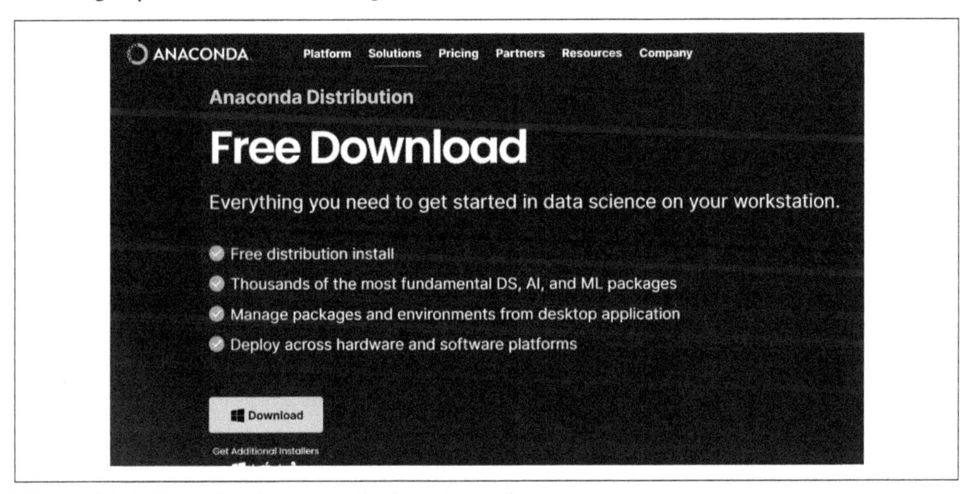

Figure 13-1. Download Anaconda from its website

Once the download is complete, find the application and launch it. You should see the welcome screen, as shown in Figure 13-2.

As it suggests, shut down any applications that are running before starting the setup. This will make it go smoothly and reduce the possibility of having to reboot your computer. From this screen, click Next to proceed. The next screen will have you review and agree to the terms and conditions of the tool, as shown in Figure 13-3.

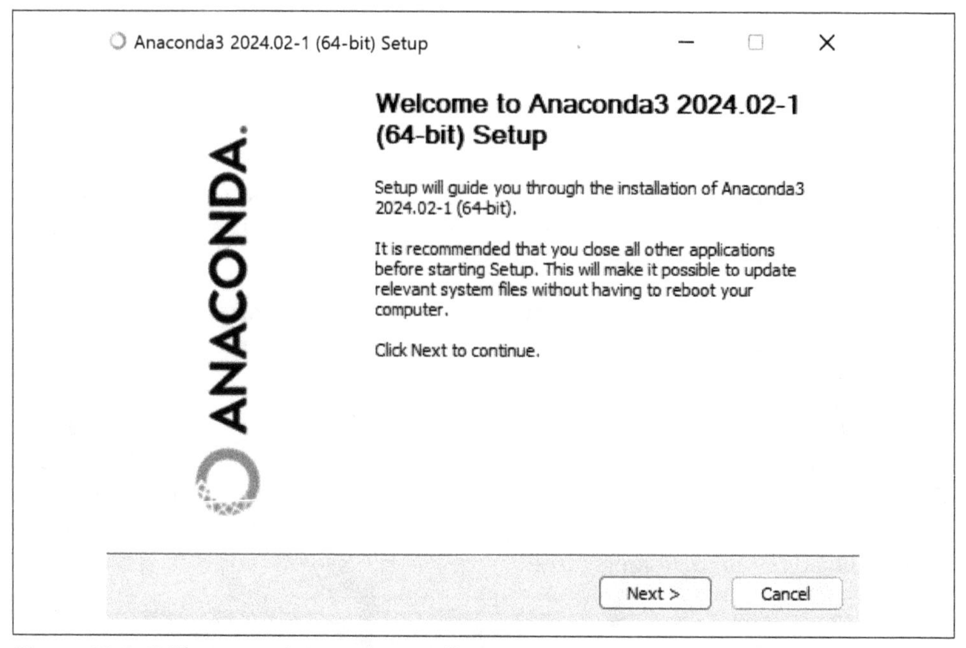

Figure 13-2. Welcome to Anaconda installation start menu

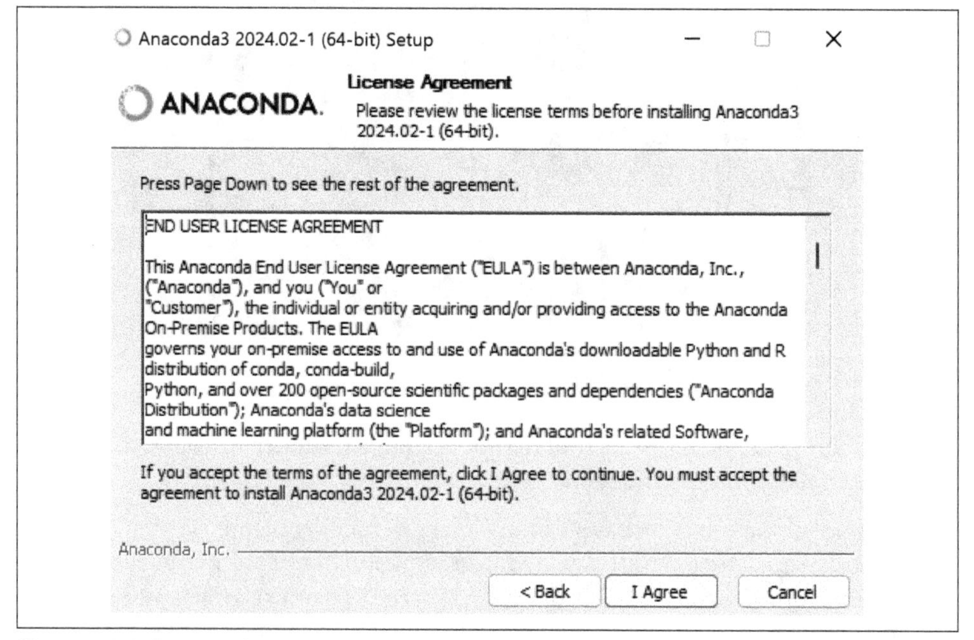

Figure 13-3. License Agreement

As always, read through the terms and agreements and then select the I Agree button to continue. The next screen will ask which installation type you are doing, as shown in Figure 13-4.

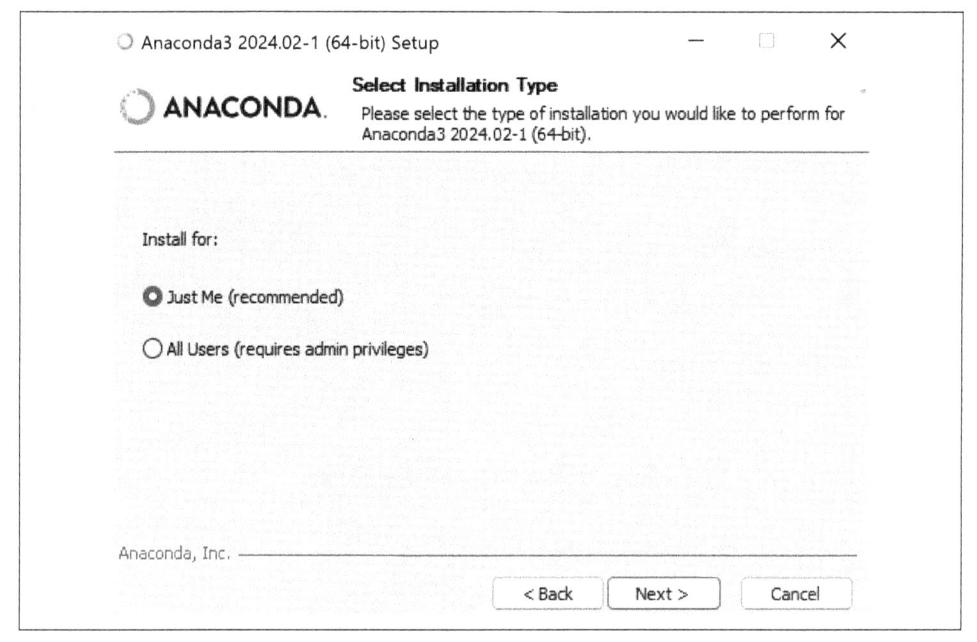

Figure 13-4. Select an installation type

In most cases, you will select "Just Me (recommended)." This will proceed with the installation for the username you are logged into on your machine. The All Users option is primarily for IT and may require you to authenticate as the administrator on your computer. Proceed with the Just Me option by clicking Next.

The next step is to choose a destination folder on your machine, as shown in Figure 13-5.

You can browse your computer's files and choose a destination if you want. Or you can leave it as the default, which will create a new directory for you on your computer. Select your chosen destination and then click Next.

On the next screen are some advanced installation options, as shown in Figure 13-6.

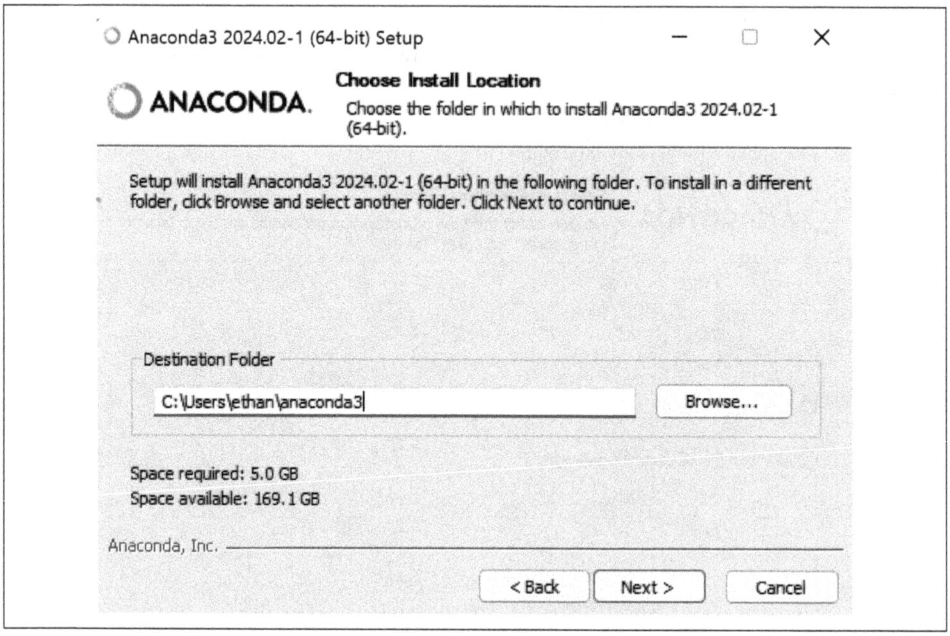

Anaconda3 2024.02-1 (64-bit) Setup
ANACONDA.
Choose Install Location
Choose the folder in which to install Anaconda3 2024.02-1 (64-bit).
Setup will install Anaconda3 2024.02-1 (64-bit) in the following folder. To install in a different folder, click Browse and select another folder. Click Next to continue.
Destination Folder
C:\Users\ethan\anaconda3
Browse...
Space required: 5.0 GB
Space available: 169.1 GB
Anaconda, Inc.
< Back
Next >
Cancel

Figure 13-5. Choose the installation location

Anaconda3 2024.02-1 (64-bit) Setup
ANACONDA.
Advanced Installation Options
Customize how Anaconda3 integrates with Windows
Create start menu shortcuts (supported packages only).
Add Anaconda3 to my PATH environment variable
NOT recommended. This can lead to conflicts with other applications. Instead, use the Commmand Prompt and Powershell menus added to the Windows Start Menu.
Register Anaconda3 as my default Python 3.11
Recommended. Allows other programs, such as VSCode, PyCharm, etc. to automatically detect Anaconda3 as the primary Python 3.11 on the system.
Clear the package cache upon completion
Recommended. Recovers some disk space without harming functionality.
Anaconda, Inc.
< Back
Install
Cancel

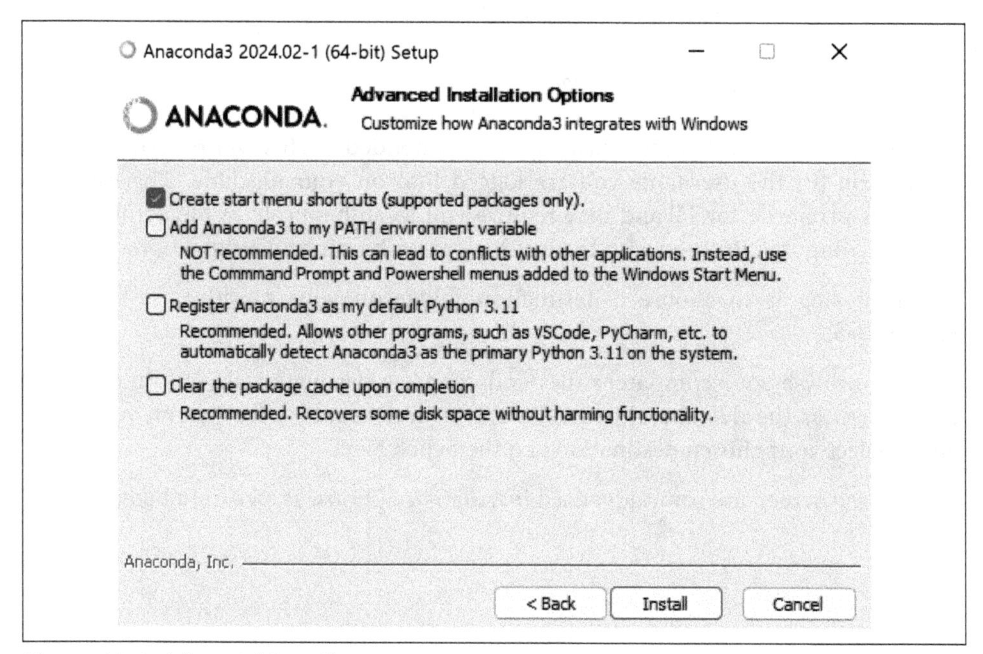

Figure 13-6. Advanced installation options

By default, this will create a start menu shortcut for you. However, there are two checkbox options that you should also select. The "Register Anaconda3 as my default Python 3.11" option will allow some of the applications within Anaconda Navigator to launch and detect Anaconda as the source Python. The next one that is recommended is the "Clear the package cache upon completion" checkbox. This is just a healthy option to have in place for your computer's longevity.

After you make those selections, you end up with the options shown in Figure 13-7.

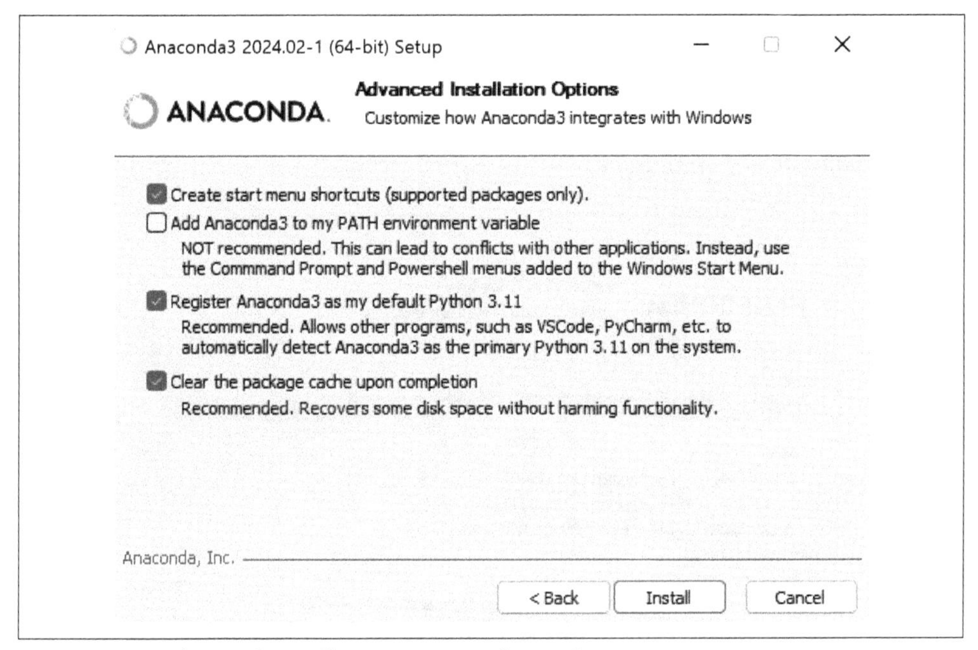

Figure 13-7. Advanced installation options after updates

Next, click Install and you will see a loading bar appear, as shown in Figure 13-8.

The actual installation will take a few minutes. Once the installation is complete, you will see a screen similar to Figure 13-9.

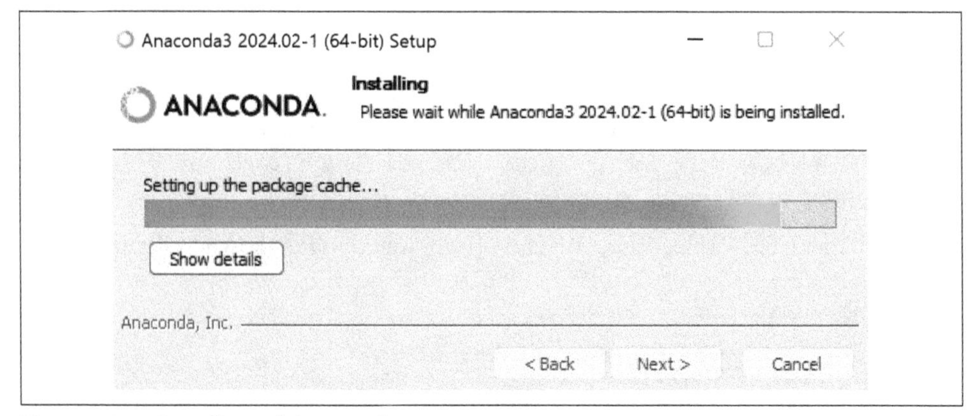

Figure 13-8. Installing of Anaconda

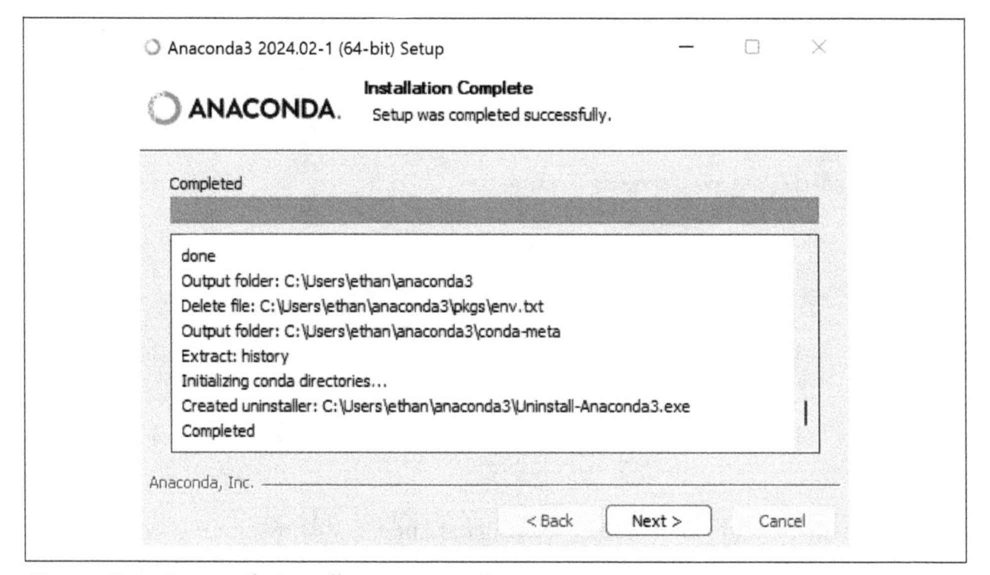

Figure 13-9. Anaconda installation is complete

From this screen, click Next. On the next screen, you should see a prompt to continue to get Jupyter Notebook, which is a cloud-based coding application, as shown in Figure 13-10.

This step is completely optional. If you desire to learn more about Jupyter Notebook, click the hyperlink in blue. You will not be using that particular coding application in this book, so you can click Next to finalize the setup, as shown in Figure 13-11.

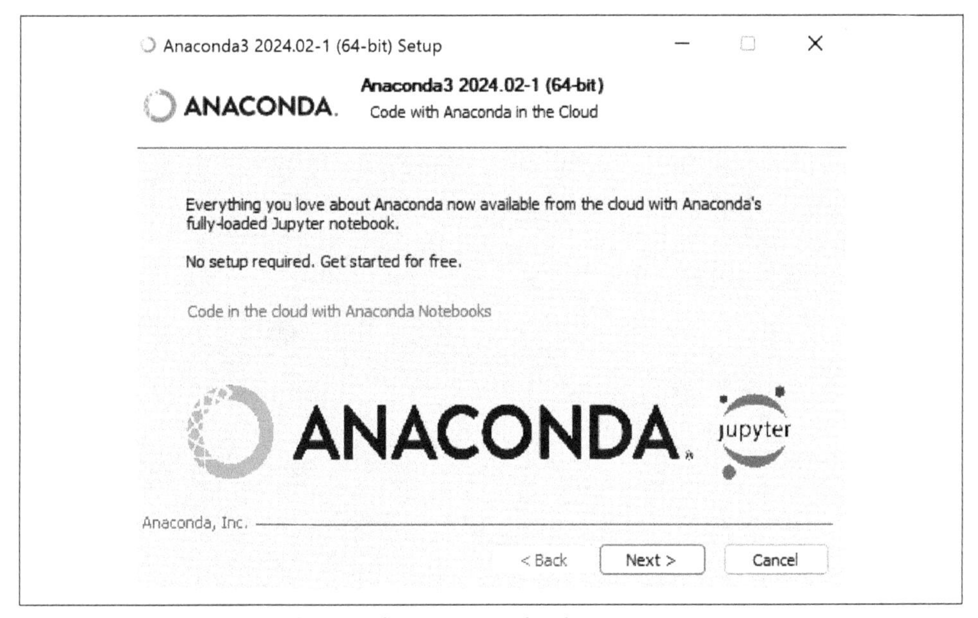

Figure 13-10. Begin installation of Jupyter Notebook

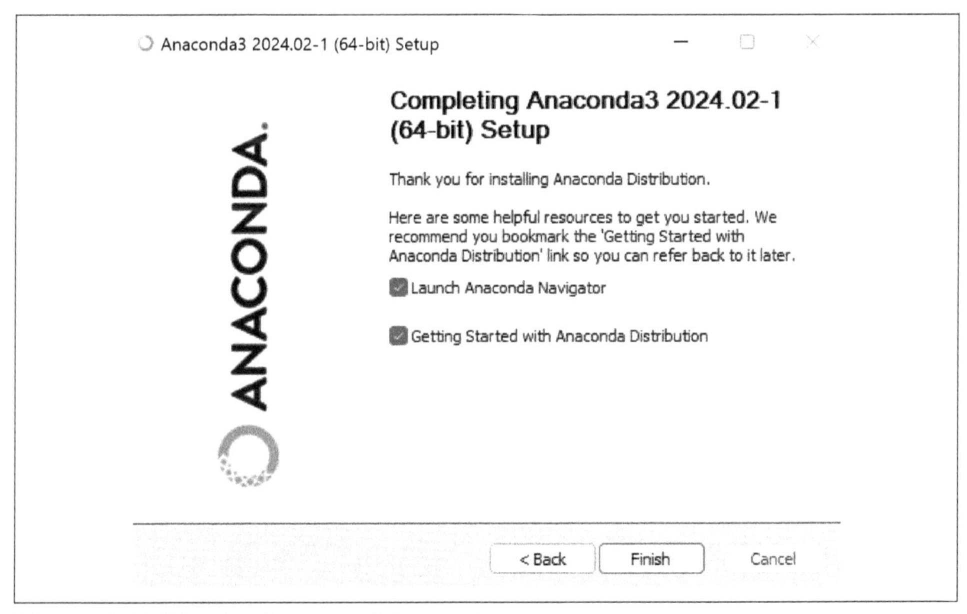

Figure 13-11. Finalize the setup of Anaconda.

Click Finish to finalize the setup, and Anaconda Navigator will launch.

Establishing an External Connection to Python in Tableau

You should now have Anaconda installed on your computer and are ready to begin setting up a connection to Tableau. Before you begin, here is a quick overview of the home screen of Anaconda Navigator, as shown in Figure 13-12.

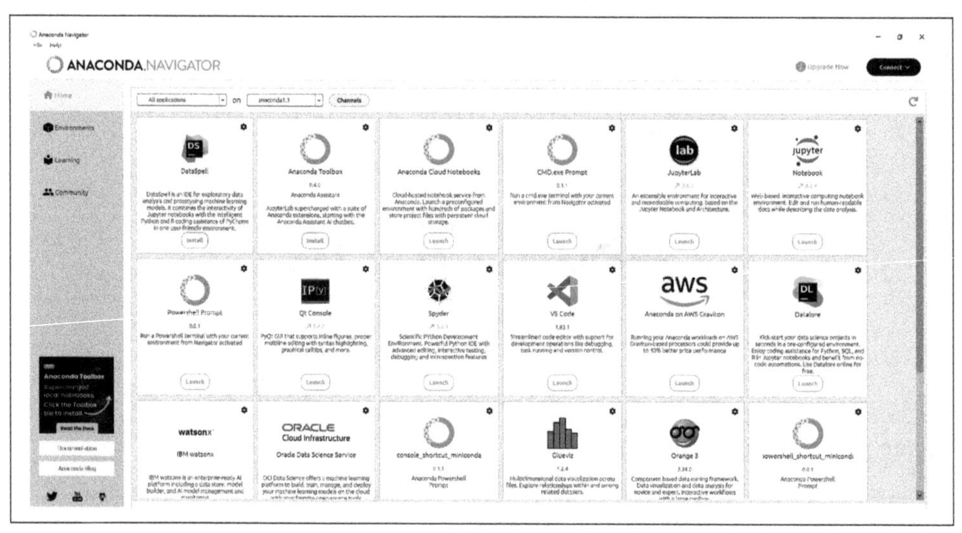

Figure 13-12. Anaconda Navigator home screen

You can see in the middle that there are many different coding applications you can launch from Anaconda. I'll be primarily using Spyder for my demonstrations; however, feel free to play around and choose one that suits you and your organization. They all have their pros and cons.

At the top left, there are some different panes you can toggle among. By default, you start on the Home pane. I will cover the Environments pane in more detail later in this section. The other two panes are Learning and Community, which give you great resources for both beginners and advanced users.

To continue connecting to Tableau, toggle to the Environments pane, as shown in Figure 13-13).

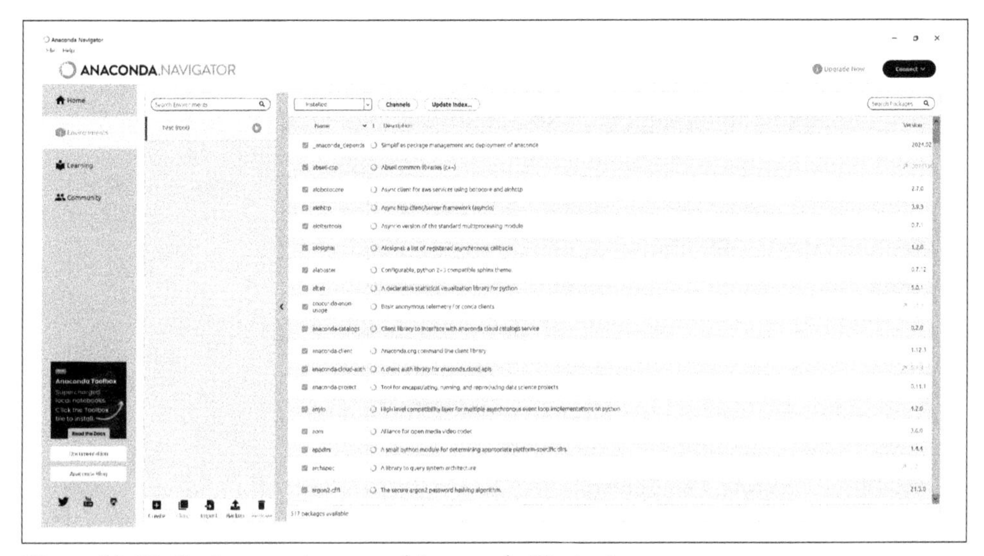

Figure 13-13. Environments pane of Anaconda Navigator

You will see all the packages that are installed in the base environment in the main screen on the right. This is where you can install more packages and create new environments. Let's start by creating a new environment you can use throughout the rest of this book. First, in the bottom left you will see a button that says Create, as shown in Figure 13-14.

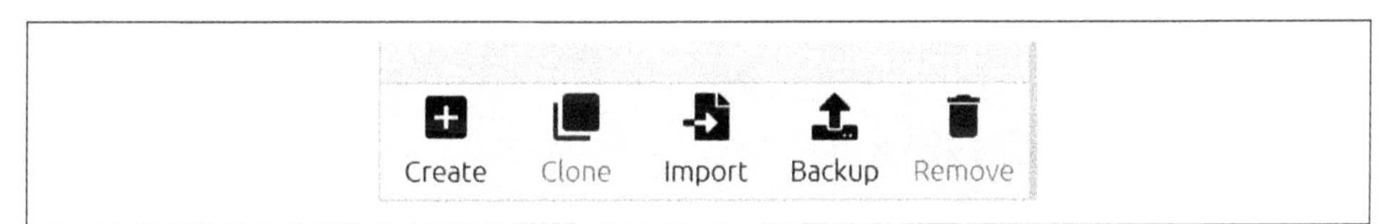

Figure 13-14. Create environment button

Click Create, and a menu titled "Create new environment" will appear. Title your environment "StatisticalTableau" and then select version 3.9.19 of Python within the drop-down, as shown in Figure 13-15.

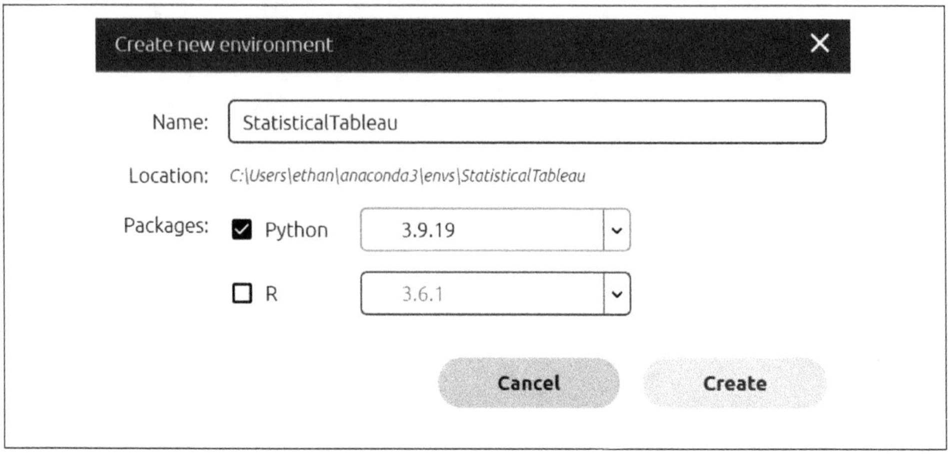

Create new environment

Name: StatisticalTableau

Location: C:\Users\ethan\anaconda3\envs\StatisticalTableau

Packages: Python 3.9.19

R 3.6.1

Cancel Create

Figure 13-15. Creating a new environment

At the time of this writing, TabPy, which is the package needed to connect to Tableau, supports Python versions 3.7, 3.8, and 3.9. Always use the most recent update adopted by your organization.

After your selections, click Create. You will see Anaconda Navigator start to create this environment and install packages in it. This process will take a few minutes, but in the end you will see your new environment light up green with a play button, as shown in Figure 13-16.

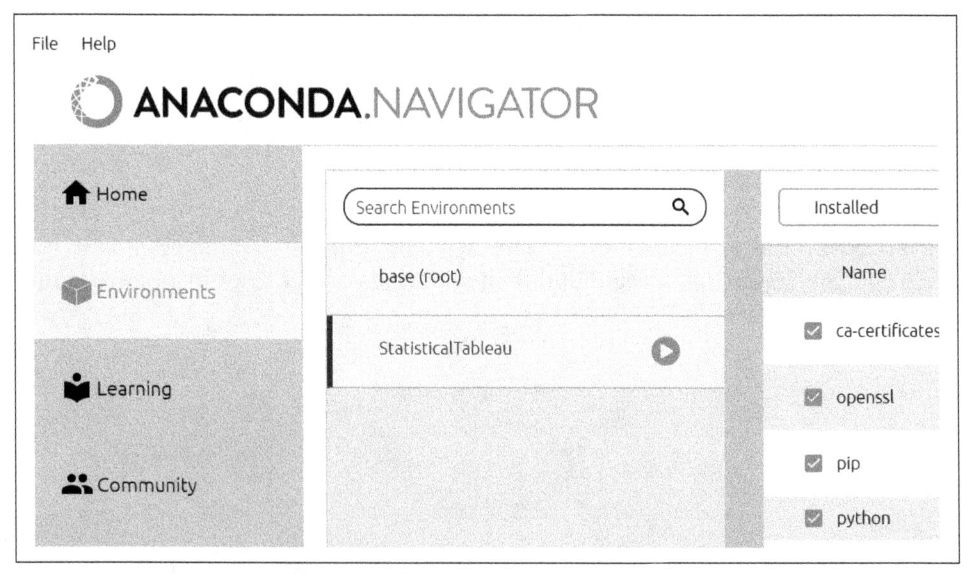

Figure 13-16. New environment ready to go

Now that you have your environment created, you need to install a package called TabPy, a framework for working with Python code. The easiest way to install it is through a terminal command prompt. Click the play button next to the Statistical-Tableau environment and select Open Terminal, as shown in Figure 13-17.

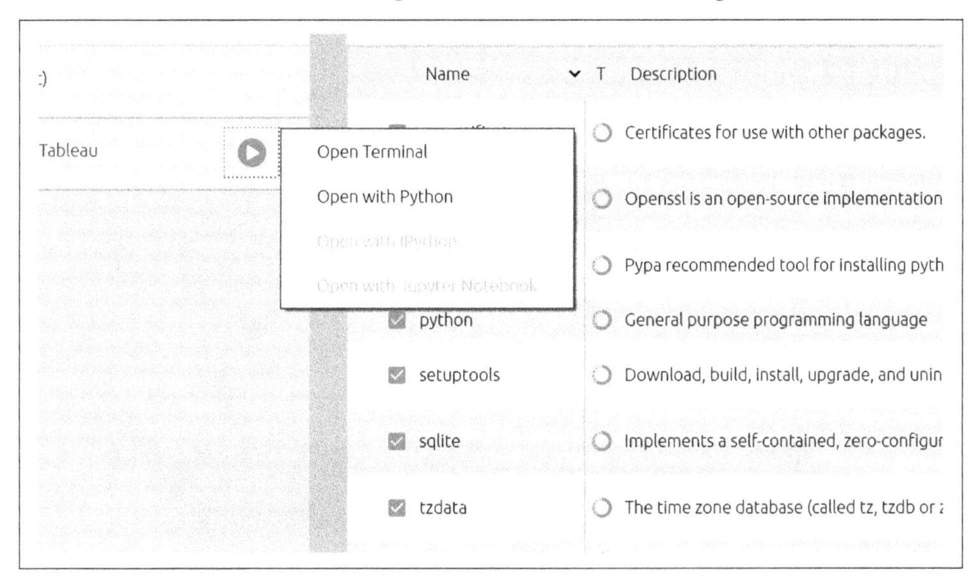

Figure 13-17. Opening a terminal using the new environment

You will see a command prompt appear. The first step is to install and update pip to the environment. To do this, type the following command (see Figure 13-18) and press Enter on your keyboard:

```
python -m pip install --upgrade pip
```

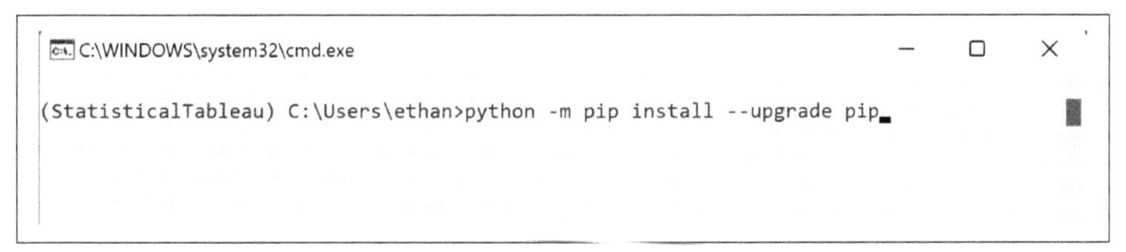

Figure 13-18. Installing and updating pip

You will see several things happen as it executes, and it should tell you that pip was successfully installed. Now you need to call on the pip package manager and install TabPy. To do this, type the following command (see Figure 13-19) and press Enter:

```
pip install tabpy
```

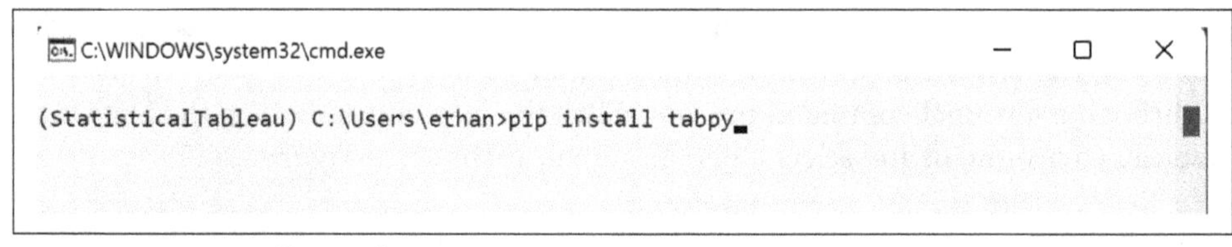

Figure 13-19. Installing TabPy

This takes a few minutes because you are not installing just TabPy but also any dependencies that package has on other packages. Once this is done, you will get a message that it successfully installed all of those packages, and the command prompt will reappear.

You are ready to start the local server and allow a connection to open to your machine. To start the Python server for Tableau, enter the following command (see Figure 13-20) and press Enter:

```
tabpy
```

```
(StatisticalTableau) C:\Users\ethan>tabpy
```

Figure 13-20. Launching TabPy

You will get a warning that you are enabling the TabPy server without authentication configured properly. It will ask if you want to proceed (y/N), as shown in Figure 13-21. You can configure a username and password if you choose. That process is beyond the scope of this writing, but there is clear documentation available (*https://oreil.ly/ndP_o*) if you want to enable that.

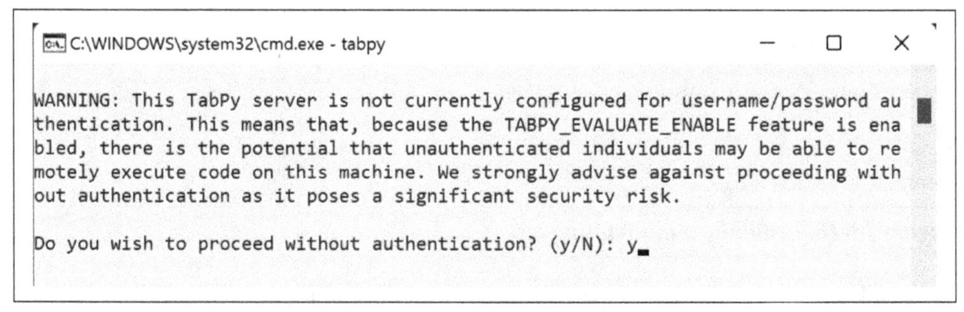

Figure 13-21. Warning to authenticate

For now, enter "y" and then click Enter on your keyboard. This will activate TabPy, and you will see some information appear about the server. Most importantly is the port that the web service is listening on. By default it is port 9004. This is important to note once you transition to Tableau.

With the server running in the command terminal, open Tableau and establish a connection to the external resource. To do this, click Help from the top navigation, hover over "Settings and Performance," then click Manage Analytics Extension Connection, as shown in Figure 13-22.

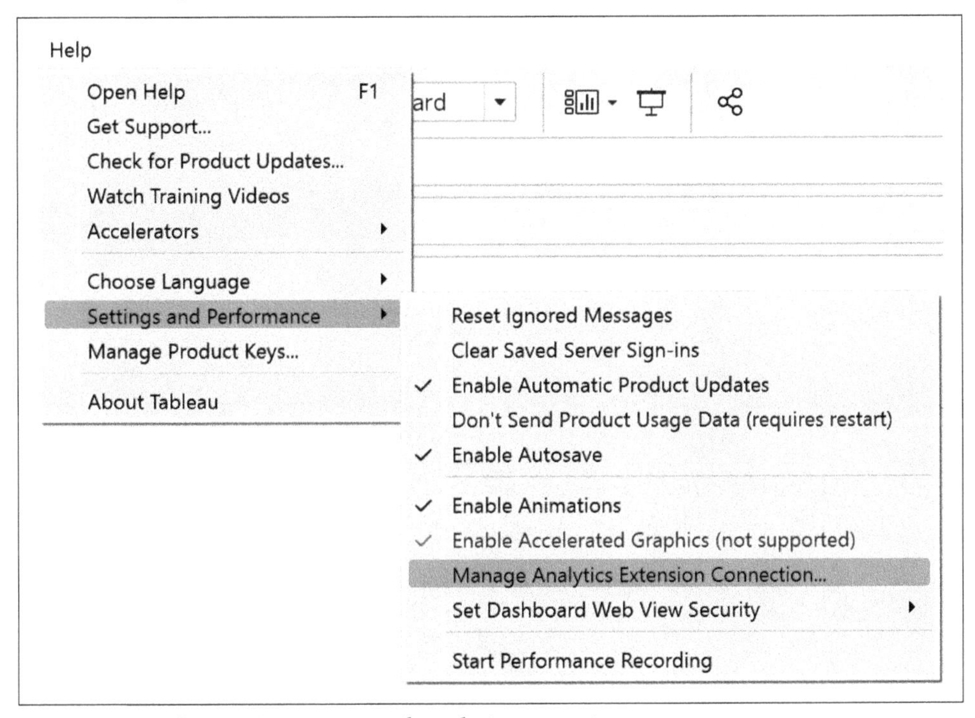

Figure 13-22. Connecting to external analytics extension

You will see a window appear, as shown in Figure 13-23, that will ask you to select a connection type. You will choose TabPy.

On the next screen, enter "localhost" as the Hostname and then enter 9004 as the Port (or whichever port you configured), as shown in Figure 13-24; then click Test Connection.

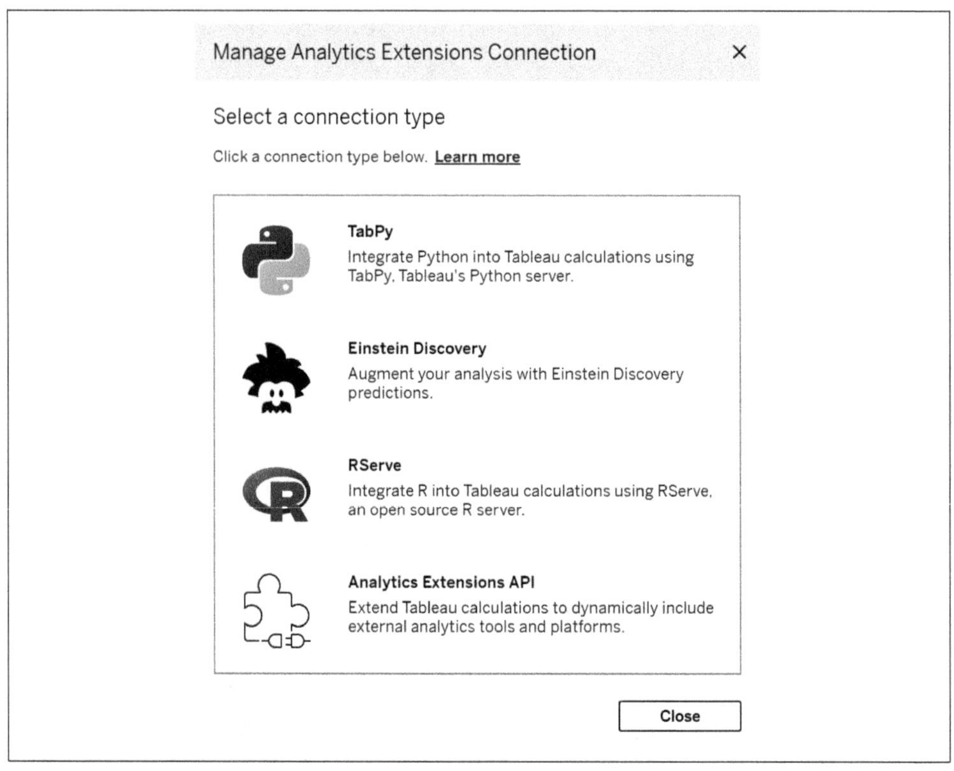

Figure 13-23. Select a connection type

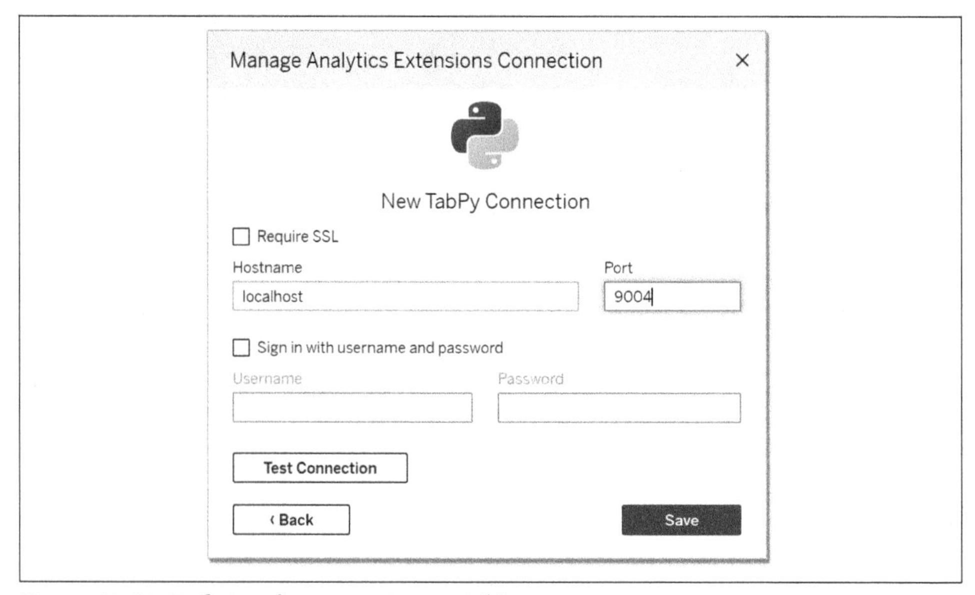

Figure 13-24. Defining the connection variables

If the information is right and the server is running from the command prompt, you will see a message appear that says Tableau Desktop successfully connected to the analytics extension, as shown in Figure 13-25.

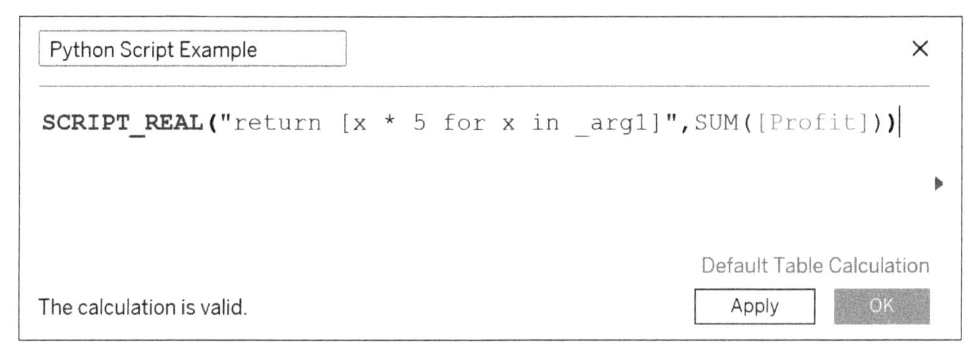

Figure 13-25. Successfully connected to the TabPy extension

With the connection successful, click Save to close the menu. Now you can write Python script within Tableau Desktop. Those scripts will run in the environment you created and then output back into Tableau Desktop as new calculated fields.

As a test, write a very basic equation that will take the profit values from the data in Tableau Desktop and multiply it by 5 in Python. It will return the result back from Python as a calculated field you can use in Tableau Desktop.

To start, create a new calculated field and enter the following calculation (see Figure 13-26):

```
SCRIPT_INT("return [int(x * 5) for x in _arg1]",SUM([Profit]))
```

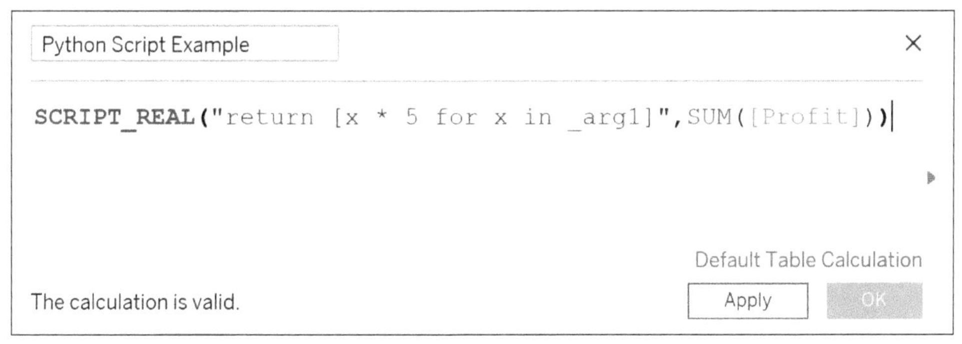

Figure 13-26. Python script example

Now create a small crosstab you can check to see that the values are correct. Add Category to the Rows shelf and then SUM(Profit) and Python Script Example to the Text property of the Marks card. You should see results similar to Figure 13-27.

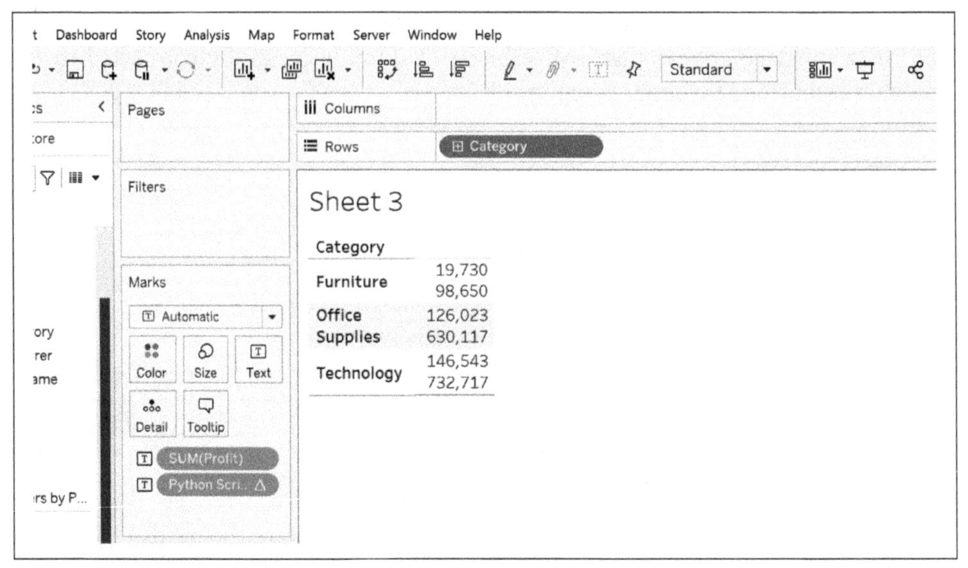

Figure 13-27. Example of the Python script

This is a very basic example to demonstrate everything is working. You will learn a more complex example of these functions in Chapter 15.

After you are done, it is always important to disconnect from the server connection in both Tableau Desktop and from the terminal. In Tableau Desktop, go back to the Manage Analytics Extension Connection menu and select Disconnect, as shown in Figure 13-28.

Then jump back to the terminal screen and press Ctrl + C. You will see a message display that says, "Shutting down TabPy," and then your command line will reappear.

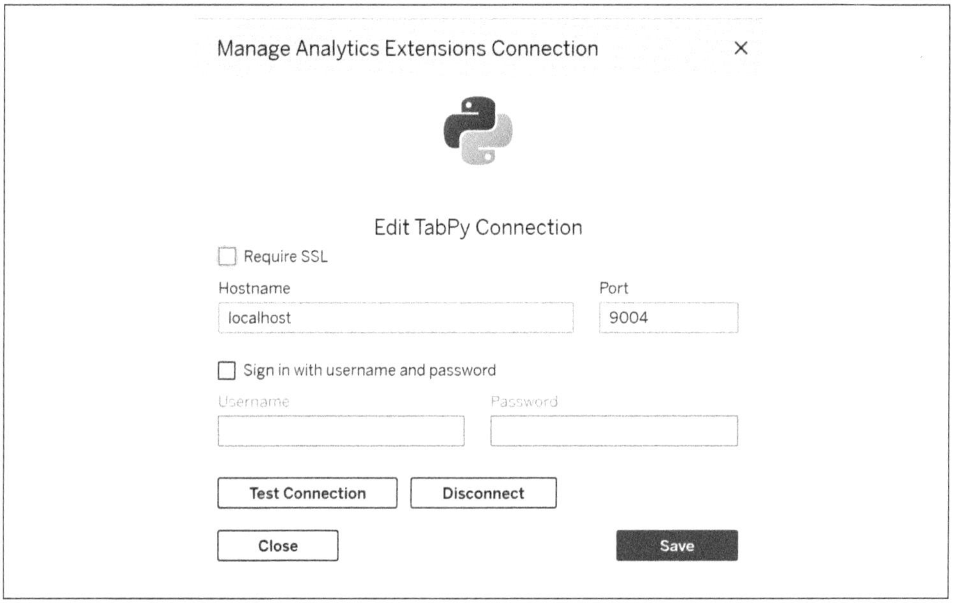

Figure 13-28. Disconnect from TabPy in Tableau Desktop

Summary

In this chapter, you learned how to download and install Anaconda. Anaconda is a great resource to work with and has a lot of benefits. You also learned how to install TabPy and run the TabPy server on your machine. Using this Python connection, you also learned how to connect to Tableau Desktop and run Python scripts there.

CHAPTER 14

Understanding Multiple Linear Regression in R and Python

In Chapter 8, you were introduced to simple linear regression. This model is used to analyze the relationship between an independent variable and a single dependent variable. However, there are many times when several variables together may cause an effect, but individually, those effects may not be seen.

In this chapter, you will learn what multiple linear regression is and its equation, and you will implement some high-level examples. You will also use R and Python scripts to learn how to implement the model in both languages. In Chapter 15, you will practice using these scripts to begin implementing more advanced models in Tableau using external connections.

What Is Multiple Linear Regression?

Multiple linear regression is a statistical method used to analyze the relationship between multiple independent variables and a dependent variable. In simple linear regression, there is only one independent variable used to predict the dependent variable. However, in multiple linear regression, two or more independent variables are employed to forecast or understand changes in the dependent variable.

Multiple linear regression has many applications in business when you are trying to use multiple inputs to understand something. A few examples of its application are as follows:

Sales forecasting
: Businesses often use multiple linear regression to predict sales based on various factors such as advertising expenditure, seasonality, pricing, and other market

conditions. By analyzing historical data, companies can predict future sales and adjust their strategies accordingly.

Marketing ROI analysis
: Multiple linear regression can help in assessing the effectiveness of marketing campaigns. It can be used to understand the impact of multiple marketing channels, such as social media, TV ads, online ads, and email campaigns, on sales or customer acquisition.

Financial analysis
: In finance, multiple linear regression can be employed to understand the relationship between various financial indicators and stock prices. Analysts might use regression to predict stock returns based on factors such as interest rates, company earnings, or market volatility.

Employee performance analysis
: Companies can use multiple linear regression to understand the relationship between various factors (such as training, experience, education, etc.) and employee performance or productivity. This analysis can help in making informed decisions related to training programs or hiring processes.

Operational efficiency
: Multiple linear regression can be used to optimize operational processes in business. For instance, in manufacturing, it could help in understanding how different factors such as machinery, labor, and raw materials impact the production output or quality. By identifying significant variables, companies can streamline processes for greater efficiency.

Multiple linear regression has the same assumptions as simple linear regression:

Linearity
: The relationship between the dependent variable and the independent variable is assumed to be linear. This means that a change in the independent variable is associated with a constant change in the dependent variable.

Independence
: The residuals (the differences between the observed and predicted values) should be independent of each other. In other words, the value of the dependent variable for one observation should not be influenced by the value of the dependent variable for any other observation.

Homoscedasticity
: The variance of the residuals should be constant across all levels of the independent variable. In simpler terms, the spread of the residuals should be roughly the same for all values of the independent variable.

Normality of residuals

The residuals should follow a normal distribution. This assumption is more critical with smaller sample sizes. If the residuals are approximately normally distributed, it suggests that the statistical inferences made using the model are more reliable.

No perfect multicollinearity

In the context of simple linear regression, multicollinearity is not a concern as there is only one independent variable. However, in multiple regression (with more than one independent variable), it is assumed that there is no perfect linear relationship among the independent variables.

Another thing to consider: you need a decent sample size to get a less volatile prediction model.

Multiple Linear Regression Equation

Similar to simple linear regression, the idea is to minimize the distance between the predicted value and the actual values, as shown in Figure 14-1.

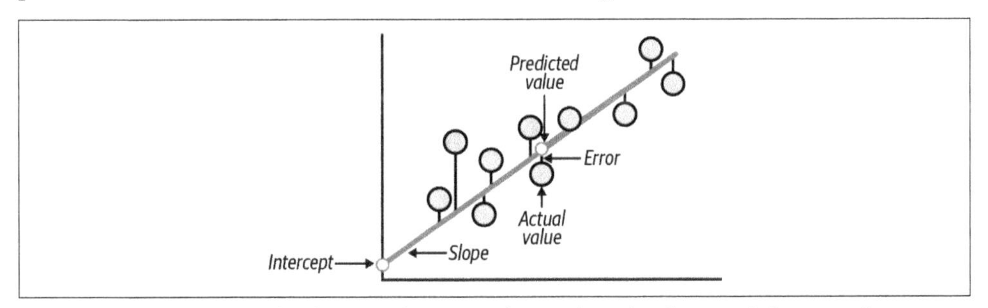

Figure 14-1. Minimize the distance between the predicted values and the actuals

What makes it different from simple linear regression is that you are using multiple inputs to create the line. This can be expressed mathematically. Simple linear regression uses the following formula:

$$Y_i = \beta_0 + \beta_1 X_i + \epsilon_i$$

where

Y_i = the value of the dependent value

β_0 = the intercept

β_1 = the coefficient calculated by the model

X_i = a known value from the dataset at the ith point

ϵ_i = a random error that occurs

Multiple linear regression is very similar and is expressed with the following formula:

$$Y_i = \beta_0 + \beta_1 X_1 + \beta_2 X_2 + \beta_i X_i + \epsilon_i$$

where

Y_i = the value of the dependent value

β_0 = the intercept

β_1 = the coefficient calculated by the model

β_2 = the coefficient calculated by the model

β_i = the ith coefficient calculated by the model

X_i = a known value from the dataset at the ith point

ϵ_i = a random error that occurs

You can see that the model pretty much stays the same except that you are adding new beta coefficients for each variable you include. In the next sections, I will show you how to code a multiple linear regression model in R and Python. To follow along, you will need to have all the required software installed on your computer. The complete steps to get the software and a brief overview of each can be found in Chapters 12 and 13.

How to Implement Multiple Linear Regression in R

In this section, you will learn to code a multiple linear regression model in R that predicts profit based on discount and quantity. You will need to know how to code in R to implement this model in Tableau. In Chapter 15, you will use this code to write the calculated field in Tableau using the SCRIPT functions you were introduced to in Chapter 12.

The first step is to launch RStudio and connect to the Sample - Superstore dataset. Once RStudio is launched, navigate to the window in the lower righthand corner. This is where you can select a directory, view visuals you have plotted, view packages installed, etc. On the Files tab, select the following directory:

Home > My Tableau Repository > Datasources > 2023.2 > en_US-US

Please note that 2023.2 is the version of Tableau at the time of this writing. You can select whichever version of Tableau you have on your machine. Depending on the version, the results may vary slightly from what appears in the figures, but the code will be the same.

Once you are in the correct directory path, you should see the same files, as shown in Figure 14-2.

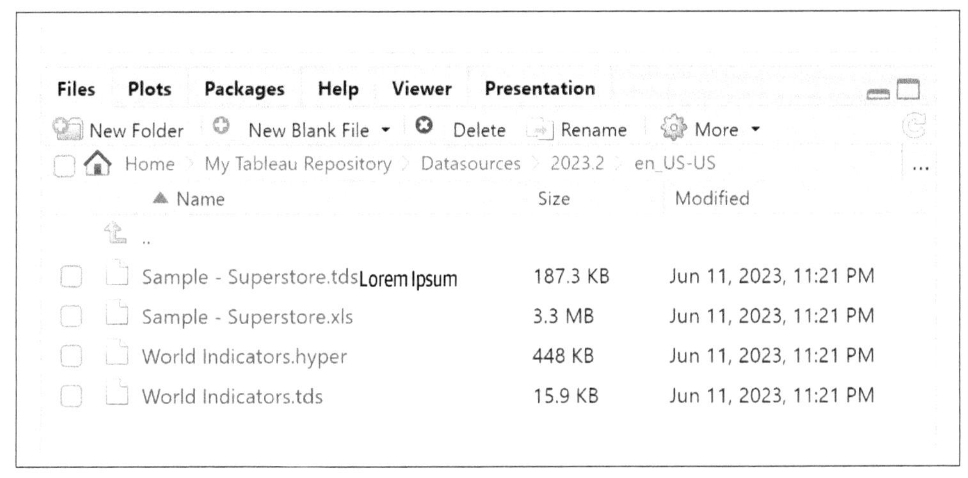

Figure 14-2. RStudio files: Sample - Superstore dataset directory

There are several ways to connect to data within RStudio. For this example click on *Sample - Superstore.xls* and then select Import Dataset, as shown in Figure 14-3.

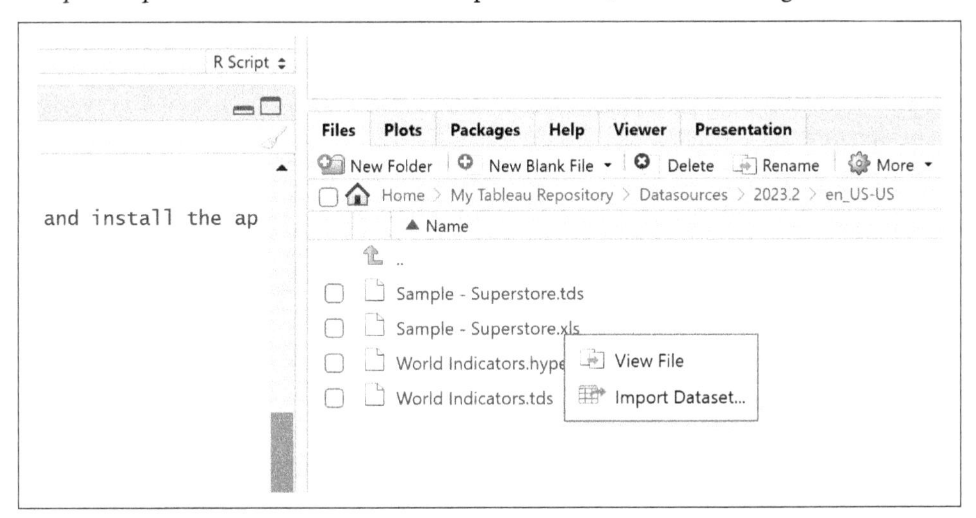

Figure 14-3. Import Dataset from file directory in RStudio

At this point, you may get a pop-up that says you need to install specific packages that are used to import the *.xls* file type, as shown in Figure 14-4.

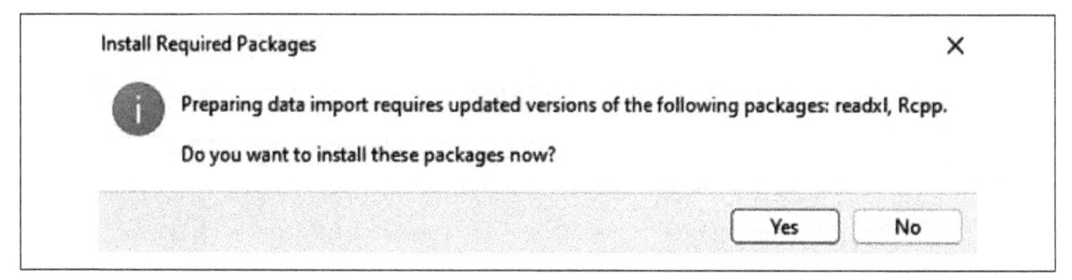

Figure 14-4. Install required packages to import data

If the packages need to be installed, you will see that job being executed in the window to the bottom left. Once it is done, or if you already had the packages installed, there will be a large window that appears. This is the import data screen and will give you a preview of the file you are importing, as shown in Figure 14-5.

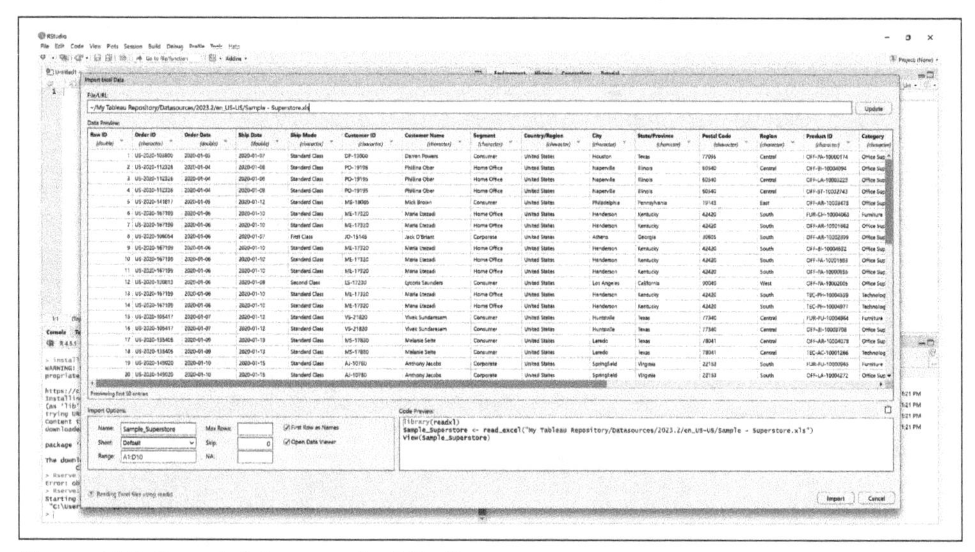

Figure 14-5. Import data screen in RStudio

From this menu, you can do a little data cleaning if needed. There are options to skip certain rows, drop columns, handle NA values, etc. The Sample - Superstore dataset is structured and doesn't require any additional cleansing. Leave all the settings as is and click Import in the bottom-right corner of the menu.

You will see a new tab appear in the upper-left quadrant of RStudio, as shown in Figure 14-6.

You can tab between the data and the code editor from the tabs at the top left. In Figure 14-6, the file is not yet saved, so it is named *Untitled1*. Click on that and start working on a very basic example of multiple linear regression.

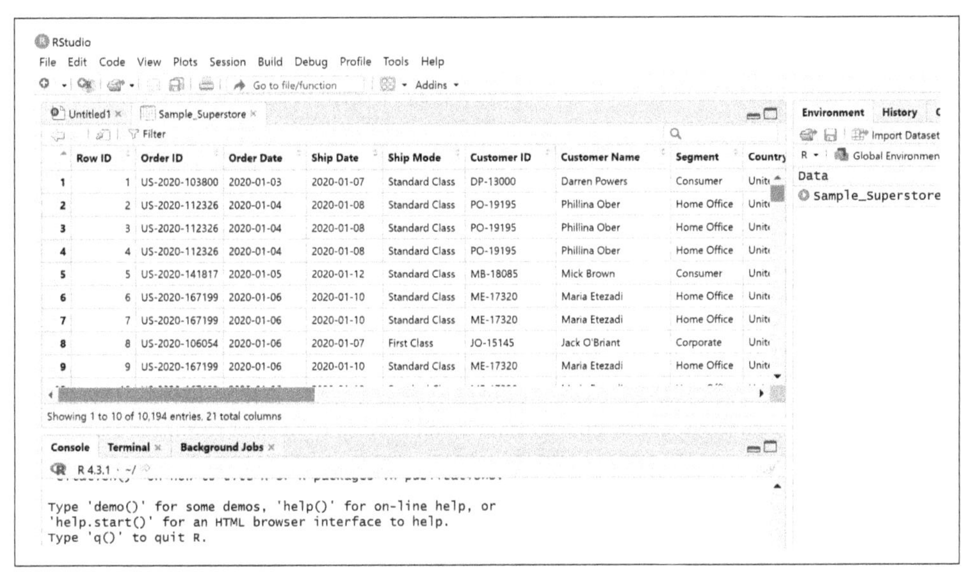

Figure 14-6. Sample_Superstore dataset tab

In this example, you're trying to predict profit based on discount and quantity. In R, the command to call a linear regression is `lm`. If you type `lm()` into the editor, RStudio will provide you with a help list of calls you can make in that command, as shown in Figure 14-7.

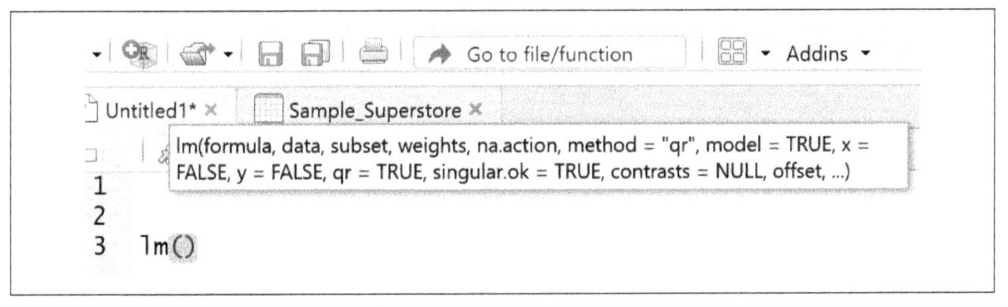

Figure 14-7. Helpful function definitions

This list of commands within the `lm` function is important to understand, and you can pass these parameters from Tableau to R once we begin that process. The great thing about R is that there is a lot of documentation and resources available to understand what each of those parameters does. If you simply type `?lm` into the editor and run that line of code, you will see the documentation appear for all of those parameters within this function at the bottom-left quadrant of RStudio. In this example, leave everything the same and just define the dependent and independent variables.

To run the model and see the results, enter and run the following code shown in Figure 14-8.

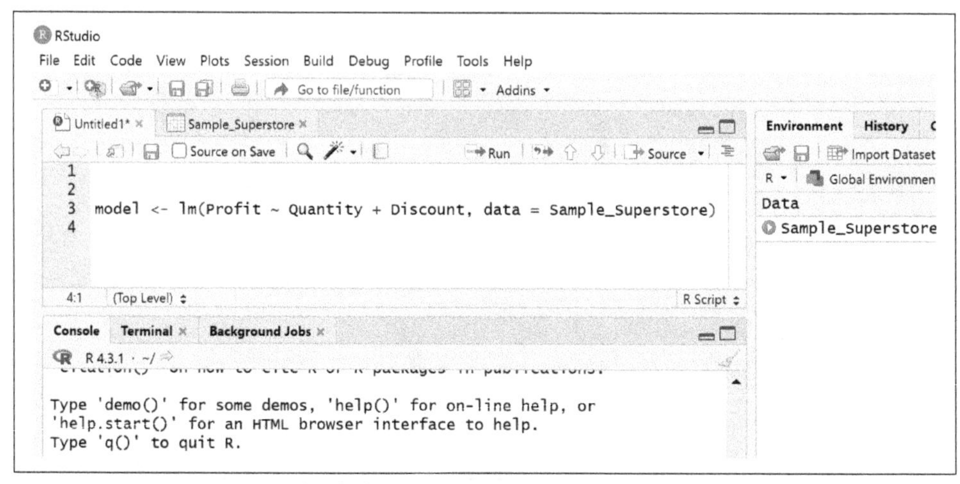

Figure 14-8. Running a multiple linear regression model in RStudio

When you enter this code and click Run, it saves the results of the multiple linear regression to a new object that you can define as a model. Within the parentheses, define Profit as the dependent variable and separate it with a tilde (~) from the independent variables. In this example, use Quantity and Discount as the independent variables you want to predict Profit with. At the end, use a comma to end the formula parameter within the `lm` function and then define the data parameter, which is the data source you connected to at the beginning of this section.

Save the results to the new object, and you will see `model` appear in the upper-right window of the interface as a new object, as shown in Figure 14-9.

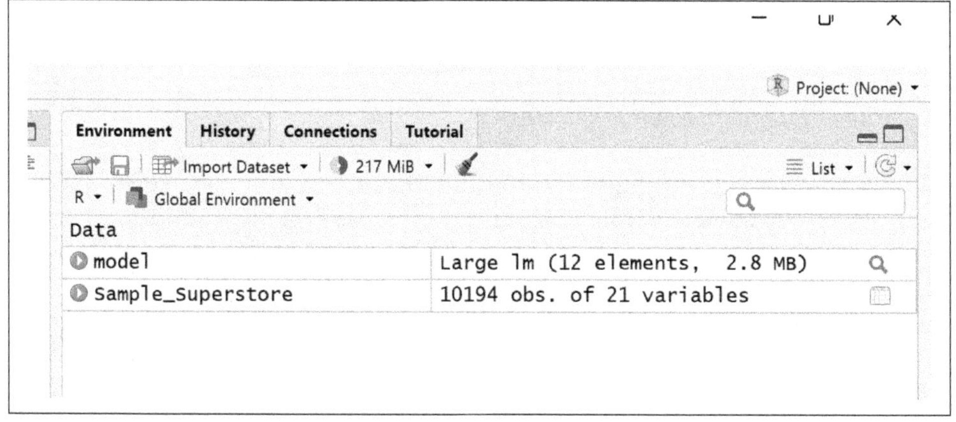

Figure 14-9. Objects within the environment

To view the results of the model, enter `summary(model)` into the code editor and run that line, as shown in Figure 14-10.

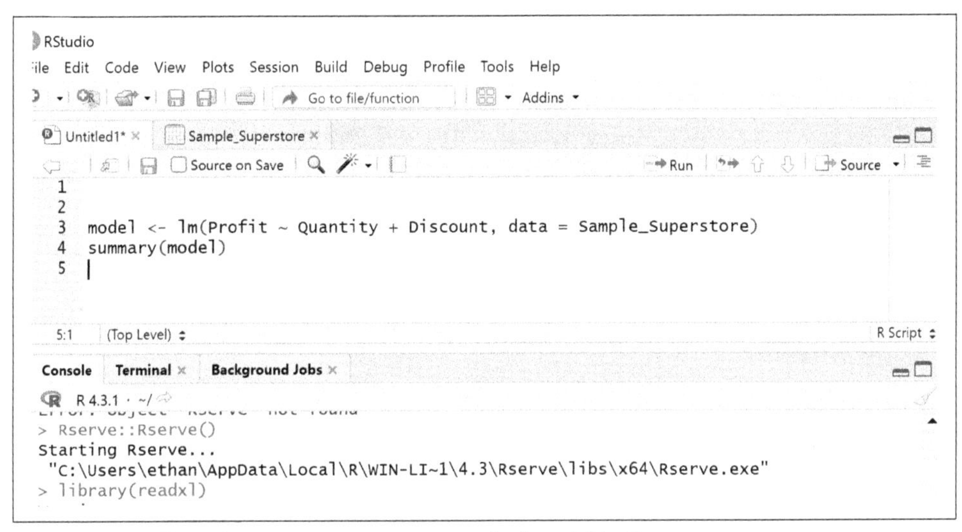

Figure 14-10. Running a summary command of model in R

In the console window at the bottom left of the interface, you will see a table of summary statistics appear, as shown in Figure 14-11.

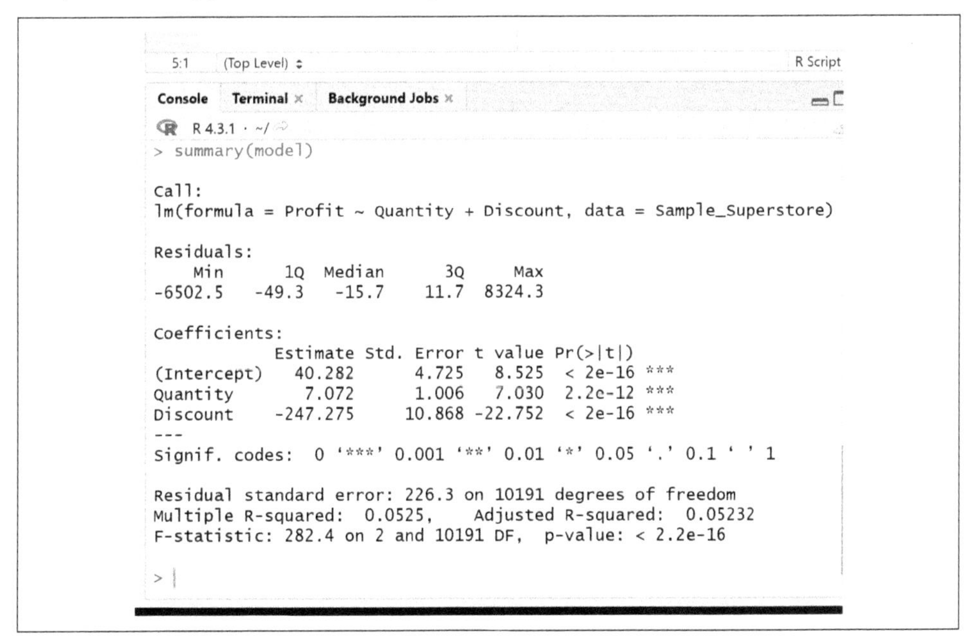

Figure 14-11. Summary statistics of multiple linear regression in R

In this table, you will see the coefficients, intercept, and the p-values of each variable. In Chapter 15, you will create scripts that will pass the data from Tableau to R, run this model, and then return the coefficient values so you can use them to build a predictive model.

There is one final piece to the code that you will use in Chapter 15. In Tableau, the functions you will use return a single value. For this example, you want to return the coefficients of the model that you can use to build a dynamic predictive model in Tableau.

To isolate the coefficient values, you simply need to call on the object you created. You previously named that object "model." As you can see in Figure 14-12, you will call the `model` object and then reference the coefficient array to isolate the intercept or the two coefficients.

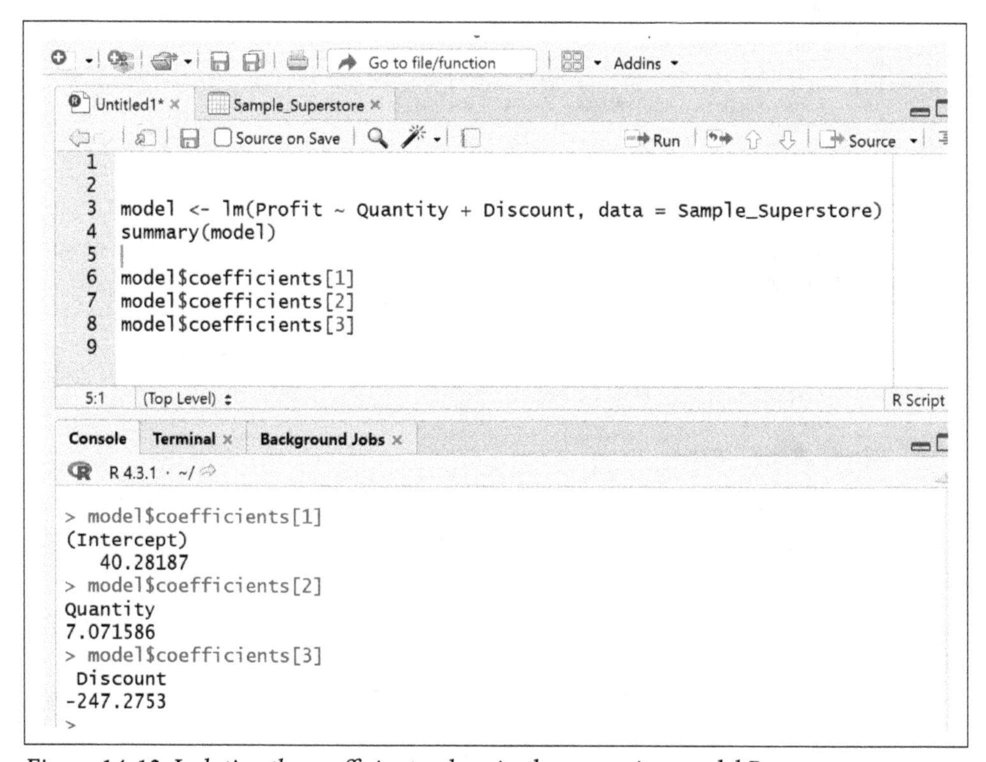

Figure 14-12. Isolating the coefficient values in the regression model R

Since the values are held in an array, you can use integers to reference the position of each value in the array. The first number in the array is the intercept, followed by your coefficient values. That is all you need to know for now to run this simple model. Knowing R code and the different models allows you to incorporate anything into Tableau. Save this code so you can reference it in Tableau in Chapter 15.

Now that you have explored building this model in R, you will repeat this process using Python in the next section.

How to Implement Multiple Linear Regression in Python

In this section, you will learn how to code a multiple linear regression model in Python. To begin, launch the Anaconda Navigator, as shown in Figure 14-13.

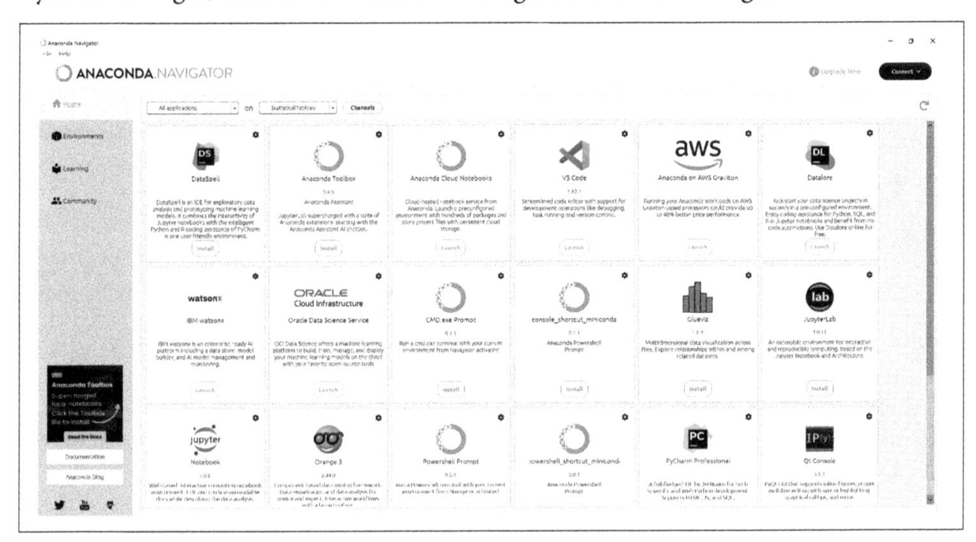

Figure 14-13. Launching Anaconda Navigator

In Chapter 12, you created an environment that references the correct version of Python for Tableau to connect to. However, there are a few more packages or libraries that you need to install to run a multiple linear regression model effectively in Python. As shown in Figure 14-14, toggle to the correct environment. For this example, use the environment you created in Chapter 13, titled "StatisticalTableau."

The first library you are going to install is called *xlrd*. This library was created to help developers extract data from Microsoft Excel and store that data in a data frame in Python. Since the Sample - Superstore dataset that you are using is stored as an Excel file, use this library to extract the data.

Just as a note, if you have a different file type you are trying to connect to, there are other libraries you can use. It's out of the scope of this book to cover all the possibilities but search the repository in Anaconda and you will find what you are looking for. Most libraries have great documentation that you can reference with example code.

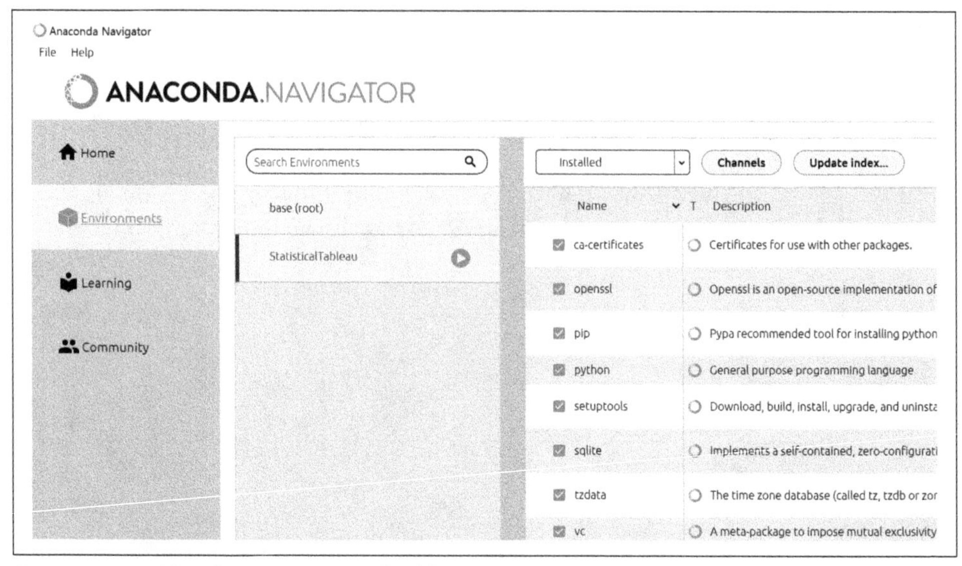

Figure 14-14. Toggling to StatisticalTableau environment to install new libraries

As shown in Figure 14-15, search for the xlrd library. Then click the checkbox next to the library name and click the Apply button in the bottom-right corner of the interface.

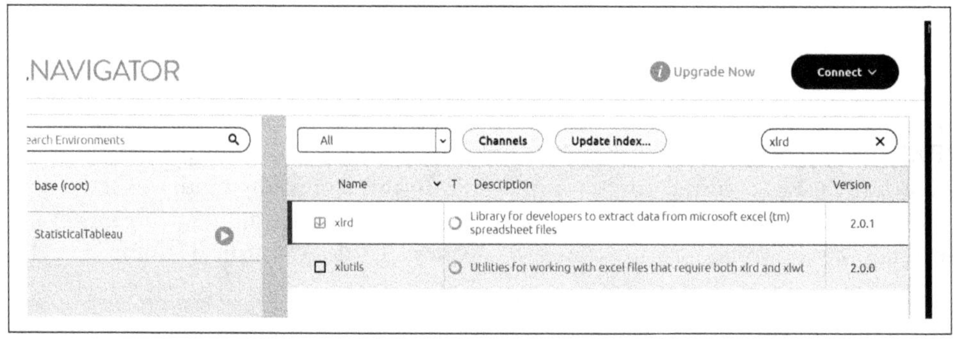

Figure 14-15. Downloading the xlrd library in Anaconda Navigator

After clicking the Apply button, there will be a loading bar on the interface. Give it some time to load the library and look for other dependencies. Eventually, you should see a window appear that will have a list of the libraries that will be installed. As you can see in Figure 14-16, this was the only package required. Click Apply on the new window to begin the installation.

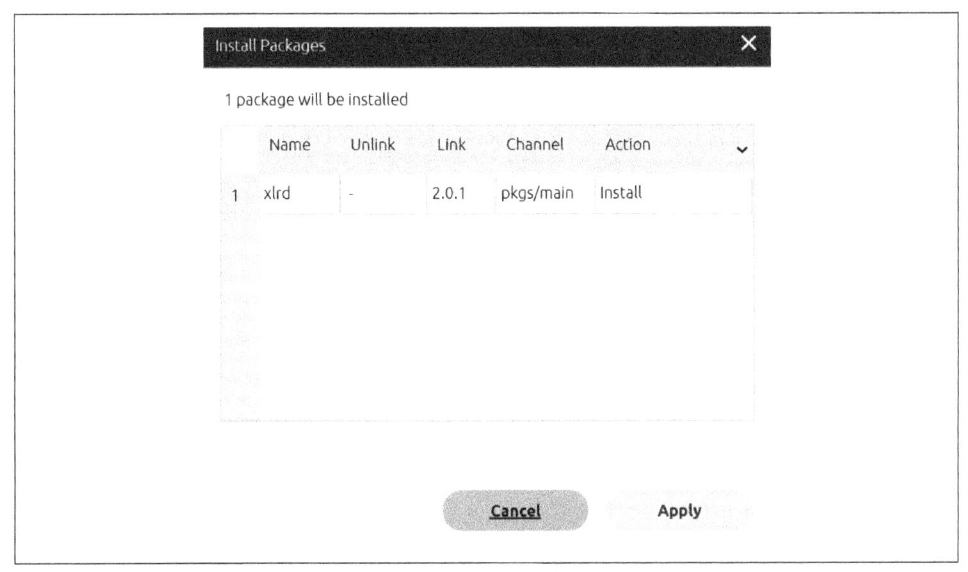

Figure 14-16. Install Packages window in Anaconda Navigator

After the installation process, you should see a green checkmark appear next to the package name. This indicates that it is installed to your environment.

The next package you need to install is called *statsmodels*. This library has a lot of different statistical models that you can call on in Python and is very useful. Just like the last package, search for it, as shown in Figure 14-17, click the checkbox, then click Apply at the bottom-right corner of the interface.

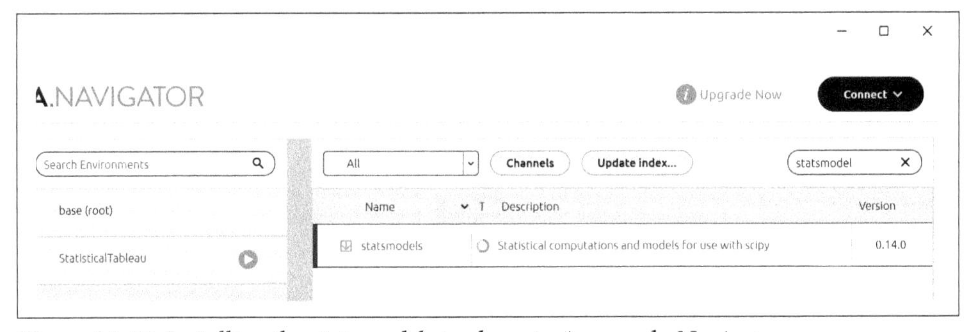

Figure 14-17. Installing the statsmodels package in Anaconda Navigator

Once you click Apply, the Install Packages window will appear. As you can see in Figure 14-18, this package has a lot of dependencies. Clicking Apply will install 17 packages to the environment.

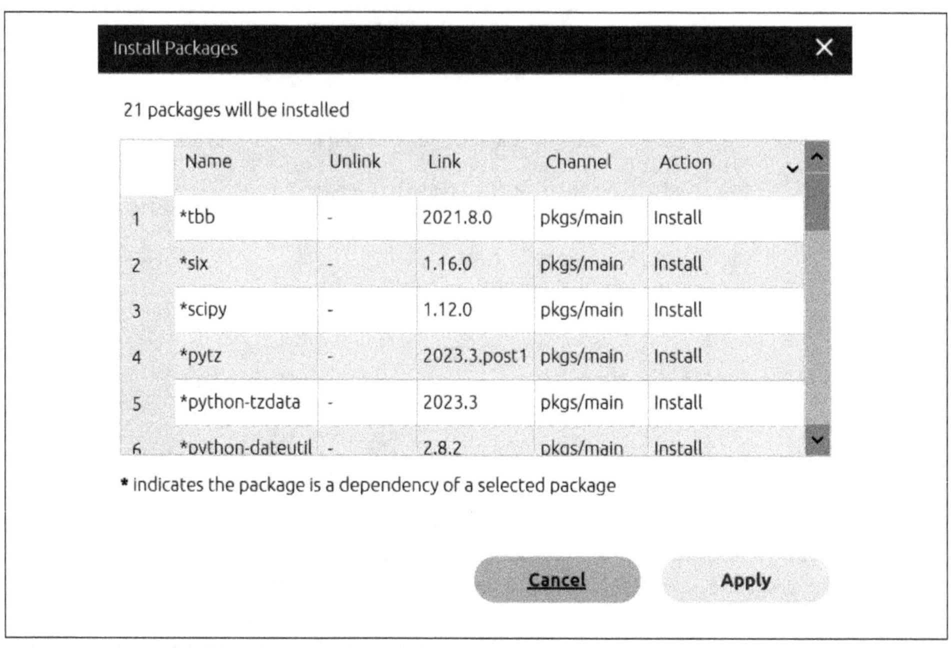

Figure 14-18. Installing statsmodels library and dependencies in Anaconda Navigator

After reviewing the list of dependencies, click Apply to begin the installation process. This install may take a little longer than the last because of the dependencies. In the end, you should see a green checkmark appear next to statsmodels in the search results.

Now that the correct libraries are installed, select the correct environment, navigate back to the home screen of Anaconda Navigator, and launch Spyder, as shown in Figure 14-19.

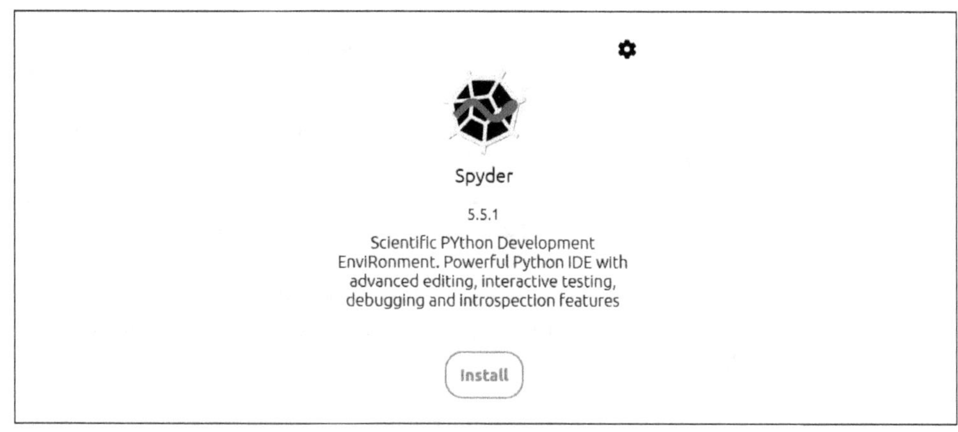

Figure 14-19. Launching Spyder from Anaconda home screen

After Spyder launches, import the libraries you will be using and assign them a reference. You need to import two packages in particular: pandas and statsmodels. As shown in Figure 14-20, enter the following code and Click the run button at the top of the interface.

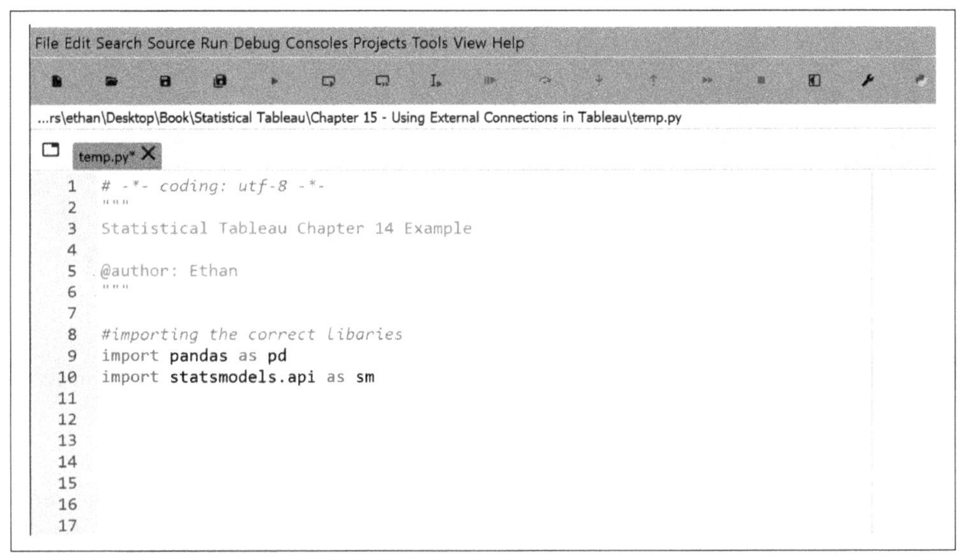

Figure 14-20. Importing the libraries in Python

Now create a data frame that you will reference to run the code on. This is similar to the importing step you did in R. However, in Python, you are going to find the location of the file you are referencing and then use code to save that data to a data frame.

Label the data frame `df` and then use the `read_excel` function from pandas to read in the Excel file of the Sample - Superstore dataset that comes with Tableau. This file is located in a specific directory that can be copied and pasted into the function.

There is one important thing to note in this step. You will need to add an extra backslash in the directory text, as you can see in Figure 14-21.

```
df = pd.read_excel('C:\\Users\\ethan\\Documents\\My Tableau Repository
            \\Datasources\\2023.2\\en_US-US\\Sample - Superstore.xls')
```

With the data frame created, run the model. In Python, you need to declare which variable from the data frame you will use as the y-intercept and which variables you will use as the beta coefficients.

```
# -*- coding: utf-8 -*-
"""
Statistical Tableau Chapter 14 Example

@author: Ethan
"""

#importing the correct libaries
import pandas as pd
import statsmodels.api as sm

df = pd.read_excel('C:\\Users\\ethan\\Documents\\My Tableau Repository\\Datasources\\2023.2\\en_US-US\\Sample - Superstore.
```

Figure 14-21. Read in the data from a local Excel file in Python

Use Discount and Quantity as the dependent variables and the Profit as the independent variable. To keep it easier to follow, title these variables "x" and "y," respectively, as shown at the bottom of Figure 14-22.

```
# -*- coding: utf-8 -*-
"""
Statistical Tableau Chapter 14 Example

@author: Ethan
"""

#importing the correct libaries
import pandas as pd
import statsmodels.api as sm

df = pd.read_excel('C:\\Users\\ethan\\Documents\\My Tableau Repository\\Datasources\\2023.2\\en_US-US\\Sample - Superstore.x

x = sm.add_constant(df[['Discount','Quantity']])
y = df['Profit']
```

Figure 14-22. Declaring the variables that will be used in the regression model

Next, run the multiple linear regression model and print the summary statistics, as shown in Figure 14-23.

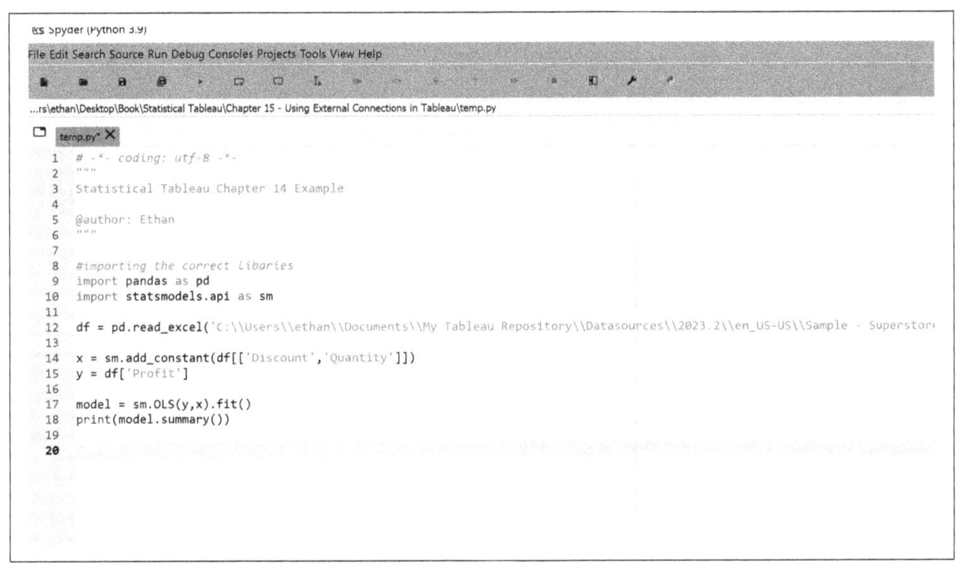

Figure 14-23. Running the model and printing the results

In the console at the bottom right of the interface, you will see a summary statistics table print, as shown in Figure 14-24. This has all the detailed information similar to what you have seen throughout this book.

```
Help  Variable Explor
Console 1/A
Using External Connections in Tableau')
                            OLS Regression Results
==============================================================================
Dep. Variable:                 Profit   R-squared:                       0.053
Model:                            OLS   Adj. R-squared:                  0.052
Method:                 Least Squares   F-statistic:                     282.4
Date:                Wed, 03 Apr 2024   Prob (F-statistic):          4.47e-120
Time:                        20:38:08   Log-Likelihood:                -69734.
No. Observations:               10194   AIC:                         1.395e+05
Df Residuals:                   10191   BIC:                         1.395e+05
Df Model:                           2
Covariance Type:            nonrobust
==============================================================================
                 coef    std err          t      P>|t|      [0.025      0.975]
------------------------------------------------------------------------------
const         40.2819      4.725      8.525      0.000      31.019      49.544
Discount    -247.2753     10.868    -22.752      0.000    -268.579    -225.971
Quantity       7.0716      1.006      7.030      0.000       5.100       9.043
==============================================================================
Omnibus:                    15319.872   Durbin-Watson:                   1.994
Prob(Omnibus):                  0.000   Jarque-Bera (JB):         77773675.550
Skew:                           8.301   Prob(JB):                         0.00
Kurtosis:                     430.585   Cond. No.                         22.0
==============================================================================

Notes:
[1] Standard Errors assume that the covariance matrix of the errors is correctly specified.

In [2]:
```

Figure 14-24. Summary statistics of regression model in Python

The last step I will show in Python is how to isolate the coefficient values. Again, you will need to know how to code this for Chapter 15. You will learn how to return those values into Tableau to use in calculated fields for a dynamic predictive model.

To isolate the coefficients, call on the `model` object and return them using the "param" variable. Just like in R, these values are stored in an array, so you can use integers to reference which variable you want to isolate, as shown in Figure 14-25.

That is the final touch to code this model in Python. As you can see, the code can be similar but the syntax and the approach used in both languages are different.

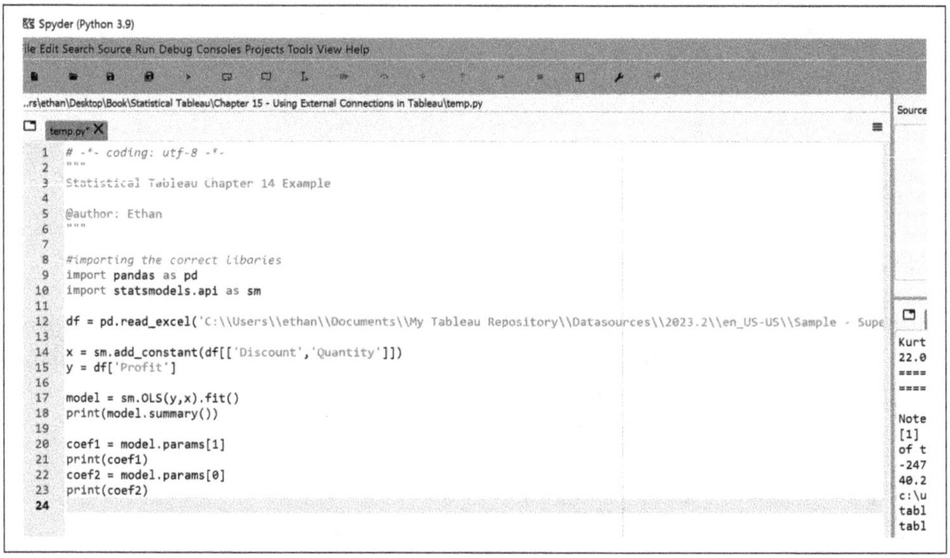
Spyder (Python 3.9)
ile Edit Search Source Run Debug Consoles Projects Tools View Help
..rs\ethan\Desktop\Book\Statistical Tableau\Chapter 15 - Using External Connections in Tableau\temp.py
temp.py*
-*- coding: utf-8 -*-
"""
Statistical Tableau Chapter 14 Example
@author: Ethan
"""
#importing the correct libaries
import pandas as pd
import statsmodels.api as sm
df = pd.read_excel('C:\\Users\\ethan\\Documents\\My Tableau Repository\\Datasources\\2023.2\\en_US-US\\Sample - Supe
x = sm.add_constant(df[['Discount','Quantity']])
y = df['Profit']
model = sm.OLS(y,x).fit()
print(model.summary())
coef1 = model.params[1]
print(coef1)
coef2 = model.params[0]
print(coef2)
Source

Figure 14-25. Isolating the coefficients

Summary

In this chapter, you have learned how to code a multiple linear regression model in R and Python. You will use these examples and build on them in Chapter 15 when you learn how to implement multiple linear regression in Tableau.

CHAPTER 15

Using External Connections in Tableau

There are many use cases where you would want to rely on the external connections to R or Python in Tableau. The most obvious one, from the content in this book, is to rely on R and Python to run advanced statistical models and incorporate them into a dashboard or data visualization. However, if you allow your imagination to guide you for a moment, you'll discover additional potential use cases. Here is a list of other things you could do using external connections that I haven't covered in this book:

Advanced chart types
: Wrangling the data to develop advanced visualizations can be a challenge that R and Python can support. This opens a lot of new chart types that are harder to develop in Tableau without support.

Custom calculations and data manipulation
: Ideally, you would prep your data before loading it into Tableau. However, these external connections can be used to shape your data in different ways and return new values.

Advanced user experience
: By combining parameters and external connections, you could create new and exciting ways to receive input from your user in Tableau and return values back to them or write the input to a database.

Integration with external APIs and web services
: R and Python have extensive library ecosystems, including modules for working with external APIs and web services. Tableau users can leverage scripts to connect to external data sources, pull in real-time data, and incorporate that data into Tableau dashboards for dynamic and up-to-date visualizations.

Geospatial analysis and mapping
: R and Python have powerful libraries for geospatial analysis, such as GeoPandas and Folium. By connecting Tableau to external resources, users can enhance their geospatial visualizations by leveraging other platforms for specialized mapping tasks, allowing for more advanced geographic analysis and visualization within Tableau.

These are just a few examples of other use cases. Tableau is very flexible, and the potential use cases are almost limitless.

In this chapter, you will learn about script functions in Tableau. These functions are what you will use in Tableau calculated fields to send scripts to R or Python. You will also learn how to use these functions to implement a multiple linear regression model in Tableau and how to extract the coefficients of the model.

Script Functions in Tableau

Script functions are used in Tableau to write R or Python code in and send via an external connection to one of these services. In Tableau, there are four functions that you can use in calculated fields to interact with external connections. These functions are `SCRIPT_BOOL`, `SCRIPT_REAL`, `SCRIPT_STR`, and `SCRIPT_INT`. The difference between these functions is the data type that the script will return to Tableau. To effectively use these functions, you will have to establish an external connection like R or Python, which you learned how to do in Chapters 12 and 13.

In Tableau, if you search for "script" in the calculated field editor you will find each of these functions, as shown in Figure 15-1.

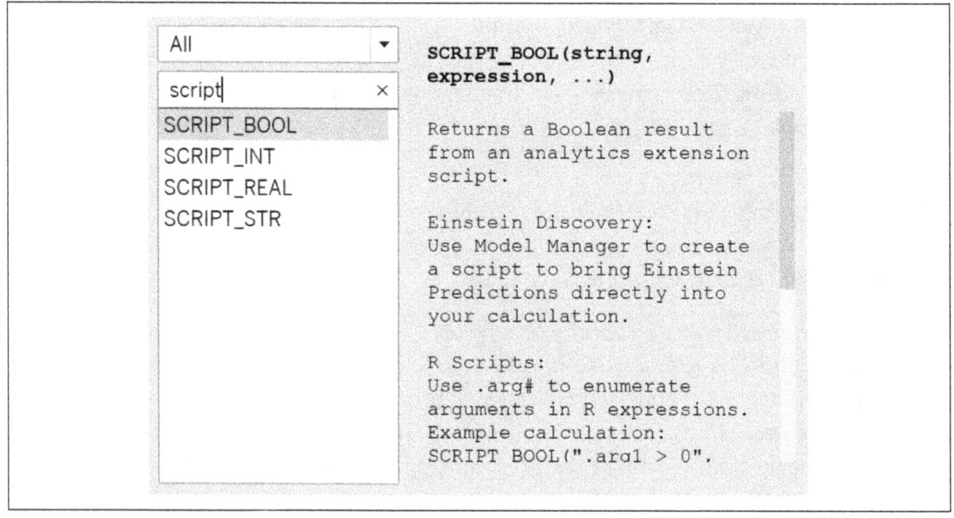

Figure 15-1. Script functions in Tableau

If you select a function in the data dictionary to the right of the calculated field editor window, you will see what data type will be returned and also some simple examples of code for Python and R. Writing the scripts in Tableau is very similar to what you did in Chapter 14. However, there are some nuances that are important to understand. The two most important ones are the concept of passing arguments in the code and the fact that these functions are table calculations.

Passing Arguments in a Tableau Script Function

An argument is basically a placeholder for a value that you will list at the end of the script. Within the code, instead of using `SUM([Profit])` throughout, you would use a numbered argument and then define what that argument is. Look at an example from Tableau's documentation for `SCRIPT_INT`, shown in Figure 15-2.

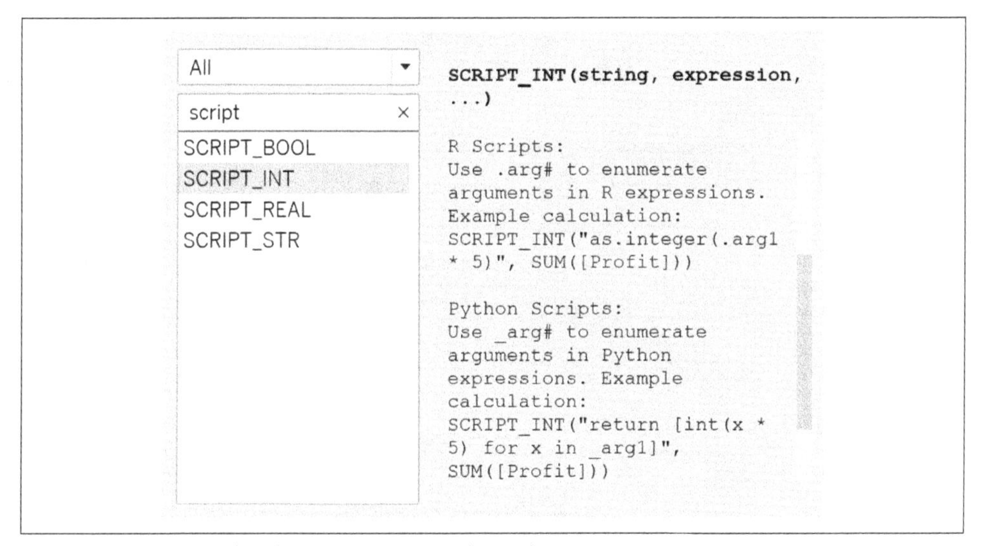

Figure 15-2. Example of an argument in the code

You can see that you write the script using `.arg#` or `_arg#`, depending on whether you are connecting to R or Python, respectively. Then you define what that argument's value is at the end of the function. You will see more examples of this in the next two sections as I show you how to write the multiple linear regression formula in R and Python.

Table Calculations and Knowing the Level of Detail

When using these script functions in Tableau, it is important to know that they are table calculations. That means they are going to run on anything that is in the view, so you must know what level of detail you will be working with and establish that within the view.

As an example, if you are working with the Sample - Superstore dataset provided by Tableau, its lowest level of detail is Order ID and Product ID. That combination will show you all the products that were ordered and separate them out by the Order ID. In this case, that is the exact level of detail you used in Chapter 14 when you learned to code in each language.

However, the advantage of Tableau's table calculations is that you can dynamically change the level of detail. Let's say you wanted to roll it up a level to predict profit at the order level. In Tableau, you can simply remove Product ID from the view. If you were working in R or Python, you would have to manually prep the data—either in the code or in the source file. To summarize, Tableau is actually creating the data frames for the external connections to execute from.

Another thing to be aware of with table calculations in Tableau is where this falls in Tableau's order of operations, as shown in Figure 15-3.

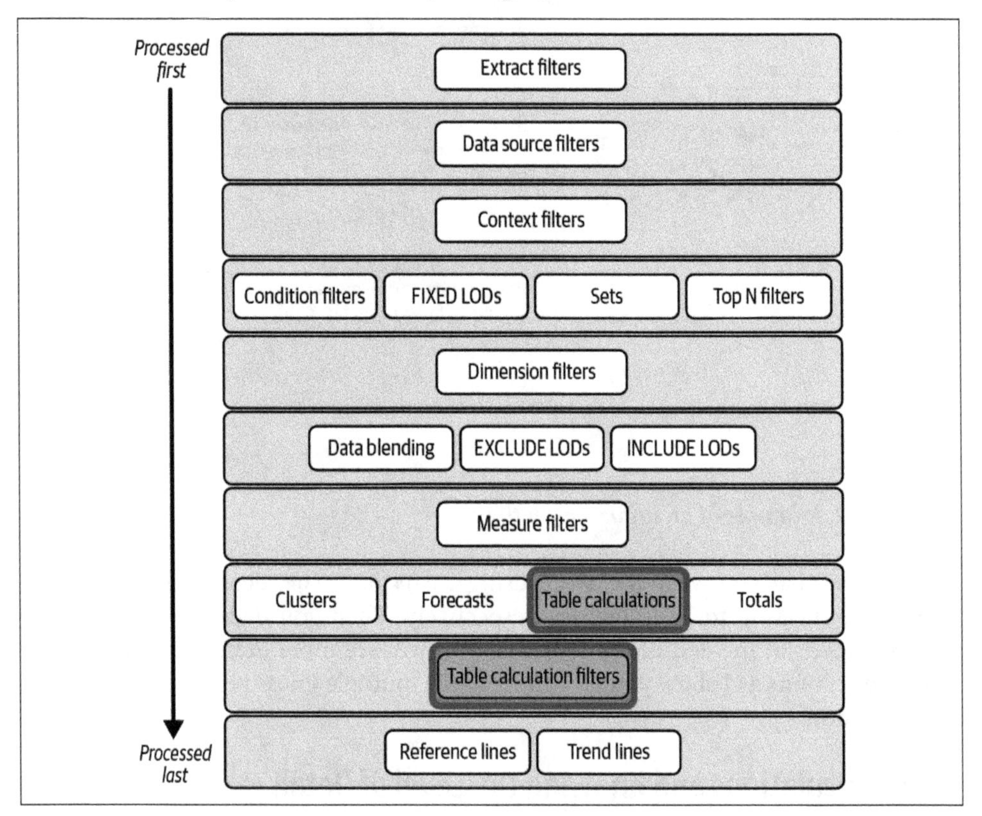

Figure 15-3. Tableau's order of operations for table calculations

It is important to note because they are one of the last things processed by Tableau and everything processed before that will affect the outcome of your models.

Example of Using an External Connection to R from Tableau

Let's take a look at an example where you want to predict profit by discount and quantity. You already learned how to code this model in R from Chapter 14. In this section, you will learn to code that model in Tableau using a script function and then pass that to R using an external analytics connection.

To get started, launch RStudio and turn on the Rserve connection. To do this, simply click in the console window at the bottom left of the interface, type `Rserve::Rserve()`, and then click Enter (see Figure 15-4).

```
5:1    (Top Level) ÷

Console   Terminal ×   Background Jobs ×

R 4.3.1 · ~/
> Rserve::Rserve()
Starting Rserve...
 "C:\Users\ethan\AppData\Local\R\WIN-LI~1\4.3\Rserve\libs\x64\Rserve.exe"
> |
```

Figure 15-4. Launching Rserve from the console window

You will see a return stating that Rserve is starting. After setting up the Rserve, you will want to connect to the same data source that you used in Chapter 14. If you recall, you used your machine's directory to find the Sample - Superstore data from the *Datasources* folder within the *My Tableau Repository* folder. To do this, open Tableau and choose Microsoft Excel from the connection pane. You can locate the same directory by navigating to Documents > My Tableau Repository > Datasources > 2023.2 > en_US-US > Sample - Superstore.xlsx, as shown in Figure 15-5.

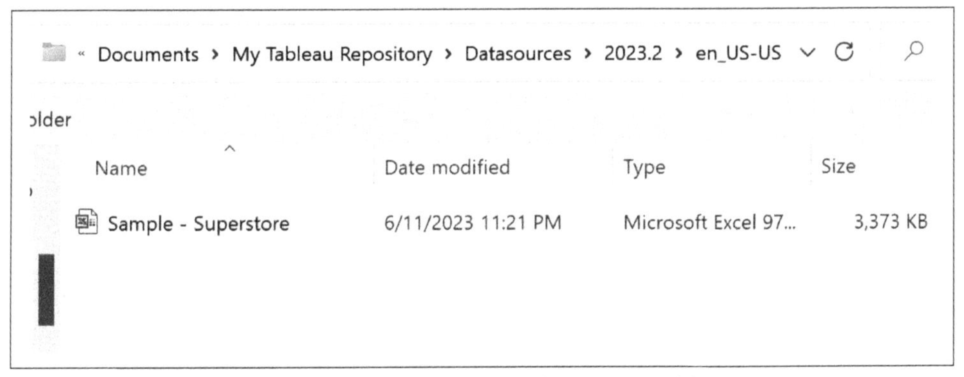

Figure 15-5. Connecting the Sample - Superstore dataset in Tableau

Since this is the exact file you used in Chapter 14, it is important to connect to the same file so you can compare the results. It is also important to know that, depending on the version of Tableau you have, the region where you live, and the language of your version of Tableau Desktop, the actual file path will be different.

Now that you are connecting to the data, launch Tableau and connect to the analytics extension connection. As shown in Figure 15-6, you will click Help from the top navigation and then click "Settings and Performance" > Manage Analytics Extension Connection.

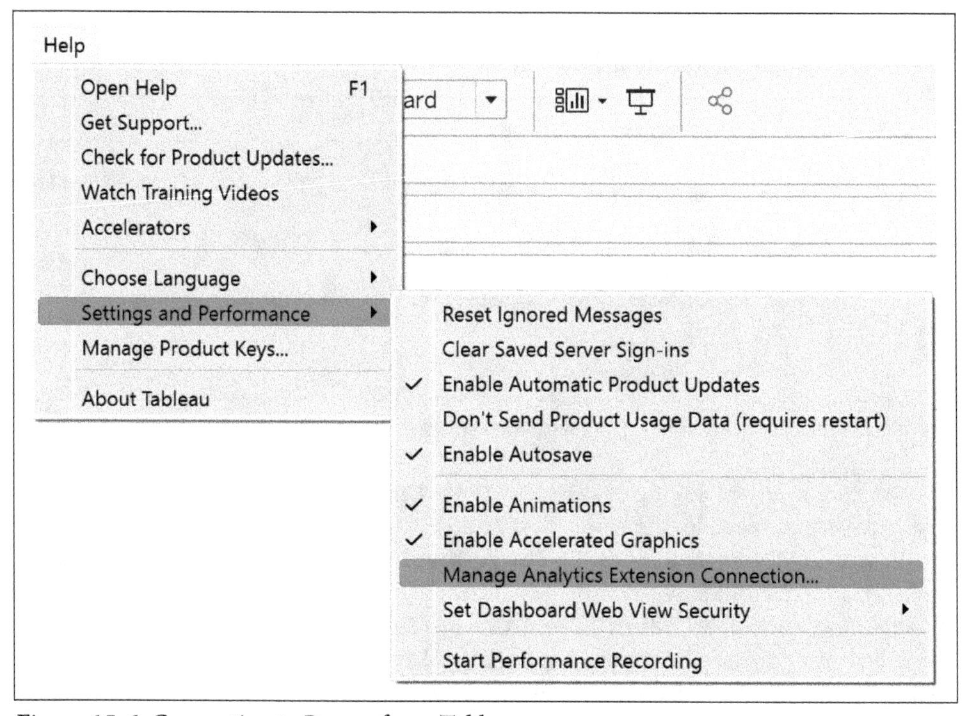

Figure 15-6. Connecting to Rserve from Tableau

You will see a new window appear that allows you to choose the connection type, as shown in Figure 15-7.

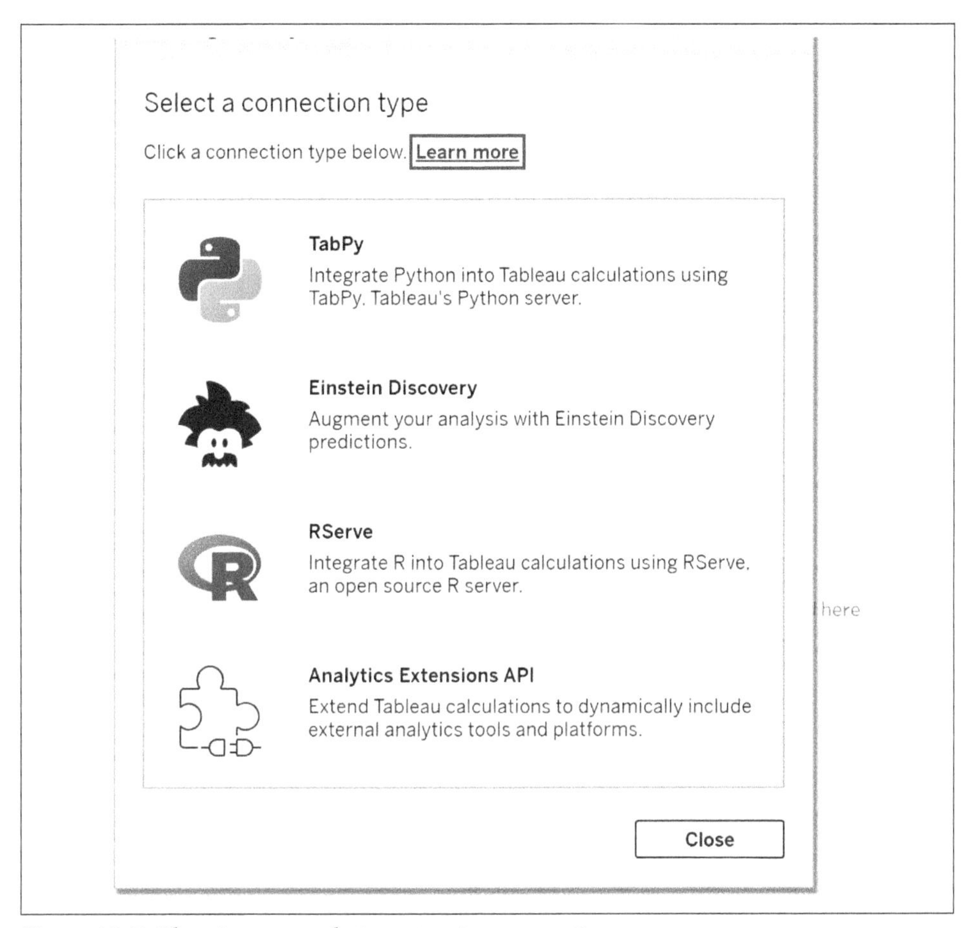

Figure 15-7. Choosing an analytics extension connection

In the next menu, you will define your connection. If you are following along, the connection will be the same as in Chapter 13. For Hostname, type in "localhost" and for Port, type in 6311, as shown in Figure 15-8. If you are using a more defined connection to a specific environment, enter the necessary information and then click Test Connection.

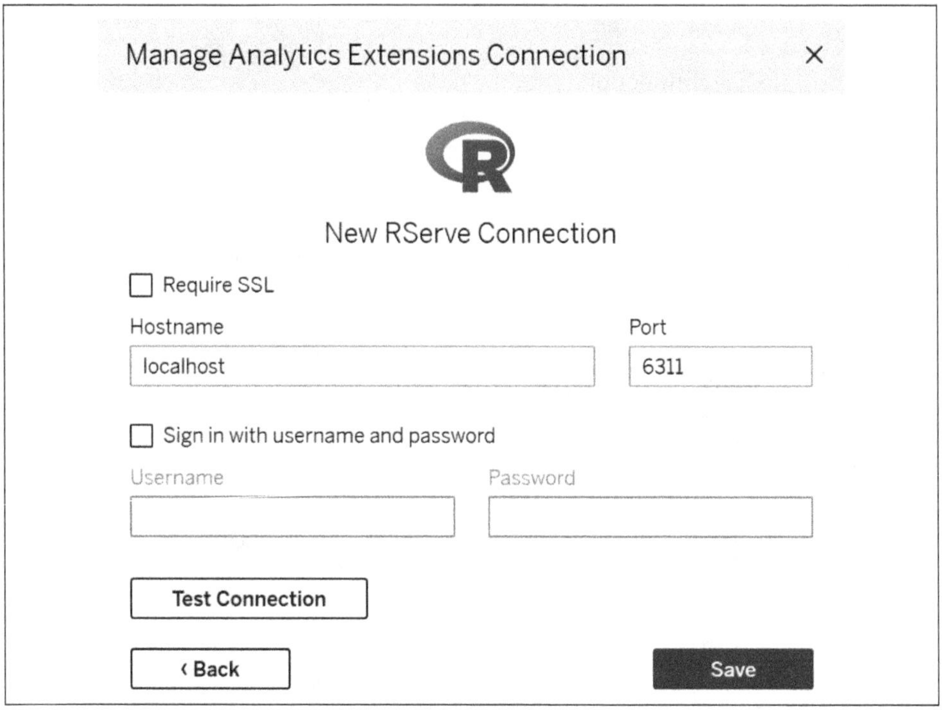

Figure 15-8. Connecting to Rserve

After clicking Test Connection, you should see a pop-up that says, "Successfully connected to the analytics extension." Click OK to close that pop-up and then click Save. That will complete the connection to Rserve.

Before you create the calculated fields for the script, you need to set up the view to the specific level of detail you want the model to run. To match the data from the previous example in Chapter 14, add Order ID and Product ID to the Rows shelf on a new sheet. Your view should look similar to Figure 15-9.

Now create a new calculated field and begin writing the script for a multiple linear regression model. The script needs to return the intercept and coefficients back so that you can use them in Tableau to write a predictive model. Since the values you want to return are decimal numbers, use the `SCRIPT_REAL` function.

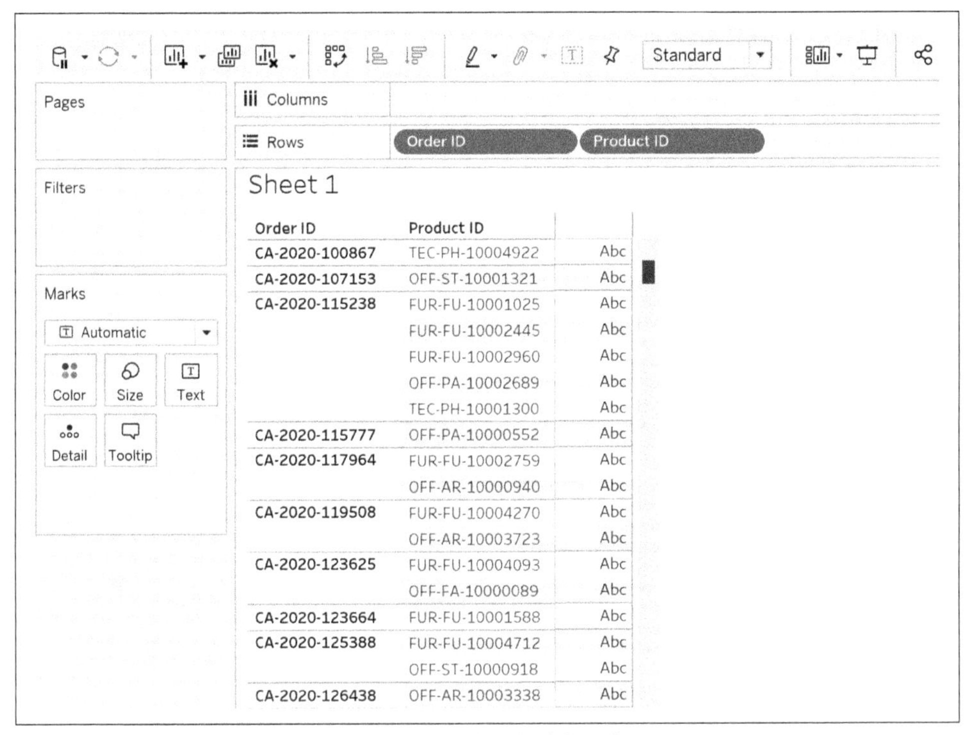

Figure 15-9. Creating a sheet at the correct level of detail

The first value you want is the intercept. To make this easy, use the R code from Chapter 14 and copy and paste it into the calculated field. Afterward, switch each of the variables in the code for the corresponding arguments in Tableau. Let's work through it step by step:

1. Copy and paste the code from R into the calculated field editor, as shown in Figure 15-10.

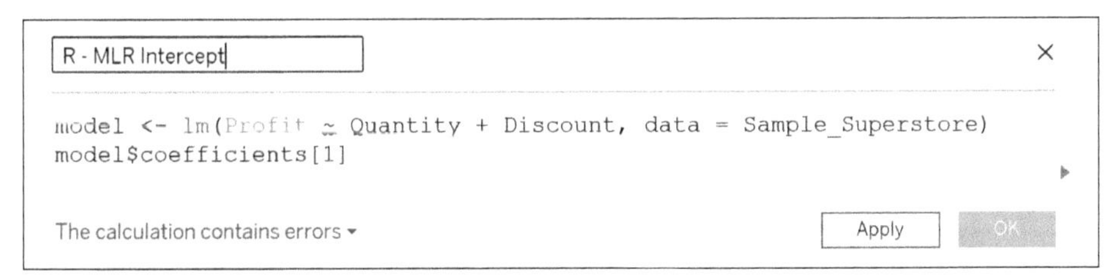

Figure 15-10. Copy and paste the R code into the editor

2. Add the `SCRIPT_REAL` function and wrap the code in quotation marks, as shown in Figure 15-11.

R - MLR Intercept

```
SCRIPT_REAL(
model <- lm(Profit ~ Quantity + Discount, data = Sample_Superstore)
model$coefficients[1]
)
```

The calculation contains errors ▾ | Apply | OK

Figure 15-11. Adding `SCRIPT_REAL` and wrapping the code in quotations

3. Add a comma and list the measures you want to use as the arguments in the code. For this example, argument 1 is Profit, argument 2 is Quantity, and argument 3 is Discount, as shown in Figure 15-12.

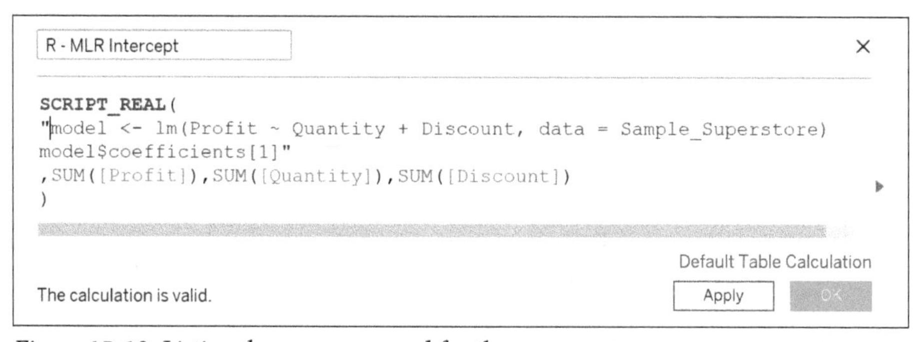

Figure 15-12. Listing the measures used for the arguments

4. The last step is to replace the variables in the code with `.arg#`. So, Profit will change to `.arg1`, Quantity will change to `.arg2`, and Discount will change to `.arg3`, as shown in Figure 15-13. Again, it is extremely important that you assign the variables in the correct order, especially for the dependent variable in the case `SUM([Profit])`. Also, remove the `data = Sample_Superstore` parameter from the code. Since you are using Tableau as the medium for the data, this isn't necessary.

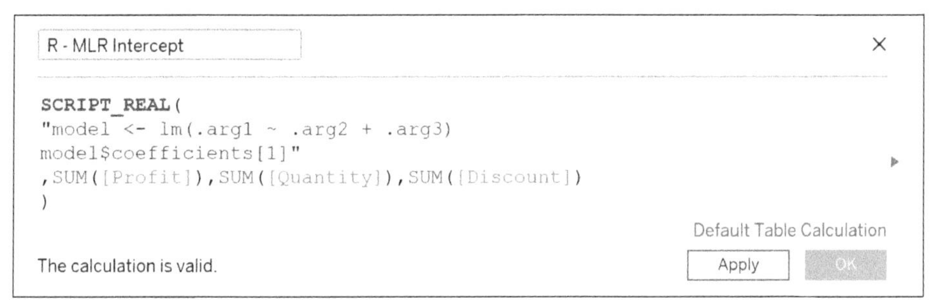

Figure 15-13. Final script for the intercept

This calculation is going to pass the data in the view from Tableau to R, run the model, and then return the intercept coefficient as a value to use in Tableau. Click OK to create the calculated field and then add it to the sheet you created earlier, as shown in Figure 15-14.

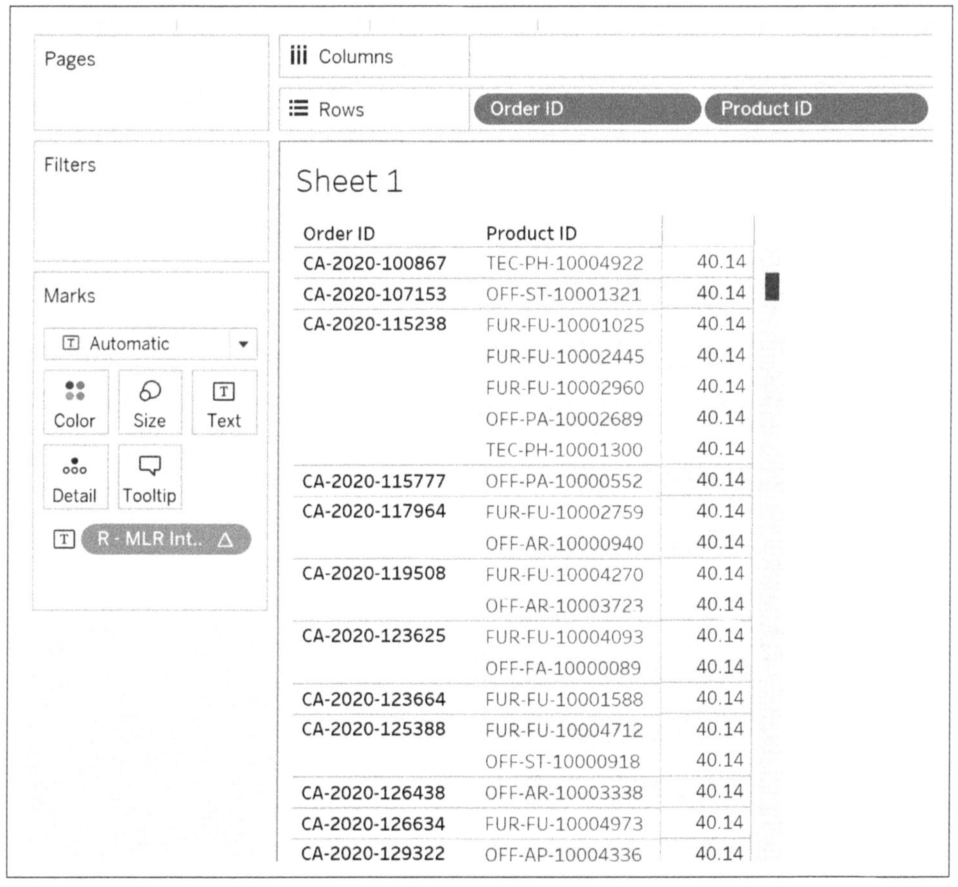

Figure 15-14. Adding the intercept calculated field to the view

Now duplicate that calculation and change the code to return the coefficients for the dependent variables. The only thing you have to change in the code is the second line when you call the `model` object to return a coefficient. If you change from 1 to 2 and from 2 to 3, you will return the next coefficient in the `model` object array. The final code can be seen in Figures 15-15 and 15-16.

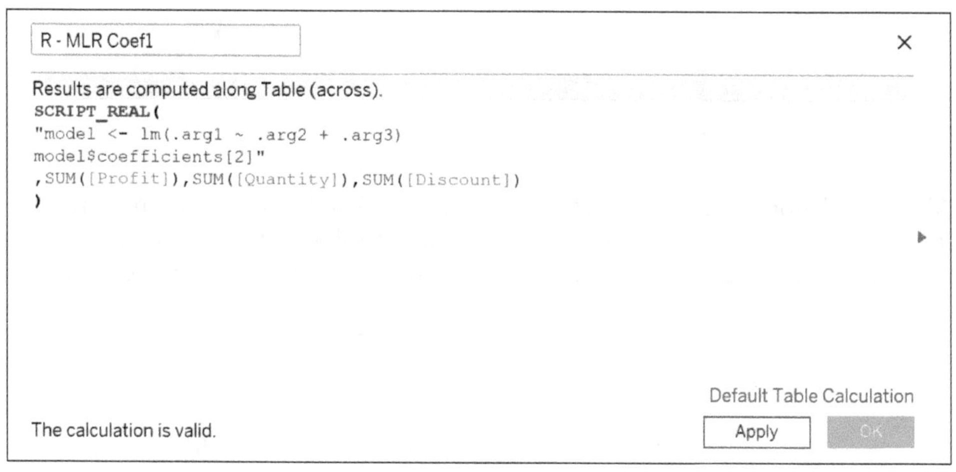

Figure 15-15. Coefficient 1 calculated field

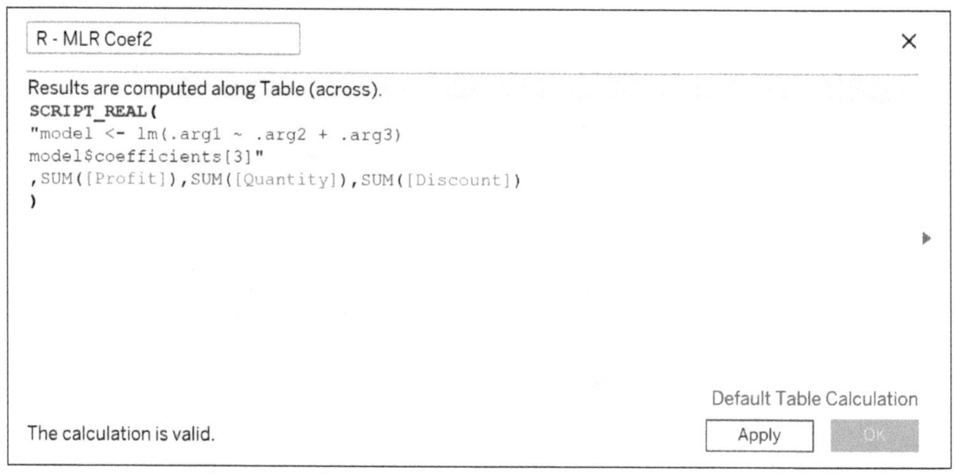

Figure 15-16. Coefficient 2 calculated field

Now add those calculated fields to the view, as seen in Figure 15-17.

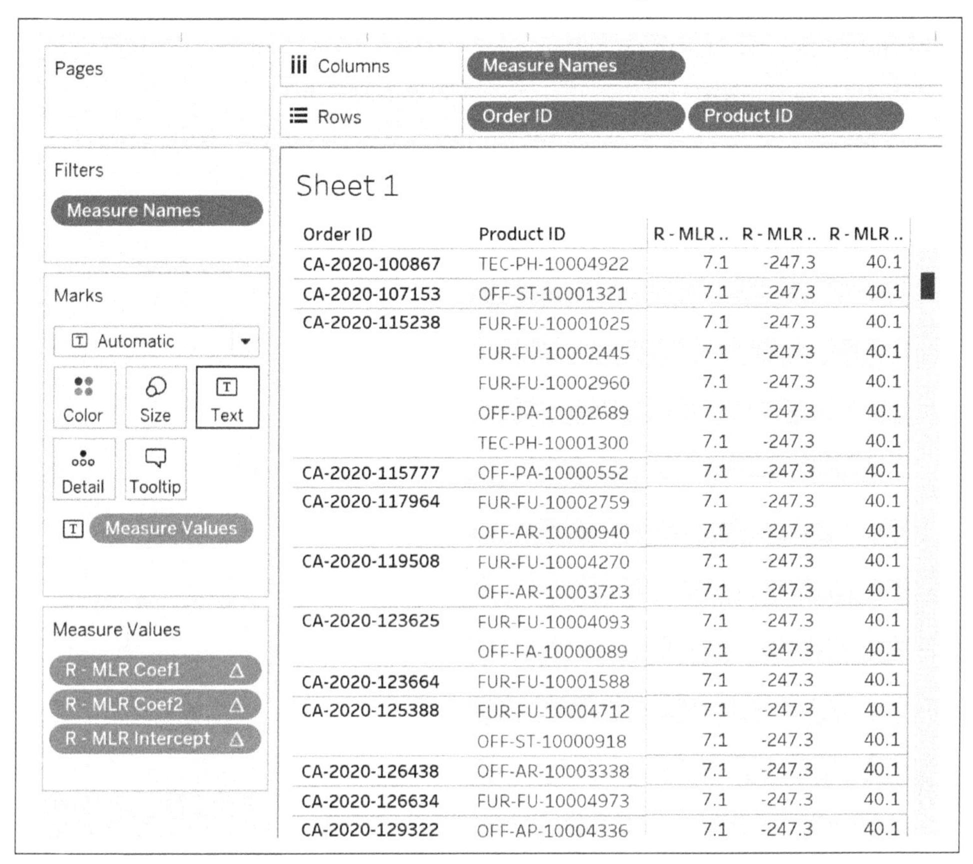

Figure 15-17. Adding the fields to view

Now that you have the intercept and coefficients in Tableau, you can create calculations to turn those values into a predictive model. All you need to do is plug them into the multiple linear regression model equation.

Multiple linear regression is very similar to simple linear regression where you add another beta coefficient for each independent variable you add. Multiple linear regression is expressed with the following formula:

$$Y_i = \beta_0 + \beta_1 X_1 + \beta_2 X_2 + \beta_i X_i + \epsilon_i$$

where

Y_i = the value of the dependent value

β_0 = the intercept

β_1 = the coefficient calculated by the model

β_2 = the coefficient calculated by the model

β_i = the ith coefficient calculated by the model

X_i = a known value from the dataset at the ith point

ϵ_i = a random error that occurs

With that said, plug the values into a calculated field, as shown in Figure 15-18.

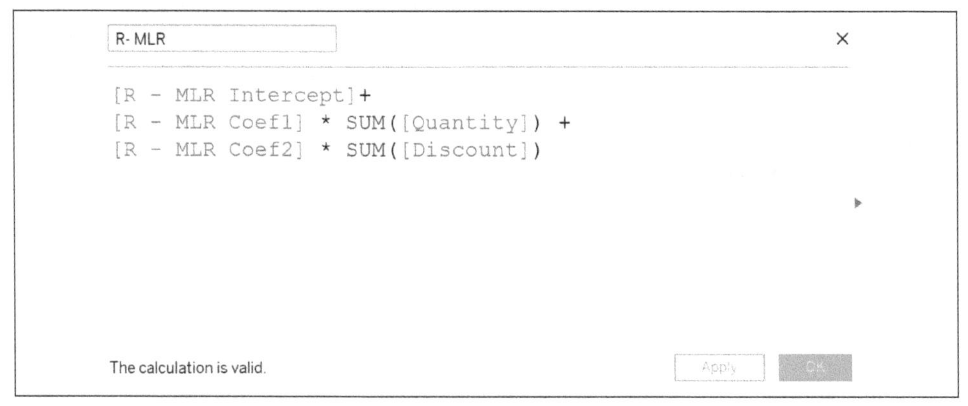

Figure 15-18. Full multiple linear regression model equation

You can now add this new calculated field and Profit to the view for a comparison of the predicted values versus the actuals, as shown in Figure 15-19.

This is a very simple example, but the capabilities span a lot further than this. Try new things and explore R code. The better you become with coding in R, the more effective this tactic will become for you in Tableau.

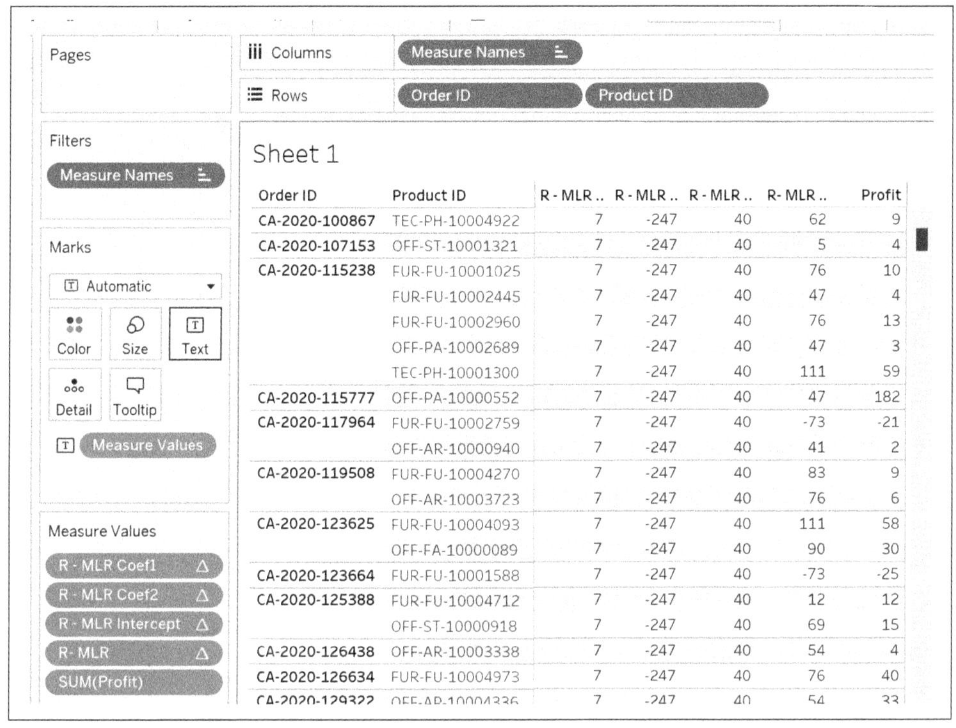

Figure 15-19. Predicted versus actuals in Tableau

When you're done, be sure to close RStudio and disconnect from Rserve. It is important to note that when you do so, any field that was using the external connection will no longer function. With that being said, you can record your intercept and coefficients and implement them in other ways. The simplest way to do this is to hardcode those values, as shown in Figure 15-20.

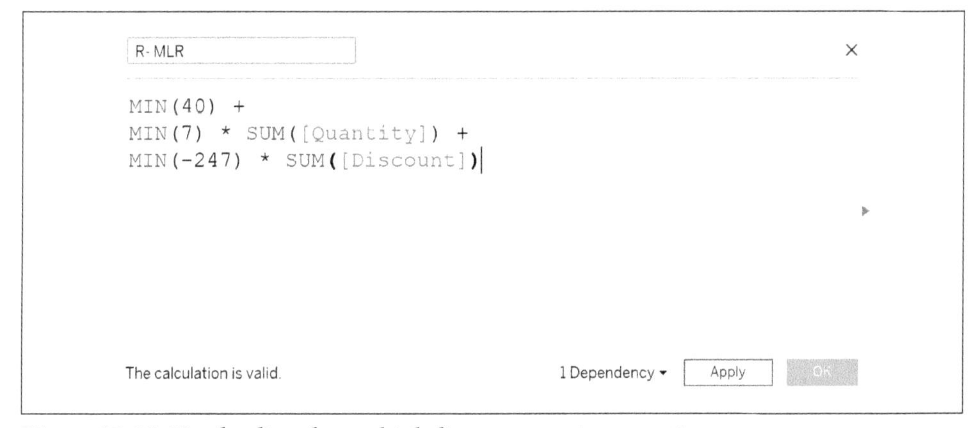

Figure 15-20. Hardcoding the multiple linear regression equation

That wraps up a simple example of using R in Tableau to implement multiple linear regression. Next, we will walk through the same example in Python.

Example of Using an External Connection to Python from Tableau

Let's take a look at the same example in Python. To get started, launch Anaconda Navigator and turn on the TabPy connection. To do this, simply click on Environments, select StatisticalTableau, then open a terminal window, as shown in Figure 15-21.

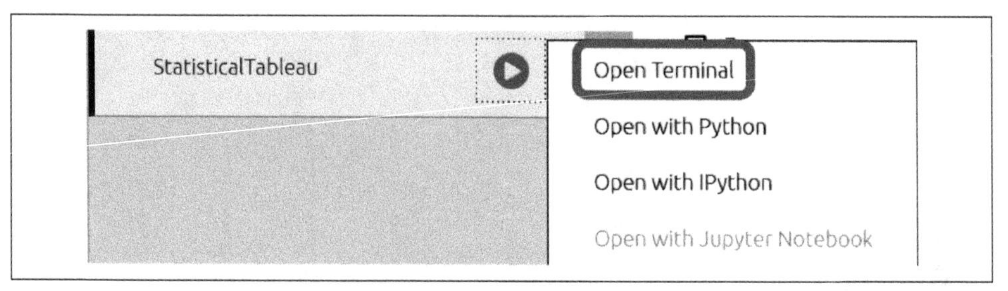

Figure 15-21. Launching the terminal to activate the local server

Once the terminal is open, enter TabPy and click Enter. You will get a warning asking if you want to proceed. Enter "y" into the terminal and click Enter again. This will activate the local server, which is running on port 9004, as shown in Figure 15-22. If you are getting any kind of errors, I would recommend following the instructions in Chapter 12.

With TabPy running, launch Tableau and make the connection to the local server. To do this, click Help in the main navigation at the top of the authoring interface, then click "Settings and Performance" > Manage Analytics Extension Connection, as shown in Figure 15-23.

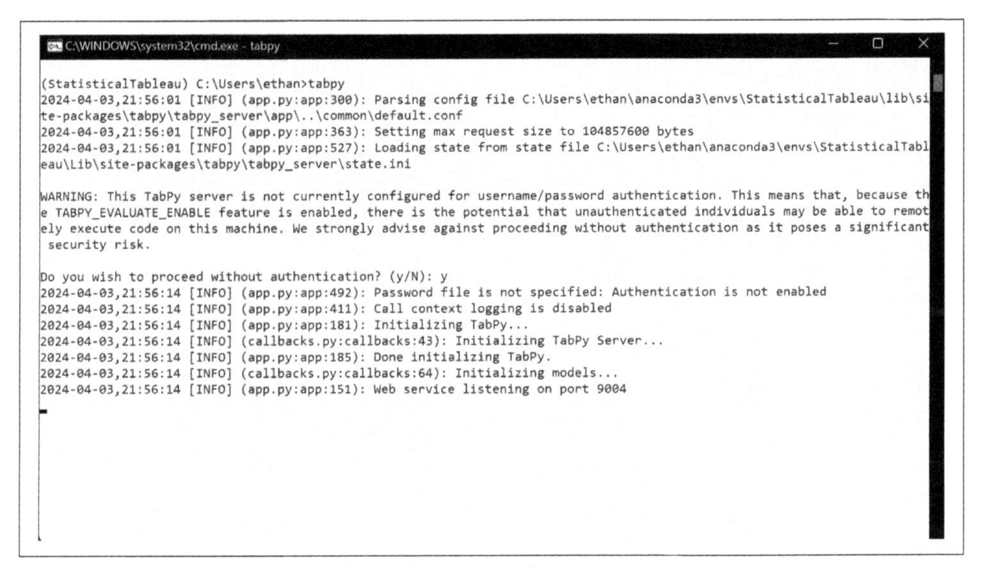

Figure 15-22. Running TabPy from the terminal

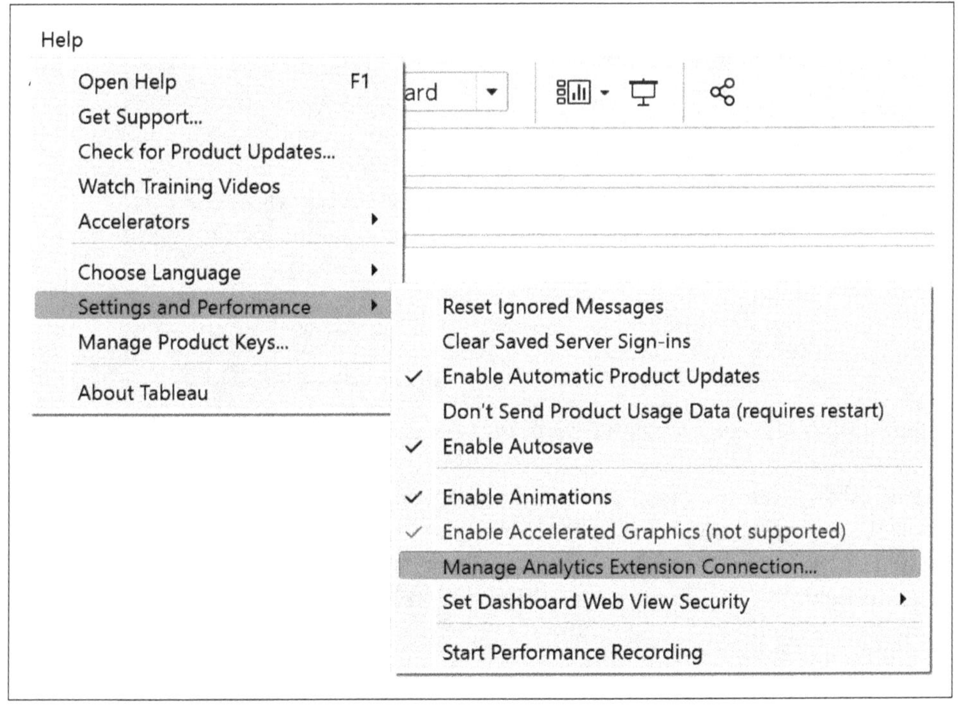

Figure 15-23. Connecting to the Python server

A new menu opens that asks what external connection you would like to use, as shown in Figure 15-24.

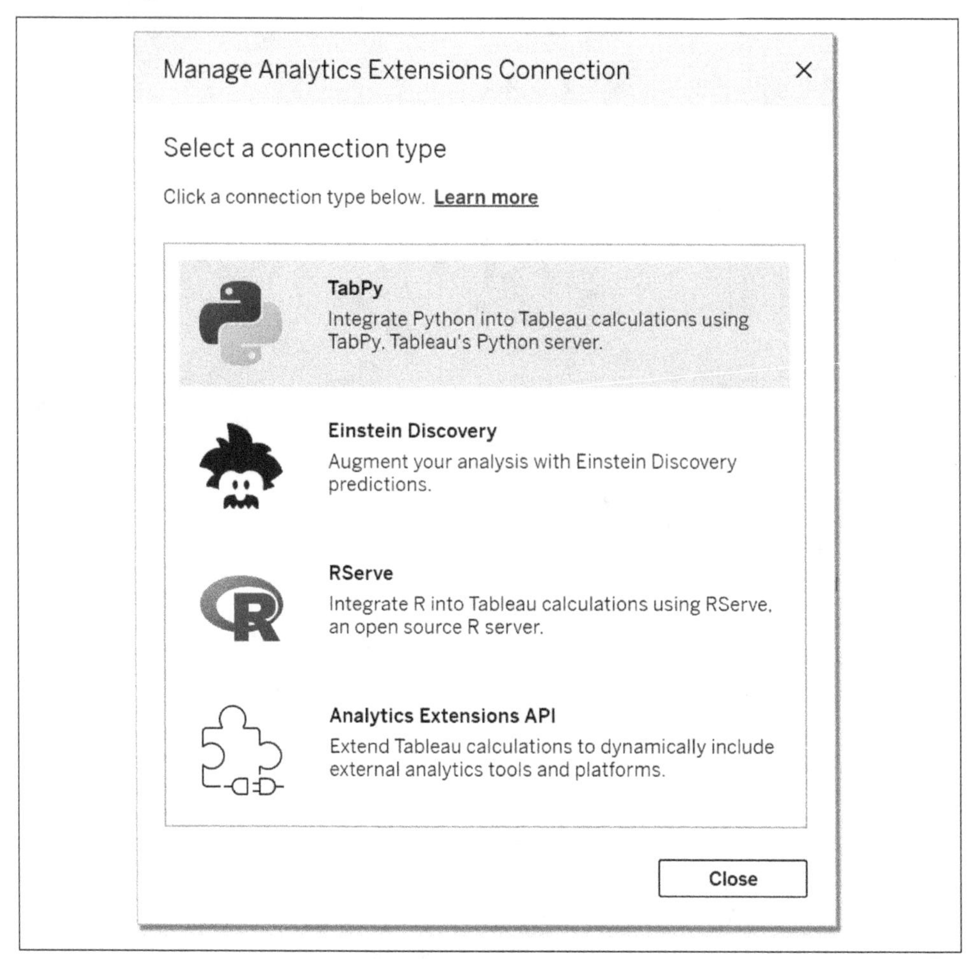

Figure 15-24. Manage Analytics Extensions Connection

Select TabPy from the menu and another pop-up window will appear asking you to enter the authentication information. Since you are likely connecting to a local server hosted on your own computer, enter "localhost" into the Hostname field and 9004 in the Port field. You should have something similar to Figure 15-25.

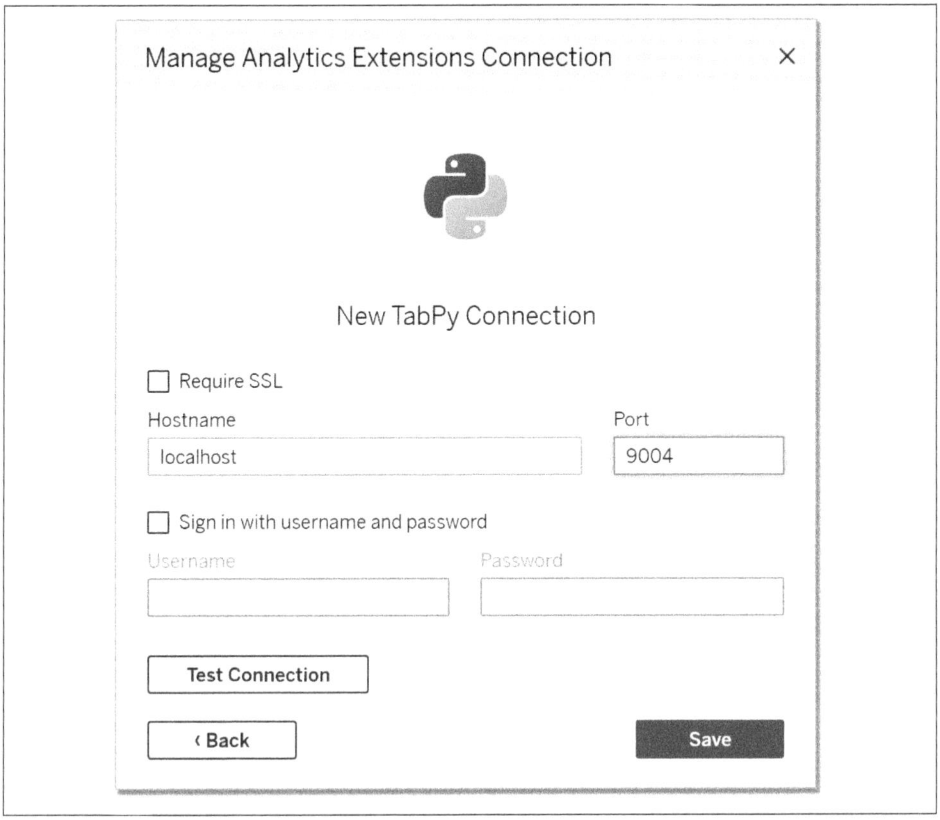

Figure 15-25. New TabPy connection

Click Test Connection; if everything is working properly, you will see a message appear that says, "Successfully connected to the analytics extension." You can close this window by clicking OK and then click Save to connect to the server.

Before you create the calculated fields for the script, you need to set up the view to the specific level of detail you want the model to run. To match the data from the previous example in Chapter 14, add Order ID and Product ID to the Rows shelf on a new sheet. Your view should look similar to Figure 15-26.

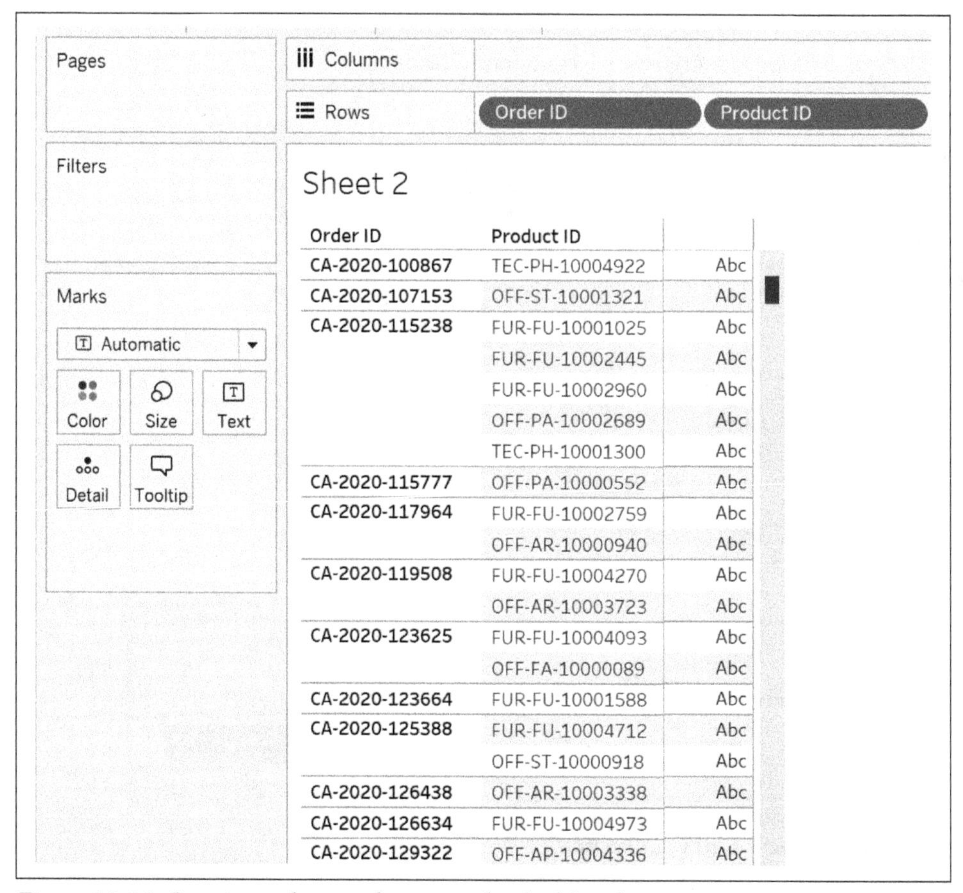

Figure 15-26. Creating a sheet at the correct level of detail

Now create a new calculated field and begin writing the script for a multiple linear regression model. The idea for this model is to return the intercept and coefficients back so that you can use them in Tableau to write a predictive model. Since the values you want to return are decimal numbers, use the `SCRIPT_REAL` function.

The first value you want is the intercept. To make this easy, use the Python script from Chapter 14 and copy and paste it into the calculated field. Afterward, switch each of the variables in the code for the corresponding arguments in Tableau. Let's work through it step by step:

1. Copy and paste the script from Chapter 14 into the calculated field, as shown in Figure 15-27.

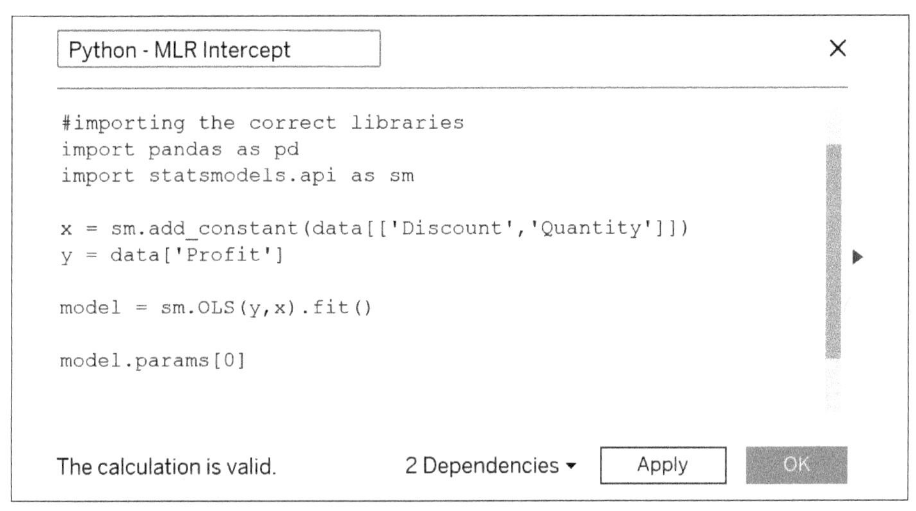

Figure 15-27. Copy and paste code from Python to Tableau

2. Add the `SCRIPT_REAL` function and add quotation marks around the code, as shown in Figure 15-28.

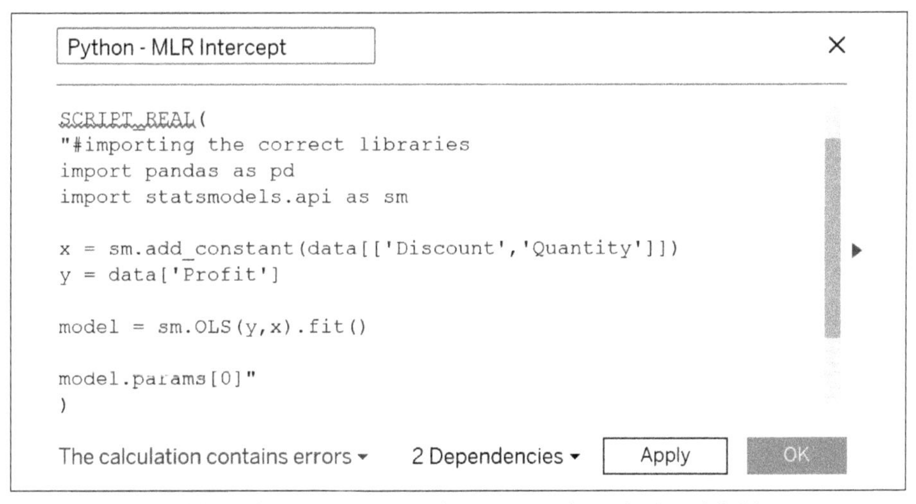

Figure 15-28. Adding the `SCRIPT_REAL` function to the calculated field

3. Add the list of measures you want to pass from Tableau to Python as arguments, as shown in Figure 15-29.

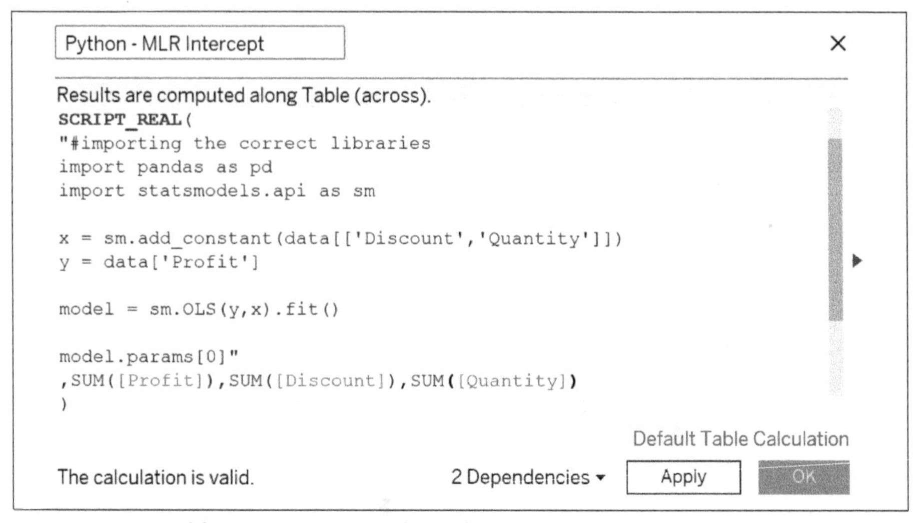

Figure 15-29. Adding arguments to the code

4. In Python, you now have to pass the arguments into a data frame. Add a line of code just below the importing libraries step that creates a new data frame and assigns the arguments to their corresponding values. You should also add `return` to the last line of code. This is a function in Python that is a signal that you want to receive `model.params[0]` as the final output, as shown in Figure 15-30.

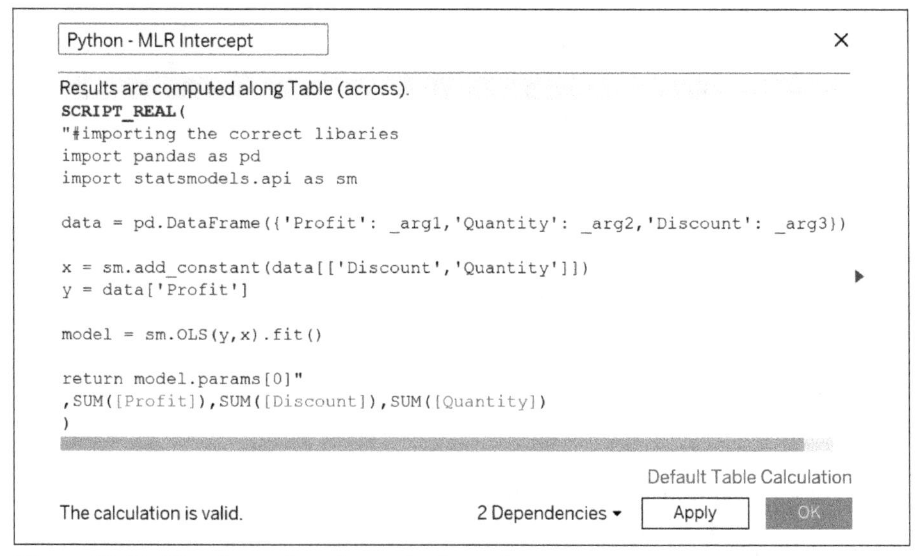

Figure 15-30. Creating a data frame in the code

That should finalize the calculated field. Click OK to save the field and add it to the sheet you created, as shown in Figure 15-31.

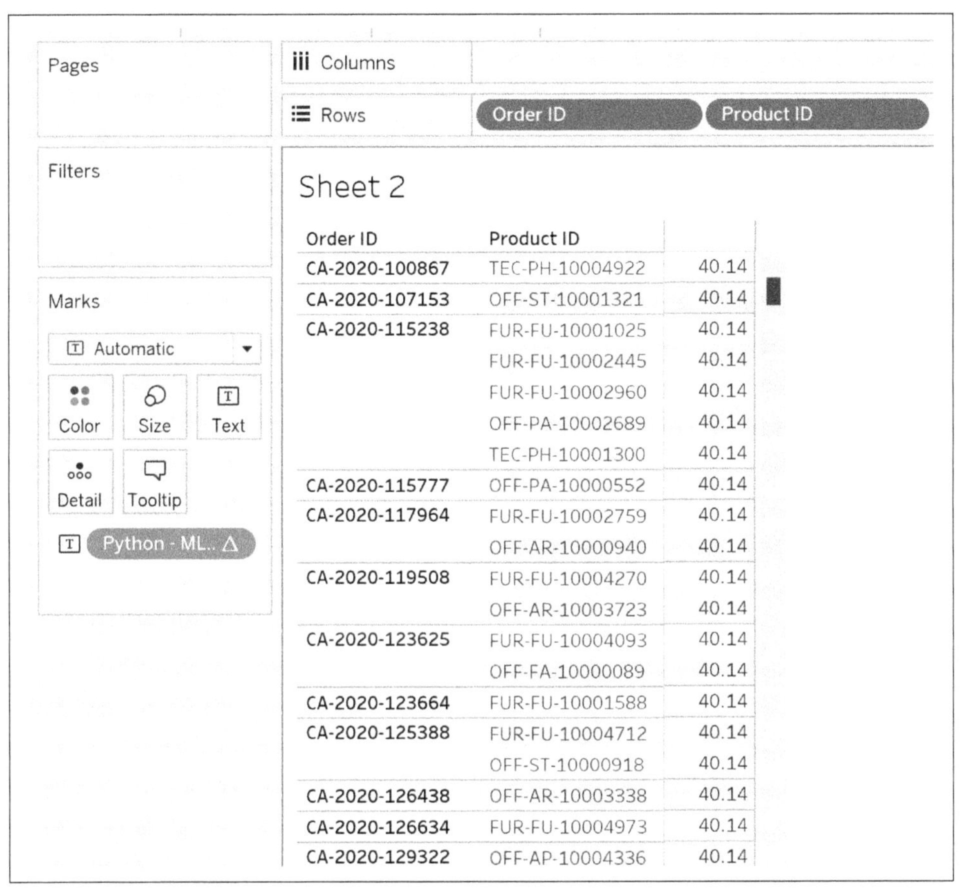

Figure 15-31. Adding the intercept value to the table

Now duplicate that calculation and change the code to return the coefficients for the dependent variables. The only thing you have to change in the code is the last line when you return `model.params[0]`. If you change that number from 0 to 1 and then from 1 to 2, you will return the next coefficient in the `model` object array. The final code can be seen in Figures 15-32 and 15-33.

Python - MLR Coef1

Results are computed along Table (across).

```
SCRIPT_REAL(
"#importing the correct libraries
import pandas as pd
import statsmodels.api as sm

data = pd.DataFrame({'Profit': _arg1,'Quantity': _arg2,'Discount': _arg3})

x = sm.add_constant(data[['Discount','Quantity']])
y = data['Profit']

model = sm.OLS(y,x).fit()

return model.params[1]"
,SUM([Profit]),SUM([Discount]),SUM([Quantity])
)
```

Default Table Calculation

The calculation is valid. 2 Dependencies ▾ Apply OK

Figure 15-32. Returning the first coefficient

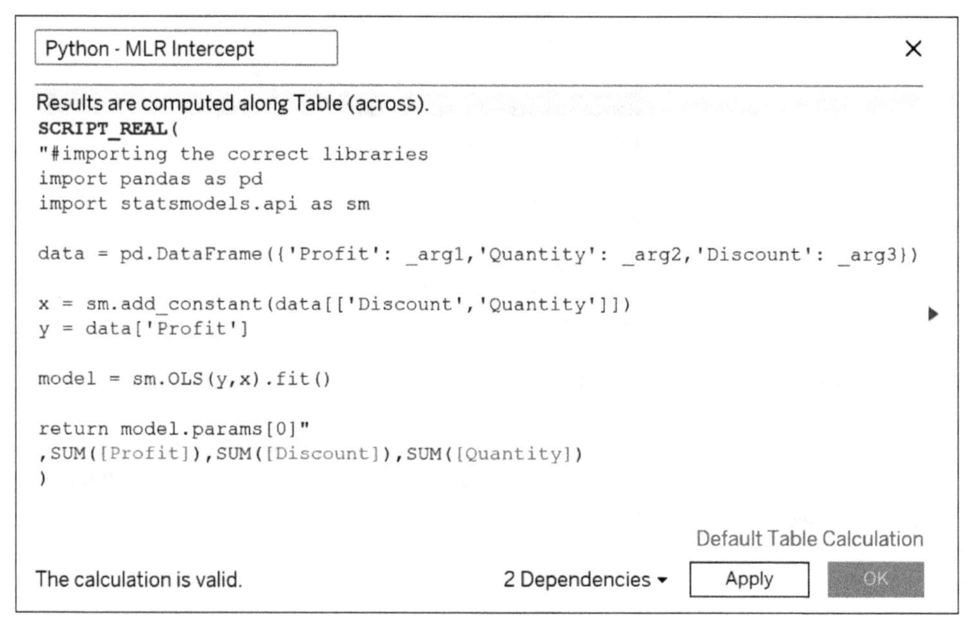

Figure 15-33. Returning the second coefficient

Now add those two new calculated fields to the view, as shown in Figure 15-34.

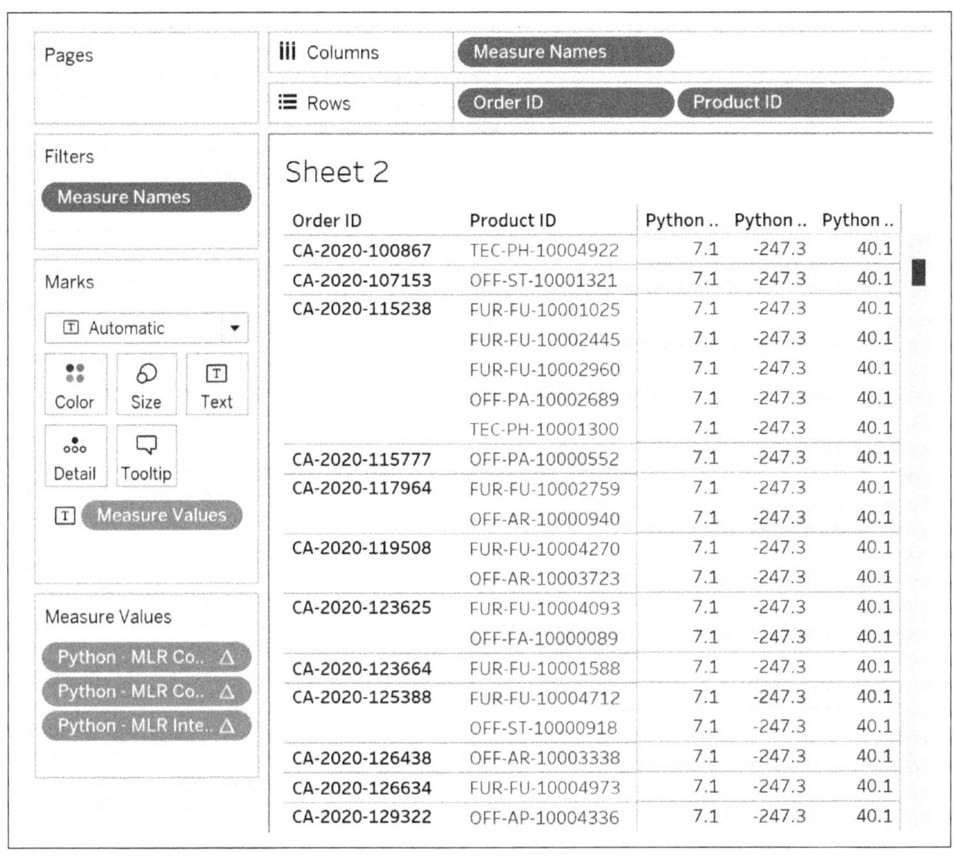

Order ID	Product ID	Python ..	Python ..	Python ..
CA-2020-100867	TEC-PH-10004922	7.1	-247.3	40.1
CA-2020-107153	OFF-ST-10001321	7.1	-247.3	40.1
CA-2020-115238	FUR-FU-10001025	7.1	-247.3	40.1
	FUR-FU-10002445	7.1	-247.3	40.1
	FUR-FU-10002960	7.1	-247.3	40.1
	OFF-PA-10002689	7.1	-247.3	40.1
	TEC-PH-10001300	7.1	-247.3	40.1
CA-2020-115777	OFF-PA-10000552	7.1	-247.3	40.1
CA-2020-117964	FUR-FU-10002759	7.1	-247.3	40.1
	OFF-AR-10000940	7.1	-247.3	40.1
CA-2020-119508	FUR-FU-10004270	7.1	-247.3	40.1
	OFF-AR-10003723	7.1	-247.3	40.1
CA-2020-123625	FUR-FU-10004093	7.1	-247.3	40.1
	OFF-FA-10000089	7.1	-247.3	40.1
CA-2020-123664	FUR-FU-10001588	7.1	-247.3	40.1
CA-2020-125388	FUR-FU-10004712	7.1	-247.3	40.1
	OFF-ST-10000918	7.1	-247.3	40.1
CA-2020-126438	OFF-AR-10003338	7.1	-247.3	40.1
CA-2020-126634	FUR-FU-10004973	7.1	-247.3	40.1
CA-2020-129322	OFF-AP-10004336	7.1	-247.3	40.1

Figure 15-34. Adding the fields to view

Now that you have the intercept and coefficients in Tableau, you can create calculations to turn those values into a predictive model by plugging them into the multiple linear regression model equation, as shown in Figure 15-35.

Python - MLR

```
([Python - MLR Intercept]) +
([Python - MLR Coef1] * SUM([Quantity])) +
([Python - MLR Coef2] * SUM([Discount]))
```

The calculation is valid.

Apply OK

Figure 15-35. Full multiple linear regression model equation

You can now add this new calculated field and Profit to the view for a comparison of the predicted values versus the actuals, as shown in Figure 15-36.

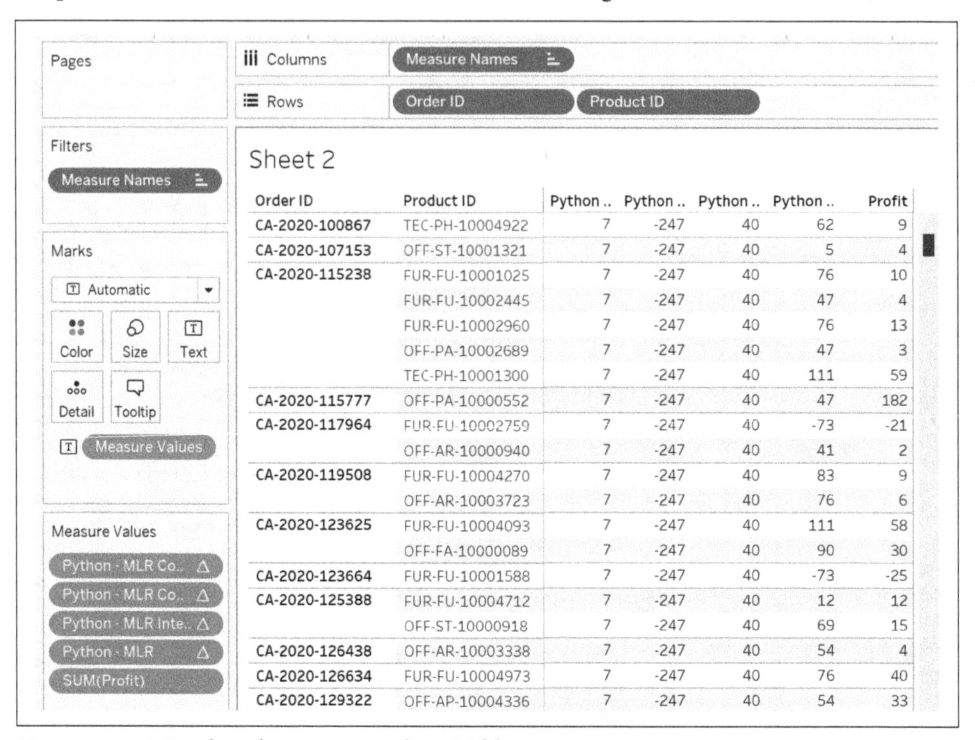

Order ID	Product ID	Python ..	Python ..	Python ..	Python ..	Profit
CA-2020-100867	TEC-PH-10004922	7	-247	40	62	9
CA-2020-107153	OFF-ST-10001321	7	-247	40	5	4
CA-2020-115238	FUR-FU-10001025	7	-247	40	76	10
	FUR-FU-10002445	7	-247	40	47	4
	FUR-FU-10002960	7	-247	40	76	13
	OFF-PA-10002689	7	-247	40	47	3
	TEC-PH-10001300	7	-247	40	111	59
CA-2020-115777	OFF-PA-10000552	7	-247	40	47	182
CA-2020-117964	FUR-FU-10002759	7	-247	40	-73	-21
	OFF-AR-10000940	7	-247	40	41	2
CA-2020-119508	FUR-FU-10004270	7	-247	40	83	9
	OFF-AR-10003723	7	-247	40	76	6
CA-2020-123625	FUR-FU-10004093	7	-247	40	111	58
	OFF-FA-10000089	7	-247	40	90	30
CA-2020-123664	FUR-FU-10001588	7	-247	40	-73	-25
CA-2020-125388	FUR-FU-10004712	7	-247	40	12	12
	OFF-ST-10000918	7	-247	40	69	15
CA-2020-126438	OFF-AR-10003338	7	-247	40	54	4
CA-2020-126634	FUR-FU-10004973	7	-247	40	76	40
CA-2020-129322	OFF-AP-10004336	7	-247	40	54	33

Figure 15-36. Predicted versus actuals in Tableau

This is a very simple example, but the capabilities span a lot further than this. Try new things and explore Python. The better you become with coding in Python, the more effective this tactic will become for you in Tableau.

When you're done, be sure to close the terminal and disconnect from TabPy in Tableau. It is important to note that when you do so, any field that was using the external connection will no longer function. With that being said, you can record your intercept and coefficients and implement them in other ways. The simplest way to do this is to hardcode those values, as shown in Figure 15-37.

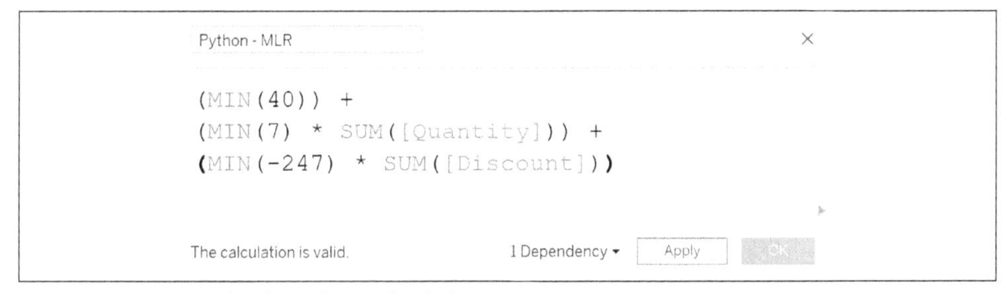

Figure 15-37. Hardcoding the multiple linear regression equation

That wraps up a simple example of using Python in Tableau to implement multiple linear regression.

Summary

In this chapter, you learned how to connect to the external connections and run a model, in both R and Python, that isn't native in Tableau. This is just one simple example of what you can do using external connections in Tableau.

I challenge you to take what you have learned in this book and apply it in new and exciting ways. Until next time! This is Ethan Lang signing off.

Index

A

B

C

D

E

F

H

I

J

K

L

M

N

O

P

R

S

T

U

V

W

X

Z

About the Author

Ethan Lang is an award-winning data visualization designer and engineer. He is the coleader of the Veterans Advocacy Tableau User Group, a Tableau User Group ambassador, and a technical reviewer on multiple best-selling Tableau books. His data visualization strategies have been shared in dozens of cities across North America including Orlando, Portland, and Kansas City.

Colophon

The animal on the cover of *Statistical Tableau* is an East African sunset moth (*Chrysiridia croesus*). As its name suggests, it is found in Kenya, Tanzania, Mozambique, and Zimbabwe.

The sunset moth was originally identified as a butterfly because of its bright colors and diurnal behavior. In contrast, most moths are shades of brown or gray and are active at night. The shape of the scales on the sunset moth's wings scatters light waves like a prism, which gives the moth its bright colors.

Sunset moths live their first week as caterpillars. Then they begin metamorphosis, which lasts another week. After that, they emerge from the cocoon and begin flying in about 10 minutes.

The caterpillars of this species eat the leaves of toxic plants and this toxicity remains in the body into adulthood, making the moth poisonous to predators. It is thought that their bright colors are a visual warning of their toxicity.

Many of the animals on O'Reilly covers are endangered; all of them are important to the world.

The cover illustration is by Karen Montgomery, based on a black-and-white engraving from *Insects Abroad*. The series design is by Edie Freedman, Ellie Volckhausen, and Karen Montgomery. The cover fonts are Gilroy Semibold and Guardian Sans. The text font is Adobe Minion Pro; the heading font is Adobe Myriad Condensed; and the code font is Dalton Maag's Ubuntu Mono.

Printed in the USA
CPSIA information can be obtained
at www.ICGtesting.com
JSHW052349110524
62932JS00005B/32